AF676462

Spectral Detection Technology, Sensors and Instruments

Spectral Detection Technology, Sensors and Instruments

Editors

Qing Yu
Ran Tu
Ting Liu
Lina Li

MDPI

Basel • Beijing • Wuhan • Barcelona • Belgrade • Novi Sad • Cluj • Manchester

Editors
Qing Yu
Huaqiao University
Xiamen
China

Ran Tu
Huaqiao University
Xiamen
China

Ting Liu
Huaqiao University
Xiamen
China

Lina Li
Huaqiao University
Xiamen
China

Editorial Office
MDPI AG
Grosspeteranlage 5
4052 Basel, Switzerland

This is a reprint of articles from the Special Issue published online in the open access journal *Sensors* (ISSN 1424-8220) (available at: https://www.mdpi.com/journal/sensors/special_issues/SDT_Sensors).

For citation purposes, cite each article independently as indicated on the article page online and as indicated below:

Lastname, A.A.; Lastname, B.B. Article Title. *Journal Name* **Year**, *Volume Number*, Page Range.

ISBN 978-3-7258-2367-3 (Hbk)
ISBN 978-3-7258-2368-0 (PDF)
doi.org/10.3390/books978-3-7258-2368-0

© 2024 by the authors. Articles in this book are Open Access and distributed under the Creative Commons Attribution (CC BY) license. The book as a whole is distributed by MDPI under the terms and conditions of the Creative Commons Attribution-NonCommercial-NoDerivs (CC BY-NC-ND) license.

Contents

About the Editors

Qing Yu

Dr. Yu is currently an associate professor at the College of Mechanical Engineering and Automation, Huaqiao University, who received his Ph.D. degree Hefei University of Technology in 2011. His research interests include Multispectral Detection, Machine Vision, and Precision Measurement.

Ran Tu

Dr. Tu is currently an associate professor at the College of Mechanical Engineering and Automation, Huaqiao University, who received his Ph.D. degree in Safety Science and Engineering from University of Science and Technology of China (USTC) in 2012. His main research interests include Safety Detection Technology, Thermal Science, Energy Conversion in EVs, Combustion Science and Technology, and Plasma Assisted Combustion.

Ting Liu

Dr. Liu obtained her Ph.D degree from Tsinghua University, China, in 2014 where she performed research on optical fiber sensing. After graduation, she joined the Measurement and Control Technology and Instrumention in the College of Mechanical Engineering and Automation of Huaqiao University, China, in 2014. Then she was appointed as an associate professor in 2020. Her research interests include optical fiber sensing, intelligent detection and intelligent manufacturing .

Lina Li

Dr. Li received her Ph.D degree in Instrument Science and Technology from Beihang University in 2011. And she is currently working in the Department of Intelligent Manufacturing at the College of Mechanical Engineering and Automation, Huaqiao University. Her main research interests include spectroscopy, Visible-NIR spectral analysis technology, photoelectric detection instrument.

sensors

MDPI

Article

Soil Organic Carbon Prediction Based on Vis–NIR Spectral Classification Data Using GWPCA–FCM Algorithm

Yutong Miao [1,†], Haoyu Wang [1,†], Xiaona Huang [2], Kexin Liu [1], Qian Sun [1], Lingtong Meng [1] and Dongyun Xu [1,*]

1 College of Resources and Environment, Shandong Agricultural University, Tai'an 271000, China; mytshcc@163.com (Y.M.); 2023120270@sdau.edu.cn (H.W.); 2023120261@sdau.edu.cn (K.L.); 2023120267@sdau.edu.cn (Q.S.); 2021110735@sdau.edu.cn (L.M.)

2 Weicheng Branch of the Weifang Natural Resources and Planning Bureau, Weifang 261000, China; hxiaona123@163.com

* Correspondence: xudongyun@sdau.edu.cn

† These authors contributed equally to this work.

Abstract: Soil visible and near–infrared reflectance spectroscopy is an effective tool for the rapid estimation of soil organic carbon (SOC). The development of spectroscopic technology has increased the application of spectral libraries for SOC research. However, the direct application of spectral libraries for SOC prediction remains challenging due to the high variability in soil types and soil–forming factors. This study aims to address this challenge by improving SOC prediction accuracy through spectral classification. We utilized the European Land Use and Cover Area frame Survey (LUCAS) large–scale spectral library and employed a geographically weighted principal component analysis (GWPCA) combined with a fuzzy c–means (FCM) clustering algorithm to classify the spectra. Subsequently, we used partial least squares regression (PLSR) and the Cubist model for SOC prediction. Additionally, we classified the soil data by land cover types and compared the classification prediction results with those obtained from spectral classification. The results showed that (1) the GWPCA–FCM–Cubist model yielded the best predictions, with an average accuracy of R^2 = 0.83 and RPIQ = 2.95, representing improvements of 10.33% and 18.00% in R^2 and RPIQ, respectively, compared to unclassified full sample modeling. (2) The accuracy of spectral classification modeling based on GWPCA–FCM was significantly superior to that of land cover type classification modeling. Specifically, there was a 7.64% and 14.22% improvement in R^2 and RPIQ, respectively, under PLSR, and a 13.36% and 29.10% improvement in R^2 and RPIQ, respectively, under Cubist. (3) Overall, the prediction accuracy of Cubist models was better than that of PLSR models. These findings indicate that the application of GWPCA and FCM clustering in conjunction with the Cubist modeling technique can significantly enhance the prediction accuracy of SOC from large–scale spectral libraries.

Keywords: soil spectroscopy; LUCAS; GWPCA; FCM

Citation: Miao, Y.; Wang, H.; Huang, X.; Liu, K.; Sun, Q.; Meng, L.; Xu, D. Soil Organic Carbon Prediction Based on Vis–NIR Spectral Classification Data Using GWPCA–FCM Algorithm. *Sensors* **2024**, *24*, 4930. https://doi.org/10.3390/s24154930

Academic Editor: Somsubhra Chakraborty

Received: 4 June 2024
Revised: 20 July 2024
Accepted: 26 July 2024
Published: 30 July 2024

Copyright: © 2024 by the authors. Licensee MDPI, Basel, Switzerland. This article is an open access article distributed under the terms and conditions of the Creative Commons Attribution (CC BY) license (https://creativecommons.org/licenses/by/4.0/).

1. Introduction

Soil organic carbon (SOC) plays a key role in terrestrial ecosystems [1] and is essential for food, soil, water, and energy security [2]. The fast and accurate determination of SOC is important for global food supply and environmental protection [3]. Traditionally, SOC measurements were based on laborious soil sampling and complicated laboratory chemical analysis, which were time–consuming, costly, and environmentally unfriendly [4–6].

As an effective alternative, visible and near–infrared reflectance (Vis–NIR) spectra have been widely used in soil property prediction [7,8] because of their fast, convenient, and inexpensive advantages [9]. To better understand and analyze soil properties using Vis–NIR spectra, researchers developed and analyzed various spectral libraries at regional, continental, national, and global scales [5,10,11]. Previous studies, such as those by Clingensmith et al. [12] and Sarkodie et al. [13], used large–sample soil spectral libraries from the U.S. to predict SOC content through direct modeling, but the results were unsatisfactory due to the diversity of spectra and the complex relationships between soil properties and soil spectra [14].

To improve the accuracy of soil property predictions from large soil databases, researchers have explored various methods. One approach is to enhance model accuracy using deep learning and machine learning algorithms [15]. For example, Wang et al. [16] combined nonlinear modeling with memory–based learning in regional spectral libraries to predict soil pH, soil organic matter, and other properties, and achieved successful results. Another approach is based on classification principles and aims to improve model accuracy through classification modeling [17,18]. For example, Ogen et al. [19] used a spectral angle mapper algorithm, spectral gradient, and fuzzy *k*–means clusters for spectral clustering, followed by modeling. Stevens et al. [20] used the Land Use and Cover Area frame Survey (LUCAS) spectral library to predict SOC content and suggested that, in subsequent studies, researchers could improve prediction accuracy by classifying samples according to the soil type or SOC content and then locally modeling the samples. Previous research has shown that local modeling through classification yields superior results. For example, Shi et al. [10] used principal component analysis (PCA) and the fuzzy *k*–means method to predict soil organic matter for 1581 soil samples from 14 provinces in China. This combination significantly improved prediction accuracy (R^2 = 0.899; RPD = 3.158) compared with using partial least squares regression (PLSR) alone (R^2 = 0.697; RPD = 1.817). Liu et al. [21] established the Chinese forest soil spectral library containing 11, 213 soil samples and combined the density peaks clustering algorithm with the Cubist model. The prediction ability of SOC content (R^2 = 0.96, RPIQ = 5.83) improved significantly compared to the traditional global PLSR modeling method (R^2 = 0.75, RPIQ = 1.95). In these studies, researchers have proved that spectral classification using soil reflectance spectral properties significantly enhances the accuracy of model predictions [22,23].

The key to effective classification lies in how the spectral features are used to classify soil samples. Because of the multidimensionality of spectral data, data dimensionality reduction is performed prior to classification. Previous studies mainly used PCA for data dimensionality reduction [10,24]. However, PCA does not consider the spatial variation of locations, which can result in less accurate principal component extraction, especially on large scales. Our study is the first attempt to apply geographically weighted principal component analysis (GWPCA) for spectral data processing. GWPCA considers the uniqueness and spatial variation of locations, overcoming the limitations of traditional PCA in determining the weights of spatially varying indicators [25,26]. Therefore, we used GWPCA to extract principal components and reduce the dimensionality of spectral data.

Fuzzy c–means (FCM) is a clustering method based on fuzzy theory that demonstrates good adaptability by assigning membership values to each sample, thereby effectively clustering the data [27]. Previous studies have shown the excellent results of FCM in the digital mapping of taxonomic soil units and the delineation of natural soil environments [28]. However, few studies have used FCM for spectral classification to predict soil properties. In this study, we propose a spectral classification method based on GWPCA–FCM to classify and localize the modeling of large–scale soil spectral libraries in a simple, fast, and accurate manner.

Various data analysis techniques have been explored continuously to build predictive models of soil properties using Vis–NIR, such as convolutional neural networks [29,30], multiple linear regression [31,32], and PLSR [8,33]. PLSR is the most widely used linear model, and for high–dimensional multicollinearity, it is more stable and has higher prediction accuracy than traditional methods [23]. However, in spectral libraries with large–scale samples, the multivariate nature of soil spectral data and the nonlinear relationship between soil properties and spectral data [14] make it difficult for the linear model to directly explain the relationship between the spectra and soil properties [34]. Cubist is an advanced, nonparametric regression tree algorithm that can handle nonlinear relationships [35]. Peng et al. [36] found that Cubist achieved the best performance in modeling when PLSR, random forest (RF), and Cubist were used to predict soil salinity. There is no consensus on the best spectroscopic calibration method. In this study, we use the classical linear model PLSR and the machine learning method Cubist to establish SOC prediction models and compare the prediction effects for linear and nonlinear models.

Therefore, we aim to integrate classification approaches and machine learning techniques to provide an effective and accurate spectral prediction method for SOC content in large–scale regions. The main objectives are as follows: (1) to evaluate the performance of GWPCA–FCM in improving SOC spectroscopic prediction based on the LUCAS large–scale spectral library; (2) to compare the prediction accuracy of spectroscopic models based on spectra classification and land cover type classification; and (3) to determine the optimal modeling strategy for SOC prediction.

2. Materials and Methods

2.1. Study Area

We obtained data from LUCAS conducted by the Statistical Office of the European Union in 2008–2012 and used indoor soil Vis–NIR (400–2500 nm) spectra for the study [37]. The study area covered 23 Member States of the European Union, including Sweden, Spain, and the Netherlands. The terrain of the study area is diverse, with plains dominating and little relief, and land cover types such as cropland, woodland, and grassland dominating, with a total area of 4.38 million km^2. The climate types are complex and diverse, covering 35 climate zones, most of which have temperate oceanic climates, with warm winters and cool summers, and a small annual temperature difference. A few of them have Mediterranean, temperate continental, and polar climates. The region is characterized by major European soil types, such as gray soil, brown soil, desert soil, charcoal soil, and chestnut calcium soil. The main crops grown in the region include wheat, maize, and sugar beet. See Figure 1.

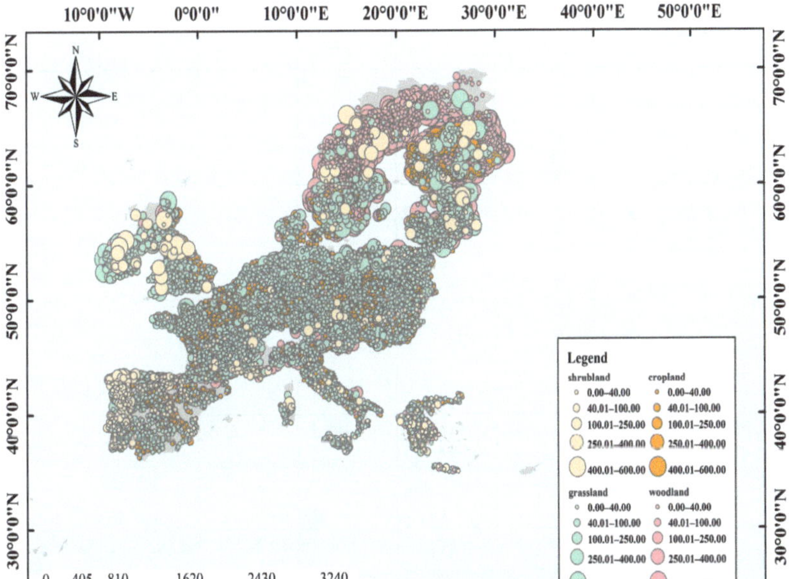

Figure 1. Location of soil samples from the Land Use and Cover Area frame Survey (LUCAS) soil spectral library. The color indicates the corresponding land cover type.

2.2. Soil Sampling and Spectra Measurement

A total of 19,967 topsoil samples (0–20 cm) were collected for different land cover types, including cropland, woodland, shrubland, and grassland, which covered all major soil types in Europe. After the sampling locations were determined using a multi–stage stratified random sampling approach, five soil samples were taken within 2 m of the sampling points using the criss–cross method. The latitude and longitude of the sampling points were recorded using GPS during sampling.

The soil samples were air–dried and sieved (2 mm) following the protocol described by the manufacturer and the Soil Spectroscopy Group [38]. The Vis–NIR soil spectra were measured using a FOSS XDS Rapid Content Analyzer (FOSS NIR Systems Inc., Laurel, MD, USA), operating in the 400–2500 nm wavelength range with a spectral resolution of 0.5 nm.

2.3. Spectral Data Preprocessing

To reduce data redundancy and improve modeling efficiency, the original spectral reflectance was first resampled with a resampling interval of 10 nm. The resampled spectra were transformed to absorbance ($\log(1/R)$), which can enlarge the spectral differences between samples and highlight the spectral characteristics of the soil. To remove background and noise effects, such as environmental factors, intrinsic factors of the samples' own reflections, and the electrical noise of the spectrometer, the spectra were smoothed using the Savitzky–Golay algorithm which was fitted using a constant with a polynomial of order 2 and a window size of 15.

2.4. GWPCA

GWPCA aims to account for certain spatial heterogeneity in data and is one of the main methods for multivariate data analysis [39]. Unlike conventional PCA, GWPCA takes into account spatial variation in the covariance structure of the variables, whose covariances are appropriately weighted using a distance function between the target and neighboring variables:

$$\sum u, v = \mathbf{X}^{\mathrm{T}}\mathbf{W}(u,v)\mathbf{X}, \tag{1}$$

where $\mathbf{X}$ is an $n \times m$ matrix, n is the number of samples, m is the number of variables, and $\mathbf{W}(u,v)$ represents the diagonal matrix of distance weights at position coordinates (u,v). At position (u_i, v_i), GWPCA defines the local feature structure as follows:

$$\mathbf{L}(u_i,v_i)\mathbf{V}(u_i,v_i)\mathbf{L}^{\mathrm{T}}(u_i,v_i) = \sum(u_i,v_i), \tag{2}$$

where $\mathbf{L}(u_i,v_i)$ is the eigenvector matrix, which represents the loading of each independent variable for each principal component. $\mathbf{V}(u_i,v_i)$ is the diagonal matrix of eigenvalues. The score of each principal component can be expressed as

$$\mathbf{Z}(u_i,v_i) = \mathbf{X}(u_i,v_i)\mathbf{L}(u_i,v_i). \tag{3}$$

Prior to the application of PCA, first, the data are standardized for the independent variables, and PCA is specified using the covariance matrix. The number of principal components is determined using the magnitude of the eigenvalues (eigenvalues greater than one are chosen for this study). The optimal bandwidth of the retained principal components is chosen based on the weighting function "bisquare". The bisquare kernel function is given by

$$w_{ij} = \begin{cases} \left(1 - \left(\frac{d_{ij}}{b}\right)^2\right)^2, \; if\left|d_{ij}\right| < 0 \\ 0, \; otherwise \end{cases} \tag{4}$$

where the bandwidth is the geographical distance b, and d_{ij} is the distance between the spatial locations of the ith and jth row in the data matrix. The final results of the principal component scores for each variable at each point are used as inputs to the FCM algorithm for cluster analysis.

The GWPCA was implemented in R using the "GW model" package [40].

2.5. Spectral Classification Methods

In this study, the FCM classification method was used to optimize the clustering results by minimizing an objective function that contained the distance between the affiliation degree and clustering centers and to provide an analytical index of the optimal number of classifications. The value function of the FCM method can be expressed as

$$J^{m}_{PCM(U,C,X)} = \sum_{i=1}^{c} \sum_{j=1}^{n} \mu_{ij}^{m} d_{ij}^{2}, \tag{5}$$

where n is the number of sample points, which is the number of principal components used in this study. C is the number of classifications; μ_{ij} indicates the degree to which x_j belongs to X_i, and must satisfy $0 \leq \mu_{ij} \leq 1$ and $\sum_{i=1}^{c} \mu_{ij} = 1, \forall_j = 1, 2, \ldots, n$; and d_{ij}^2 is equal to the square of the Euclidean distance from x_j to the center of X_i clustering. m is the degree of fuzziness, which is a parameter that controls the flexibility of the algorithm. If $m = 1$, the result of FCM is hard c–means clustering, and as m increases, the clustering results become fuzzier.

The validity of the clustering results is evaluated using a fuzzy performance index (FPI) and normalized classification entropy (NCE):

$$\mathrm{FPI} = 1 - \frac{c}{c-1}\left[1 - \frac{\sum_{k=1}^{n}\sum_{i=1}^{c}(\mu_{ik})^2}{n}\right]$$

$$\mathrm{NCE} = \frac{n}{n-c}\left[-\frac{\sum_{k=1}^{n}\sum_{i=1}^{c}\mu_{ik}\log_a(\mu_{ik})}{n}\right], \tag{6}$$

where c is the number of clusters, n is the number of samples, and μ_{ik} is the fuzzy affiliation degree. FPI represents the degree of separation between the c clusters in the data matrix and ranges from 0 to 1. The closer the value of FPI to 0, the less data are shared by the clusters, and the division of the clusters is obvious. The opposite case indicates that the division is ineffective. NCE is used to estimate the amount of decomposition of fuzzy c–partitioning. The smaller the value of NCE, the better the clustering effect.

2.6. Model Construction and Evaluation

In this study, PLSR and Cubist were used to model the inversion of SOC content and compare prediction accuracies. Before modeling, the sample soil spectra and corresponding SOC data were divided into a modeling dataset and validation dataset, and the Kennard–Stone (KS) algorithm was used to divide the data. The Euclidean distance between the spectral variables of the samples was computed. The modeling samples were selected uniformly in the feature space of the samples. The ratio of the number of modeling sets to the number of validation sets was 2:1. A log transformation was performed on the dependent variable SOC to make it conform to the normal distribution.

PLSR is a multivariate statistical analysis method with wide applicability that was proposed by Wold and Alban in 1983 and has been developed in recent years. It combines PCA, multiple linear regression, and typical correlation analysis into one regression model for solving the problem of multicollinearity encountered in multiple linear regression analysis. The PLSR algorithm integrates compression and regression steps and selects continuous orthogonal factors to maximize the covariance between the predictor and response variables [41], which are used to build a predictive model for predicting the values of the response variable. It has the advantages of simplicity and stability, easy qualitative interpretation, higher prediction accuracy, and is suitable for spectral analysis, which has more independent variables [21].

The Cubist model is a comprehensive decision tree–based learning algorithm that predicts or categorizes data by constructing multiple decision trees. A decision tree represents a segmented multivariate linear function that predicts the value of a variable through a series of independent variables. The basic concept is to create subsets of samples with similar attributes in the original data set when a variable is predicted through the constructed multivariate linear function [42], and then model each subset separately. The training rule is simple and effective and is suitable for solving the problem of the nonlinear relationship between SOC and predictor variables.

In this study, the root mean square error (RMSE), coefficient of determination (R^2), and ratio of performance to inter–quartile distance (RPIQ) were used to verify the performance of the models. The higher the R^2 and lower the RMSE, the higher the model prediction accuracy. RPIQ takes into account both the prediction error and variation in observations, thereby providing a more objective and easier approach to compare model effectiveness metrics in model validation studies. The larger the RPIQ, the better the predictive ability of the model. According to Salazar et al. [43], prediction ability can be divided into four categories based on the RPIQ value: RPIQ < 1.5 indicates that the model is very bad; if RPIQ is between 1.5 and 2.0, this indicates that the model is poor; if RPIQ is between 2.0 and 2.5, this indicates that the model is good; and RPIQ > 2.5 indicates that the model is very good.

2.7. Important Band Analysis

RF was used in this study for band importance analysis. It is an integrated machine learning algorithm that can aggregate ideas in addition to solving classification and regression problems [44]. It is applied gradually in feature importance analysis [45]. RF can quantify the degree of contribution of input features to the model output. The importance of variables is influenced by two main parameters: the size of the subset of input variables (mtry) and the number of trees in the forest (ntree).

3. Results and Discussion

3.1. Statistical Analysis of SOC Content

The histograms of raw SOC and log–transformed logSOC distributions are shown in Figure 2. The raw SOC data were positively skewed, and the logSOC data essentially conformed to a normal distribution, which satisfied the requirements of data analysis and prediction modeling [24]. Specifically, the SOC content ranged from 0 to 586.8 g kg^{-1}, with a mean, median, and standard deviation of 50 g kg^{-1}, 20.8 g kg^{-1}, and 91.3 g kg^{-1}, respectively. The skewness coefficient was 3.68, and the kurtosis coefficient was 13.46. The logSOC content ranged from 0.69 to 6.37 g kg^{-1}, with a mean value of 3.22 g kg^{-1}, standard deviation of 1.01 g kg^{-1}, and skewness coefficient and kurtosis coefficient of 0.997 and 1.13, respectively.

To enable a further comparison and analysis of the differences in the distribution of SOC content, soil samples from four land cover types were divided. The distribution characteristics of grassland, cropland, shrubland, and woodland logSOC content are shown in Figure 3b. The distribution of SOC content in different land cover types showed obvious differences. The woodland SOC content distribution was the most dispersed, with a median value of 3.77 g kg^{-1} and the highest mean value of 3.90 g kg^{-1}. This is because of the enhancement of the CO_2 absorption capacity of woodland in the atmosphere [46]. The woodland samples were mainly taken from evergreen broadleaf forests and coniferous forests, which had a higher content of organic carbon than the other forests. Additionally, woodland was less affected by human influence and had a greater accumulation than depletion of organic carbon, which led to the highest content of organic carbon. This was followed by shrubland (mean = 3.49 g kg^{-1}) and grassland (mean = 3.33 g kg^{-1}). Cropland had the lowest mean SOC content of 2.72 g kg^{-1}, which was mainly attributed to the generally higher carbon loss caused by soil disturbance that resulted from traditional farmland management compared with grassland and forest ecosystems [47].

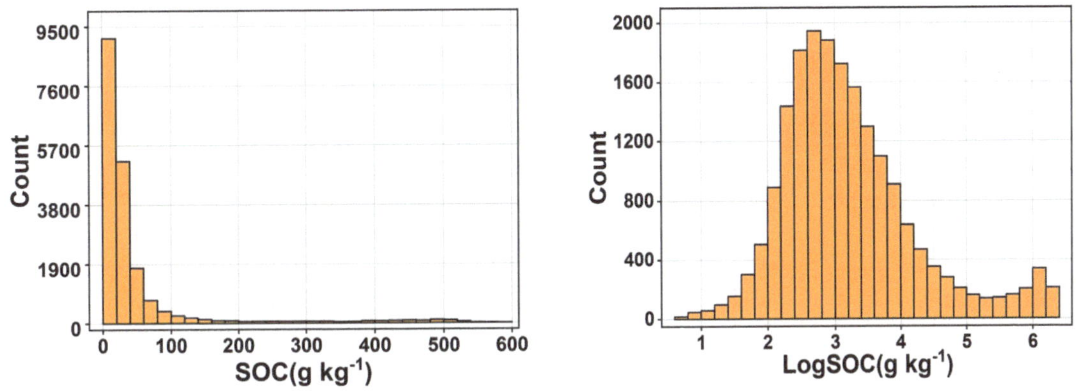

(a) (b)

Figure 2. Histogram of soil organic carbon content SOC (**a**) and log–transformed logSOC (**b**).

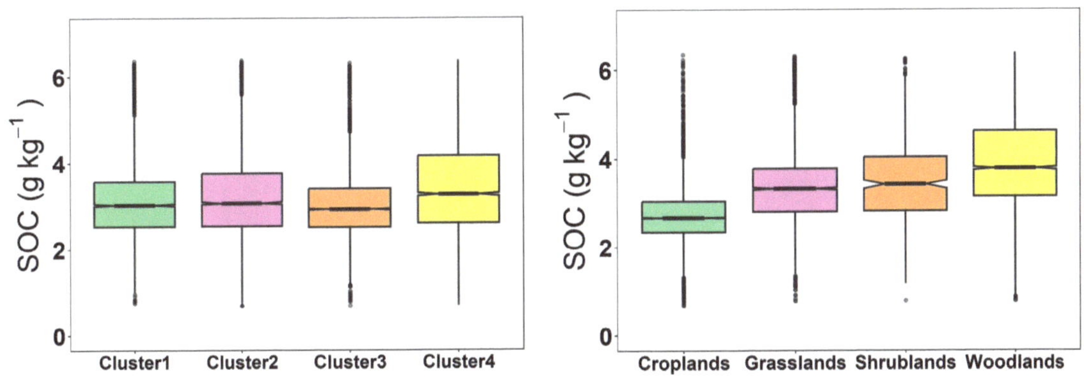

(a) (b)

Figure 3. Box line plot of SOC content for land cover type (**a**) and spectral cluster (**b**).

3.2. Spectra Classification and SOC Content

Given the considerable volume of data in the soil spectral library and the issue of multiplicity correlation between bands, all soil samples were subjected to data dimensionality reduction using GWPCA. The cumulative contribution of the first four principal components exceeded 99.45%. The FCM was then used for rational classification of soil spectra and the determination of the optimal number of classifications. The data from the first four principal components were imported for FCM cluster analysis. The maximum number of iterations was 300, the convergence threshold was 0.001, and the fuzzy weighting index was 1.5 [48]. To determine the optimal number of categories, all the samples were divided into 2, 3, 4, 5, 6, 7, 8, 9, and 10 clusters. The values of FPI and NCE for each cluster are shown in Figure 4. The optimal number of clusters in this study was finally determined to be 4 by taking into consideration the matching of the land cover types and the clustering index.

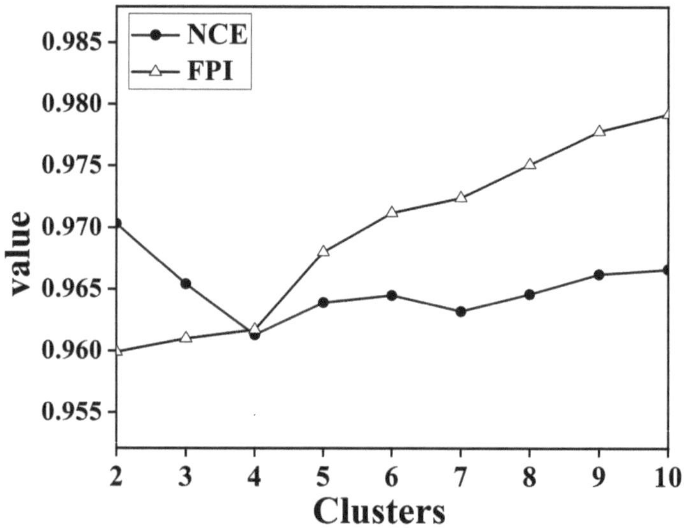

Figure 4. Plot of NCE and FPI value versus the number of clusters.

The distribution characteristics of the SOC content of soil samples for the four clusters are shown in Figure 3a. After GWPCA–FCM was used, the difference in SOC content was obviously reduced, and the distribution was essentially the same. For Cluster 3, which consisted of 5145 soil samples, 4778 soil samples were mainly from four land cover types, that is, cropland (2173 samples), grassland (1111 samples), shrubland (117 samples), and woodland (1377 samples), which accounted for 93% of the soil samples for Cluster 3. The distribution of SOC content was most concentrated, with the lowest mean value of only 2.99 g kg^{-1}. This is mainly because cropland soil samples account for 42% of the total number of soil samples, which is greatly influenced by cropland. Cluster 4 consisted of 4992 soil samples, mainly from four land cover types, that is, cropland (1605 samples), grassland (1235 samples), shrubland (113 samples), and woodland (1629 samples), for a total of 4582 samples, which accounted for 92% of the total soil samples for Cluster 4. The distribution of SOC was relatively scattered, with the highest mean value of 3.50 g kg^{-1}. This is attributed to the fact that the woodland and grassland soil samples accounted for 57% of the total soil samples and were highly influenced by woodland and grassland. The SOC contents for Cluster 1 and Cluster 2 were comparable, with average contents of 3.15 and 3.24 g kg^{-1}, respectively. This is mainly because the proportion of woodland and grassland soil samples from Cluster 2 is larger than Cluster 1.

3.3. Spectral Characteristics of Different Soil Types

The average spectral reflectance and its range of variation for each type of soil sample based on land cover type and GWPCA–FCM are shown in Figure 5. The spectral curves obtained by the two classification methods had basically the same morphology; however, the difference in spectral reflectance based on GWPCA–FCM was more obvious. Under the GWPCA–FCM classification, the slope of the curve was large, and the reflectance increased rapidly in the visible range. In the near–infrared region, the curve tended to flatten and the reflectance increased slowly; between 700 nm and 2500 nm, the difference in spectral reflectance gradually increased. The spectral curves had distinct absorption valleys near 1400 nm, 1900 nm, and 2200 nm, which were mainly caused by moisture, organic matter, iron oxides, and clay fractions [49]. The highest spectral reflectance was found in Cluster 3, followed by Clusters 1 and 2, and the lowest spectral reflectance was found in Cluster 4. The average spectral reflectance of Cluster1, Cluster2, Cluster3, and Cluster4 was 3.15 g kg^{-1}, 3.24 g kg^{-1}, 2.99 g kg^{-1}, and 3.50 g kg^{-1} respectively, which showed a significant negative correlation with the increase in SOC content. The position of spectral curves decreased, and the spectral reflectance decreased.

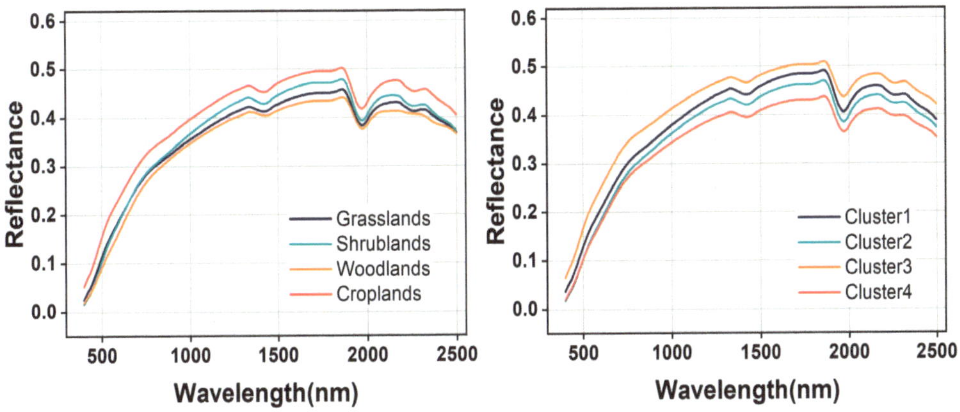

Figure 5. Comparison of mean soil reflectance curves based on the land cover type and GWPCA–FCM classification.

For the land cover type, four land cover types had a similar curve shape. In the range of 400–750 nm, reflectance values increase rapidly, while they decrease slowly in the range of 800–1800 nm. Absorption features could be identified near 1400 and 1900 nm, which are assigned to soil hygroscopic water in clay minerals [50]. The highest spectral reflectance was found in cropland, followed by shrubland and grassland, and the lowest in woodland, which is consistent with the findings of Liu et al. [51]. This is because cropland soils have a lower mean SOC content (2.72 g kg^{-1}) than woodland soils (3.94 g kg^{-1}), shrubland soils (3.49 g kg^{-1}), and grassland soils (3.33 g kg^{-1}).

3.4. Spectral Prediction of SOC

To further explore the spectral prediction effect after GWPCA–FCM classification, the KS algorithm was used to divide the datasets of the four clusters into a modeling set and a validation set in a 2:1 ratio. The SOC spectral prediction models for the clusters were established using two methods: PLSR and Cubist. Additionally, the SOC spectral prediction models of the four land cover types were also established for comparative analysis.

The introduction of GWPCA–FCM significantly improved prediction accuracy. Specifically, the use of PLSR improved the R^2 mean from 0.72 to 0.74 and the RPIQ mean from 2.36 to 2.43 compared with unclassified global modeling. Similar conclusions were reached by Ward et al. [24] and Liu et al. [52], who modeled spectral classification based on *k*–means clustering with PCA and found that the prediction of SOC improved, while significantly reducing the algorithm's run time. By contrast, the accuracy of land cover type classification did not improve but declined, with the R^2 reducing from 0.72 to 0.69 and the RPIQ reducing from 2.36 to 2.13. Using Cubist, compared to unclassified global modeling, R^2 increased by 10.33%, RMSE decreased by 17.42%, and RPIQ increased by 18.00%. However, the accuracy of land cover type classification did not improve, with R^2 decreasing from 0.75 to 0.73, RMSE improving from 0.33 g kg^{-1} to 0.39 g kg^{-1}, and RPIQ reducing from 2.50 to 2.29. Stenberg et al. [53] noted that the prediction error of spectral models increases as the standard deviation of predicted soil properties increases. Ignoring the spatial extent and distribution of samples, the large variation in SOC content across land cover types resulted in a decrease in prediction accuracy. As shown in Figures 6 and 7, after the land cover classification, the scattered points were distributed in a certain area, and the whole was more dispersed, indicating that the correlation between the predicted values and the measured values is weak. After the classification based on GWPCA–FCM, the distribution of points was more concentrated in a straight line, and the trend line was closer to a 1:1

line compared with the land cover classification, indicating that the predicted values were closer to the measured values as a whole, and the prediction effect was better.

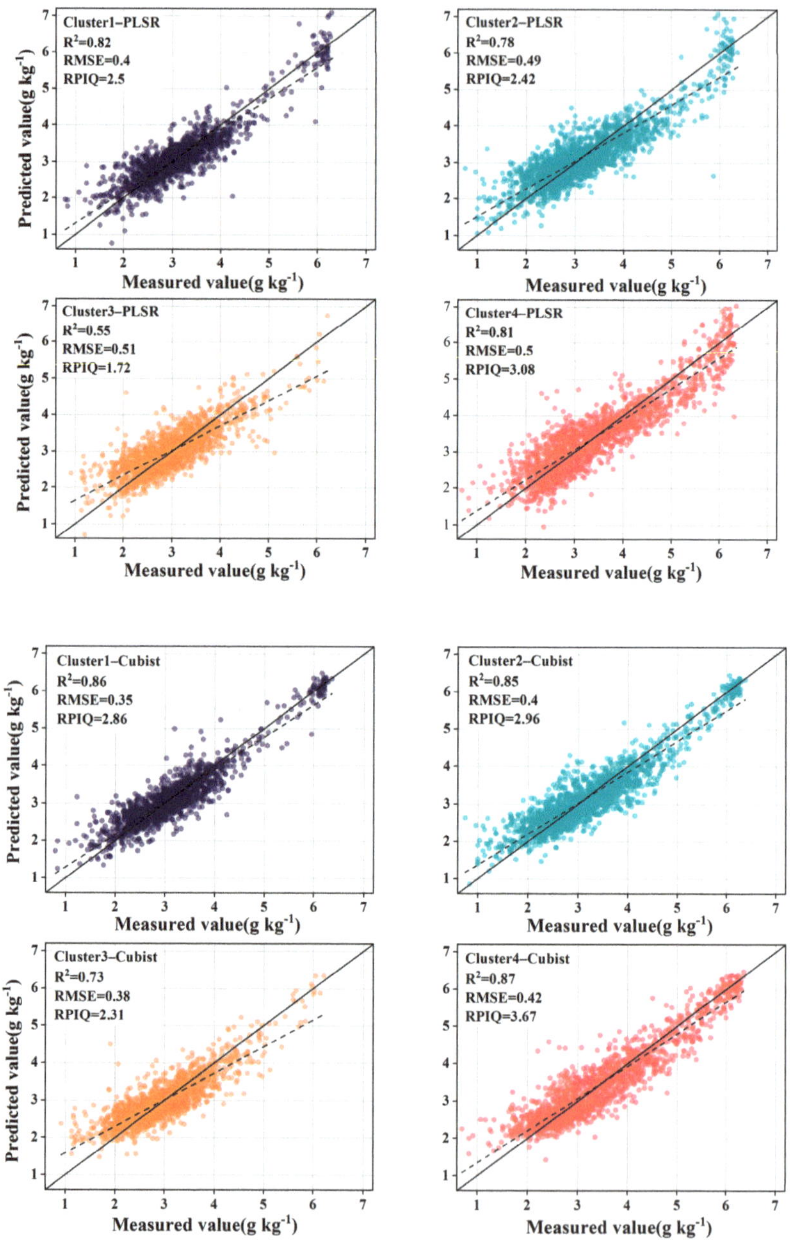

Figure 6. Observed vs. predicted SOC values of the validation samples for the GWPCA–FCM approaches.

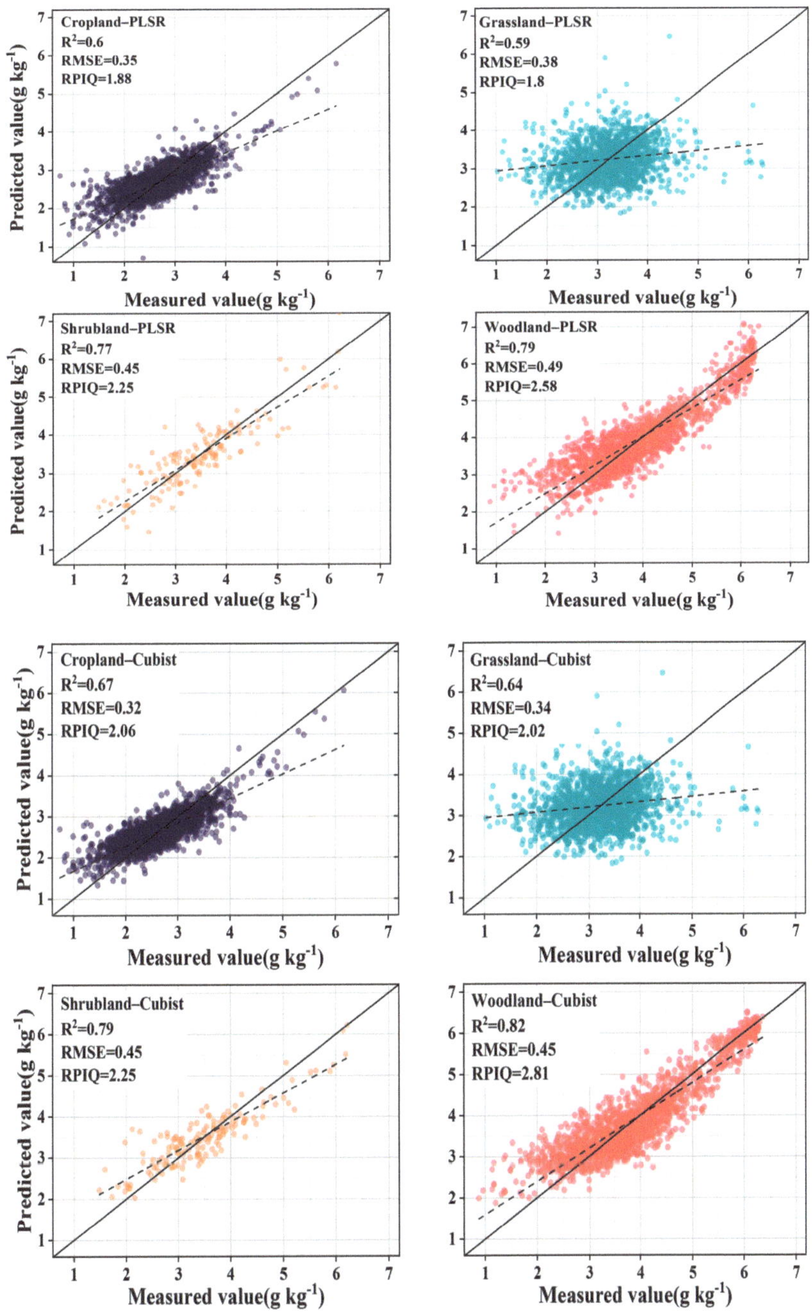

Figure 7. Observed vs. predicted SOC values of the validation samples for land cover type approaches.

The prediction performance of PLSR and Cubist was explored further. For GWPCA–FCM–Cubist compared with GWPCA–FCM–PLSR, the mean value of R^2 improved from 0.74 to 0.83, the mean value of RMSE decreased from 0.48 g kg^{-1} to 0.39 g kg^{-1}, and the mean value of RPIQ improved from 2.43 to 2.95, which is consistent with previous research results [23,54]. This is because the performance of PLSR is affected by multicollinearity [55].

Different data types, differences in dataset sizes, and the distribution of organic carbon content can have multiple effects on the prediction accuracy of the model. The results show that Cubist achieved higher accuracy when the spectral prediction of SOC content was performed in the context of large spatial and temporal variability, significant spatial heterogeneity, and the large data volume of the European LUCAS spectral library. See Tables 1 and 2.

Table 1. PLSR and Cubist predictions of SOC for different spectral clusters.

Subsets	N	Cubist			PLSR		
		R^2	RMSE	RPIQ	R^2	RMSE	RPIQ
All	18,921	0.75	0.33	2.5	0.72	0.35	2.36
Cluster1	3870	0.86	0.35	2.86	0.82	0.4	2.5
Cluster2	4914	0.85	0.4	2.96	0.78	0.49	2.42
Cluster3	5145	0.73	0.38	2.31	0.55	0.51	1.72
Cluster4	4992	0.87	0.42	3.67	0.81	0.5	3.08
Mean	4730	0.83	0.39	2.95	0.74	0.48	2.43

Table 2. PLSR and Cubist predictions of SOC for different land cover types.

Subsets	N	Cubist			PLSR		
		R^2	RMSE	RPIQ	R^2	RMSE	RPIQ
All	18,921	0.75	0.33	2.5	0.72	0.35	2.36
Cropland	7476	0.67	0.32	2.06	0.6	0.35	1.88
Grassland	4200	0.64	0.34	2.02	0.59	0.38	1.8
Shrubland	443	0.79	0.45	2.25	0.77	0.45	2.25
Woodland	5218	0.82	0.45	2.81	0.79	0.49	2.58
Mean	4334	0.73	0.39	2.29	0.69	0.42	2.13

3.5. Important Band of SOC for Each Soil Type in RF Models

The results of the land cover classification and GWPCA–FCM spectral classification of importance bands in RF models are shown in Figure 8. In Figure 8a, the curves of the four land cover types are significantly different. For grassland, the most important bands were mainly distributed in the regions of 400–550 nm and 2210–2350 nm, with obvious peaks and valleys at 540, 1480, and 1990 nm, which may be influenced by hydroxyl vibration in the samples. For woodland, the most important bands were mainly distributed in the regions of 490–590 nm and 830–1030 nm, with distinct peaks and valleys at 920, 1660, and 1850 nm. For shrubland, the most important bands were mainly distributed in the regions of 520–610 nm and 790–940 nm, with distinct peaks and valleys at 610, 1370, and 2040 nm. For cropland, the most important bands were mainly distributed in the regions of 400–570 nm and 1800–1870 nm, with distinct peaks and valleys at 480, 980, and 1560 nm.

As shown in Figure 8b, the curves did not differ much based on GWPCA–FCM, and the distributions of higher values of feature importance, peaks, and valleys were essentially the same. The most important bands were all mainly distributed in the spectral regions of 400–600 nm and 2200–2340 nm, and obvious peaks appeared near 530, 1330, and 2030 nm. There were obvious valleys near 670, 1230, and 1970 nm, which were because of the content of, for example, organic carbon, iron oxides, and clay minerals. The features in the near–infrared band were mainly caused by the multiplicative or combined frequency absorption of the molecular vibrations of C–H, N–H, C–O, O–H, and Fe–O groups in minerals [53,54,56]. The higher importance value of the features in the 400–600 nm band was mainly affected by soil carbon and iron oxides. The main moisture absorption bands were near the 1400 nm and 1950 nm bands. Moisture absorbed electromagnetic waves in this band, and the combined frequency jump of stretching vibration and corner vibration of O–H functional groups in water molecules formed the largest absorption coefficient in the near–infrared region. The absorption band of Al–OH clay minerals mainly existed near 2200 nm in the synchrotron region, and an organic matter–related C–H characteristic peak existed near 2300 nm [52].

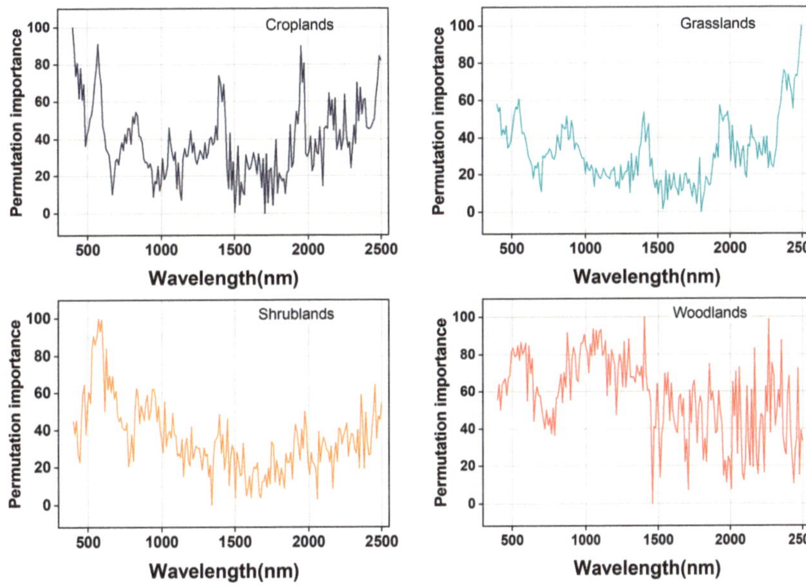

(**a**) Land cover type

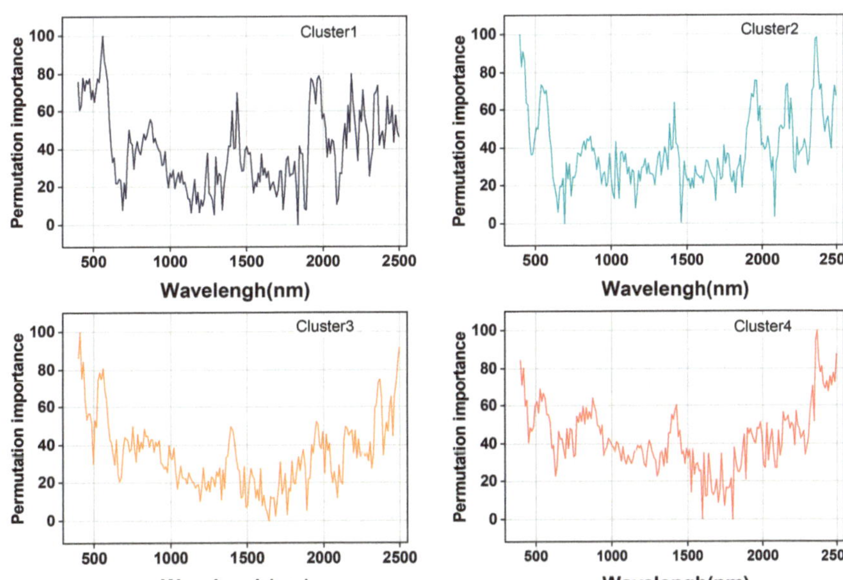

(**b**) GWPCA–FCM cluster

Figure 8. Important band diagram in RF models.

4. Conclusions

In this study, we used PLSR and Cubist models to compare SOC prediction accuracy based on full sample data, land cover classification data, and spectral classification data by GWPCA–FCM in a large spectral library. The main conclusions are as follows:

(1) The prediction accuracy of the GWPCA–FCM classification model was significantly higher than that of the unclassified global model and the land cover type classification model. This approach enhanced the accuracy of SOC predictions for large spectral libraries. (2) Among the modeling approaches, Cubist was found to be superior to PLSR, with the GWPCA–FCM–Cubist model achieving the optimal prediction results. This research underscores the potential of integrating advanced data reduction and classification techniques with robust modeling algorithms to improve the precision of SOC content prediction on a large scale.

Author Contributions: Conceptualization, D.X., Y.M. and H.W.; methodology, Y.M., H.W. and K.L.; software, K.L., Q.S. and L.M.; validation, H.W., X.H. and K.L.; formal analysis, Y.M., X.H., Q.S. and L.M.; investigation, X.H., K.L. and Q.S.; resources, D.X.; data curation, D.X. and X.H.; writing—original draft preparation, Y.M. and H.W.; writing—review and editing, D.X., X.H., K.L. and Q.S.; visualization, H.W., K.L. and L.M.; supervision, D.X.; project administration, D.X.; funding acquisition, D.X. All authors have read and agreed to the published version of the manuscript.

Funding: This research was funded by the Natural Science Foundation of Shandong Province (ZR2022QD122, ZR2023MD101) and the National Natural Science Foundation of China (42301065).

Institutional Review Board Statement: Not applicable.

Informed Consent Statement: Not applicable.

Data Availability Statement: The LUCAS topsoil dataset used in this work was made available by the European Commission through the European Soil Data Centre managed by the Joint Research Centre (JRC), https://esdac.jrc.ec.europa.eu/content/lucas-2009-topsoil-data, accessed on 1 July 2024.

Conflicts of Interest: The authors declare no conflicts of interest.

References

1. Lu, H.B.; Wang, X.Y.; Zhang, H.C.; Xie, X.H.; Nakhavali, M.; Quine, T.A.; Xu, W.F.; Xia, J.Z.; He, B.; Hao, Z.X.; et al. Soil Organic Carbon Lateral Movement Processes Integrated into a Terrestrial Ecosystem Model. *J. Adv. Model. Earth Syst.* **2024**, *16*, e2023MS003916. [CrossRef]
2. Stockmann, U.; Padarian, J.; McBratney, A.; Minasny, B.; de Brogniez, D.; Montanarella, L.; Hong, S.Y.; Rawlins, B.G.; Field, D.J. Global soil organic carbon assessment. *Glob. Food Secur.-Agric. Policy* **2015**, *6*, 9–16. [CrossRef]
3. Yousra, M.; Hussain, Q.; Khan, K.S.; Ansar, M.; Sarwar, S.; Khan, M.Z. Soil Organic Carbon Pools in Benchmark Soils of Punjab, Pakistan. *Commun. Soil Sci. Plant Anal.* **2023**, *54*, 571–585. [CrossRef]
4. Araújo, S.R.; Wetterlind, J.; Demattê, J.A.M.; Stenberg, B. Improving the prediction performance of a large tropical vis-NIR spectroscopic soil library from Brazil by clustering into smaller subsets or use of data mining calibration techniques. *Eur. J. Soil Sci.* **2014**, *65*, 718–729. [CrossRef]
5. Viscarra Rossel, R.A.; Behrens, T.; Ben-Dor, E.; Brown, D.J.; Demattê, J.A.M.; Shepherd, K.D.; Shi, Z.; Stenberg, B.; Stevens, A.; Adamchuk, V.; et al. A global spectral library to characterize the world's soil. *Earth-Sci. Rev.* **2016**, *155*, 198–230. [CrossRef]
6. Meng, X.T.; Bao, Y.L.; Zhang, X.L.; Wang, X.; Liu, H.J. Prediction of soil organic matter using different soil classification hierarchical level stratification strategies and spectral characteristic parameters. *Geoderma* **2022**, *411*, 115696. [CrossRef]
7. Ding, S.T.; Zhang, X.; Shang, K.; Xiao, Q.; Wang, W.H.; Rehman, A.U.R. Removal of environmental influences for estimating soil texture fractions based on ZY1 satellite hyperspectral images. *Catena* **2024**, *236*, 107713. [CrossRef]
8. Hou, M.J.; Ai, Z.M.; Li, X.H.; Dang, X.H.; Yao, Y.Y.; Deng, Y.; Wang, T.; Li, T.; Xiao, L. Exploring the optimal model for assessing SOC and TN in Zanthoxylum bungeanum forest on the Loess Plateau using VNIR spectroscopy. *Ecol. Inform.* **2024**, *79*, 102429. [CrossRef]
9. Omondiagbe, O.P.; Lilburne, L.; Licorish, S.A.; MacDonell, S.G. Soil texture prediction with automated deep convolutional neural networks and population-based learning. *Geoderma* **2023**, *436*, 116521. [CrossRef]
10. Shi, Z.; Wang, Q.L.; Peng, J.; Ji, W.J.; Liu, H.J.; Li, X.; Viscarra Rossel, R.A. Development of a national VNIR soil-spectral library for soil classification and prediction of organic matter concentrations. *Sci. China-Earth Sci.* **2014**, *57*, 1671–1680. [CrossRef]
11. Wang, S.; Guan, K.Y.; Zhang, C.H.; Lee, D.; Margenot, A.J.; Ge, Y.F.; Peng, J.; Zhou, W.; Zhou, Q.; Huang, Y.Z. Using soil library hyperspectral reflectance and machine learning to predict soil organic carbon: Assessing potential of airborne and spaceborne optical soil sensing. *Remote Sens. Environ.* **2022**, *271*, 116521. [CrossRef]
12. Clingensmith, C.M.; Grunwald, S. Predicting Soil Properties and Interpreting Vis-NIR Models from across Continental United States. *Sensors* **2022**, *22*, 3187. [CrossRef]
13. Sarkodie, V.Y.O.; Vasát, R.; Pouladi, N.; Srámek, V.; Sánka, M.; Fadrhonsová, V.; Hellebrandová, K.N.; Boruvka, L. Predicting soil organic carbon stocks in different layers of forest soils in the Czech Republic. *Geoderma Reg.* **2023**, *34*, e00658. [CrossRef]

14. Nocita, M.; Stevens, A.; Toth, G.; Panagos, P.; van Wesemael, B.; Montanarella, L. Prediction of soil organic carbon content by diffuse reflectance spectroscopy using a local partial least square regression approach. *Soil Biol. Biochem.* **2014**, *68*, 337–347. [CrossRef]
15. Padarian, J.; Minasny, B.; McBratney, A.B. Machine learning and soil sciences: A review aided by machine learning tools. *Soil* **2020**, *6*, 35–52. [CrossRef]
16. Wang, Z.; Chen, S.C.; Lu, R.; Zhang, X.L.; Ma, Y.X.; Shi, Z. Non-linear memory-based learning for predicting soil properties using a regional vis-NIR spectral library. *Geoderma* **2024**, *441*, 116752. [CrossRef]
17. Zeng, R.; Zhang, G.L.; Li, D.C.; Rossiter, D.G.; Zhao, Y.G. How well can VNIR spectroscopy distinguish soil classes? *Biosyst. Eng.* **2016**, *152*, 117–125. [CrossRef]
18. Xu, D.Y.; Ma, W.Z.; Chen, S.C.; Jiang, Q.S.; He, K.; Shi, Z. Assessment of important soil properties related to Chinese Soil Taxonomy based on vis-NIR reflectance spectroscopy. *Comput. Electron. Agric.* **2018**, *144*, 1–8. [CrossRef]
19. Ogen, Y.; Zaluda, J.; Francos, N.; Goldshleger, N.; Ben-Dor, E. Cluster-based spectral models for a robust assessment of soil properties. *Geoderma* **2019**, *340*, 175–184. [CrossRef]
20. Stevens, A.; Nocita, M.; Tóth, G.; Montanarella, L.; van Wesemael, B. Prediction of Soil Organic Carbon at the European Scale by Visible and Near InfraRed Reflectance Spectroscopy. *PLoS ONE* **2013**, *8*, e66409. [CrossRef]
21. Liu, S.S.; Shen, H.H.; Chen, S.C.; Zhao, X.; Biswas, A.; Jia, X.L.; Shi, Z.; Fang, J.Y. Estimating forest soil organic carbon content using vis-NIR spectroscopy: Implications for large-scale soil carbon spectroscopic assessment. *Geoderma* **2019**, *348*, 37–44. [CrossRef]
22. Heil, J.; Häring, V.; Marschner, B.; Stumpe, B. Advantages of fuzzy k-means over k-means clustering in the classification of diffuse reflectance soil spectra: A case study with West African soils. *Geoderma* **2019**, *337*, 11–21. [CrossRef]
23. Peng, L.Y.; Cheng, H.; Wang, L.J.; Zhu, D.Z. Comparisons of the prediction results of soil properties based on fuzzy c-means clustering and expert knowledge from laboratory Visible—Near-Infrared reflectance spectroscopy data. *Can. J. Soil Sci.* **2021**, *101*, 33–44. [CrossRef]
24. Ward, K.J.; Chabrillat, S.; Neumann, C.; Foerster, S. A remote sensing adapted approach for soil organic carbon prediction based on the spectrally clustered LUCAS soil database. *Geoderma* **2019**, *353*, 297–307. [CrossRef]
25. Chen, J.; Qu, M.K.; Zhang, J.L.; Xie, E.Z.; Huang, B.; Zhao, Y.C. Soil fertility quality assessment based on geographically weighted principal component analysis (GWPCA) in large-scale areas. *Catena* **2021**, *201*, 105197. [CrossRef]
26. Foroutan, E.; Hu, T.; Zhang, F.; Yu, H.B. Assessing heat vulnerability in Philadelphia using geographically weighted principal component analysis (GWPCA): A geospatial big data-driven approach. *Int. J. Appl. Earth Obs. Geoinf.* **2024**, *127*, 103653. [CrossRef]
27. Yu, B.; Zheng, Z.J.; Cai, M.J.; Pedrycz, W.; Ding, W.P. FRCM: A fuzzy rough c-means clustering method. *Fuzzy Sets Syst.* **2024**, *480*, 108860. [CrossRef]
28. Horácek, M.; Samec, P.; Minár, J. The mapping of soil taxonomic units via fuzzy clustering—A case study from the Outer Carpathians, Czechia. *Geoderma* **2018**, *326*, 111–122. [CrossRef]
29. Bai, Z.J.; Chen, S.C.; Hong, Y.S.; Hu, B.F.; Luo, D.F.; Peng, J.; Shi, Z. Estimation of soil inorganic carbon with visible near-infrared spectroscopy coupling of variable selection and deep learning in arid region of China. *Geoderma* **2023**, *437*, 116589. [CrossRef]
30. Li, H.; Ju, W.L.; Song, Y.M.; Cao, Y.Y.; Yang, W.; Li, M.Z. Soil organic matter content prediction based on two-branch convolutional neural network combining image and spectral features. *Comput. Electron. Agric.* **2024**, *217*, 108561. [CrossRef]
31. Conforti, M.; Matteucci, G.; Buttafuoco, G. Using laboratory Vis-NIR spectroscopy for monitoring some forest soil properties. *J. Soils Sediments* **2018**, *18*, 1009–1019. [CrossRef]
32. Li, H.; Wang, J.W.; Zhang, J.X.; Liu, T.Q.; Acquah, G.E.; Yuan, H.M. Combining Variable Selection and Multiple Linear Regression for Soil Organic Matter and Total Nitrogen Estimation by DRIFT-MIR Spectroscopy. *Agronomy* **2022**, *12*, 638. [CrossRef]
33. Sidike, A.; Zhao, S.H.; Wen, Y.M. Estimating soil salinity in Pingluo County of China using QuickBird data and soil reflectance spectra. *Int. J. Appl. Earth Obs. Geoinf.* **2014**, *26*, 156–175. [CrossRef]
34. Karray, E.; Elmannai, H.; Toumi, E.; Gharbia, M.H.; Meshoul, S.; Aichi, H.; Ben Rabah, Z. Evaluating the Potentials of PLSR and SVR Models for Soil Properties Prediction Using Field Imaging, Laboratory VNIR Spectroscopy and Their Combination. *CMES-Comp. Model. Eng. Sci.* **2023**, *136*, 1399–1425. [CrossRef]
35. Pouladi, N.; Moller, A.B.; Tabatabai, S.; Greve, M.H. Mapping soil organic matter contents at field level with Cubist, Random Forest and kriging. *Geoderma* **2019**, *342*, 85–92. [CrossRef]
36. Peng, J.; Li, S.; Makar, R.S.; Li, H.Y.; Feng, C.H.; Luo, D.F.; Shen, J.L.; Wang, Y.; Jiang, Q.S.; Fang, L.C. Proximal Soil Sensing of Low Salinity in Southern Xinjiang, China. *Remote Sens.* **2022**, *14*, 4448. [CrossRef]
37. Toth, G.; Jones, A.; Montanarella, L. The LUCAS topsoil database and derived information on the regional variability of cropland topsoil properties in the European Union. *Environ. Monit. Assess.* **2013**, *185*, 7409–7425. [CrossRef] [PubMed]
38. Castaldi, F.; Chabrillat, S.; Jones, A.; Vreys, K.; Bomans, B.; van Wesemael, B. Soil Organic Carbon Estimation in Croplands by Hyperspectral Remote APEX Data Using the LUCAS Topsoil Database. *Remote Sens.* **2018**, *10*, 153. [CrossRef]
39. Harris, P.; Brunsdon, C.; Charlton, M. Geographically weighted principal components analysis. *Int. J. Geogr. Inf. Sci.* **2011**, *25*, 1717–1736. [CrossRef]
40. Gollini, I.; Lu, B.B.; Charlton, M.; Brunsdon, C.; Harris, P. GWmodel: An R Package for Exploring Spatial Heterogeneity Using Geographically Weighted Models. *J. Stat. Softw.* **2015**, *63*, 1–50. [CrossRef]
41. Viscarra Rossel, R.A.; Behrens, T. Using data mining to model and interpret soil diffuse reflectance spectra. *Geoderma* **2010**, *158*, 46–54. [CrossRef]

42. Morellos, A.; Pantazi, X.E.; Moshou, D.; Alexandridis, T.; Whetton, R.; Tziotzios, G.; Wiebensohn, J.; Bill, R.; Mouazen, A.M. Machine learning based prediction of soil total nitrogen, organic carbon and moisture content by using VIS-NIR spectroscopy. *Biosyst. Eng.* **2016**, *152*, 104–116. [CrossRef]
43. Salazar, D.F.U.; Demattê, J.A.M.; Vicente, L.E.; Guimaraes, C.C.B.; Sayao, V.M.; Cerri, C.E.P.; Padilha, M.C.D.; Mendes, W.D. Emissivity of agricultural soil attributes in southeastern Brazil via terrestrial and satellite sensors. *Geoderma* **2020**, *361*, 114038. [CrossRef]
44. Wang, J.Z.; Ding, J.L.; Yu, D.L.; Ma, X.K.; Zhang, Z.P.; Ge, X.Y.; Teng, D.X.; Li, X.H.; Liang, J.; Lizag, A.; et al. Capability of Sentinel-2 MSI data for monitoring and mapping of soil salinity in dry and wet seasons in the Ebinur Lake region, Xinjiang, China. *Geoderma* **2019**, *353*, 172–187. [CrossRef]
45. Chen, R.; Wang, X.; Wang, Z.W.; Qu, H.; Ma, T.M.; Chen, Z.G.; Gao, R. Wavelength Selection Method of Near-Infrared Spectrum Based on Random Forest Feature Importance and Interval Partial Least Square Method. *Spectrosc. Spectr. Anal.* **2023**, *43*, 1043–1050.
46. Norby, R.J.; DeLucia, E.H.; Gielen, B.; Calfapietra, C.; Giardina, C.P.; King, J.S.; Ledford, J.; McCarthy, H.R.; Moore, D.J.P.; Ceulemans, R.; et al. Forest response to elevated CO2 is conserved across a broad range of productivity. *Proc. Natl. Acad. Sci. USA* **2005**, *102*, 18052–18056. [CrossRef] [PubMed]
47. Paula, R.R.; Calmon, M.; Lopes Assad, M.L.; Mendonça, E.D.S. Soil organic carbon storage in forest restoration models and environmental conditions. *J. For. Res.* **2021**, *33*, 1123–1134. [CrossRef]
48. Guo, Y.; Shi, Z.; Li, H.Y.; Triantafilis, J. Application of digital soil mapping methods for identifying salinity management classes based on a study on coastal central China. *Soil Use Manag.* **2013**, *29*, 445–456. [CrossRef]
49. Galvao, L.S.; Pizarro, M.A.; Epiphanio, J.C. Variations in reflectance of tropical soils: Spectral-chemical composition relationships from AVIRIS data. *Remote Sens. Environ.* **2001**, *75*, 245–255. [CrossRef]
50. Ben-Dor, E.; Taylor, R.G.; Hill, J.; Demattê, J.A.M.; Whiting, M.L.; Sommer, S. Imaging Spectrometry for Soil Applications. *Adv. Agron.* **2008**, *97*, 321–392.
51. Liu, L.F.; Ji, M.; Buchroithner, M. Combining Partial Least Squares and the Gradient-Boosting Method for Soil Property Retrieval Using Visible Near-Infrared Shortwave Infrared Spectra. *Remote Sens.* **2017**, *9*, 1299. [CrossRef]
52. Liu, B.Y.; Guo, B.F.; Zhuo, R.X.; Dai, F.; Chi, H.Y. Prediction of the soil organic carbon in the LUCAS soil database based on spectral clustering. *Soil Water Res.* **2023**, *18*, 43–54. [CrossRef]
53. Stenberg, B.; Viscarra Rossel, R.A.; Mouazen, A.M.; Wetterlind, J. *Visible and Near Infrared Spectroscopy in Soil Science*; Academic Press: Cambridge, MA, USA, 2010; pp. 163–215.
54. Zhang, X.L.; Xue, J.; Xiao, Y.; Shi, Z.; Chen, S.C. Towards Optimal Variable Selection Methods for Soil Property Prediction Using a Regional Soil Vis-NIR Spectral Library. *Remote Sens.* **2023**, *15*, 465. [CrossRef]
55. Xu, L.; Hong, Y.S.; Wei, Y.; Guo, L.; Shi, T.Z.; Liu, Y.; Jiang, Q.H.; Fei, T.; Liu, Y.L.; Mouazen, A.M.; et al. Estimation of Organic Carbon in Anthropogenic Soil by VIS-NIR Spectroscopy: Effect of Variable Selection. *Remote Sens.* **2020**, *12*, 3394. [CrossRef]
56. Shi, Z.; Ji, W.; Viscarra Rossel, R.A.; Chen, S.; Zhou, Y. Prediction of soil organic matter using a spatially constrained local partial least squares regression and the Chinese vis-NIR spectral library. *Eur. J. Soil Sci.* **2015**, *66*, 679–687. [CrossRef]

Disclaimer/Publisher's Note: The statements, opinions and data contained in all publications are solely those of the individual author(s) and contributor(s) and not of MDPI and/or the editor(s). MDPI and/or the editor(s) disclaim responsibility for any injury to people or property resulting from any ideas, methods, instructions or products referred to in the content.

sensors

MDPI

Article

Developing a Portable Autofluorescence Detection System and Its Application in Biological Samples

Jiaxing Zhou, Yunfei Li *, Jinfeng Zhang and Fuhong Cai

School of Biomedical Engineering, Hainan University, Sanya 572000, China; qq2628783938@outlook.com (J.Z.); 1457893655@foxmail.com (J.Z.); caifuhong@hainanu.edu.cn (F.C.)
* Correspondence: 21110810000014@hainanu.edu.cn

Abstract: Advanced glycation end-products (AGEs) are complex compounds closely associated with several chronic diseases, especially diabetes mellitus (DM). Current methods for detecting AGEs are not suitable for screening large populations, or for long-term monitoring. This paper introduces a portable autofluorescence detection system that measures the concentration of AGEs in the skin based on the fluorescence characteristics of AGEs in biological tissues. The system employs a 395 nm laser LED to excite the fluorescence of AGEs, and uses a photodetector to capture the fluorescence intensity. A model correlating fluorescence intensity with AGEs concentration facilitates the detection of AGEs levels. To account for the variation in optical properties of different individuals' skin, the system includes a 520 nm light source for calibration. The system features a compact design, measuring only 60 mm × 50 mm × 20 mm, and is equipped with a miniature STM32 module for control and a battery for extended operation, making it easy for subjects to wear. To validate the system's effectiveness, it was tested on 14 volunteers to examine the correlation between AGEs and glycated hemoglobin, revealing a correlation coefficient of 0.49. Additionally, long-term monitoring of AGEs' fluorescence and blood sugar levels showed a correlation trend exceeding 0.95, indicating that AGEs reflect changes in blood sugar levels to some extent. Further, by constructing a multivariate predictive model, the study also found that AGEs levels are correlated with age, BMI, gender, and a physical activity index, providing new insights for predicting AGEs content and blood sugar levels. This research supports the early diagnosis and treatment of chronic diseases such as diabetes, and offers a potentially useful tool for future clinical applications.

Keywords: fluorescence monitoring; advanced glycation end-products (AGEs); portable detection system; diabetes management

Citation: Zhou, J.; Li, Y.; Zhang, J.; Cai, F. Developing a Portable Autofluorescence Detection System and Its Application in Biological Samples. *Sensors* **2024**, *24*, 3351. https://doi.org/10.3390/s24113351

Academic Editor: Hisashi Satoh

Received: 24 April 2024
Revised: 14 May 2024
Accepted: 22 May 2024
Published: 23 May 2024

Copyright: © 2024 by the authors. Licensee MDPI, Basel, Switzerland. This article is an open access article distributed under the terms and conditions of the Creative Commons Attribution (CC BY) license (https://creativecommons.org/licenses/by/4.0/).

1. Introduction

Advanced glycation end-products (AGEs) are complex molecules formed through non-enzymatic reactions [1]. They can be ingested from food, as well as produced endogenously, gradually accumulating within the human body [2]. The accumulation of advanced glycation end-products (AGEs) occurs in the progression of various chronic diseases, including diabetes, renal diseases, cardiovascular diseases, and neurological disorders [3–5]. For instance, research on AGEs has uncovered their pivotal role in accelerating the development of complications associated with diabetes [6,7]. This influence extends beyond mere hyperglycemia to encompass profound damage at the microscopic molecular level [8,9]. Investigating the dynamics of advanced glycation end-products (AGEs) in the human body is essential for the early detection and intervention of chronic diseases [10]. For example, diabetes mellitus presents a significant and increasing challenge globally, affecting both developed and developing countries. As of 2021, approximately 537 million adults worldwide are living with diabetes. This number is projected to increase to 643 million by 2030 and 783 million by 2045 [11]. It is noteworthy that a substantial proportion of diabetes cases remain undiagnosed, with about half of the individuals unaware of their condition [12,13]. Although diabetes can be managed through medication, effective treatment necessitates early detection. Unfortunately, diagnosis often occurs several years after the onset of diabetes, typically at the commencement of complications [13].

AGEs are implicated in numerous chronic diseases through two primary formation pathways: exogenous and endogenous [2]. Exogenous AGEs accumulate from environmental exposures such as cigarette smoke, consumption of highly processed foods, particularly those processed with heat, and sedentary lifestyles, all of which contribute to increased formation rates of these harmful compounds [5,6]. Conversely, endogenous AGEs originate internally under conditions of elevated blood sugar, where reducing sugars like glucose, lactose, and fructose undergo non-enzymatic glycation reactions with proteins and lipids [14]. This internal process is known as the Maillard reaction, a crucial pathway for AGEs formation within the body.

AGEs are present in human blood and tissues, and their detection relies on drawing blood [15] or obtaining bodily fluids [16] for biochemical analysis [17]. This biochemical testing is not only costly, but is also time-consuming, requiring 8–9 h. Developing non-invasive, convenient, and pain-free methods of detection is essential for the practical continuous measurement of advanced glycation end-products (AGEs) in human blood and tissues, serving as a foundation for long-term monitoring and research into their role in chronic diseases. Over the past decades, a multitude of technologies and corresponding equipment have been developed to pursue more convenient and cost-effective detection methods [18–20]. Within this domain, numerous optical spectroscopic methods have attracted considerable attention due to their potential in enhancing analytical precision and reducing operational complexities [21–24]. The optical non-destructive testing method is implemented by analyzing the autofluorescence characteristics of AGEs, utilizing the unique fluorescence properties of molecules to specifically detect target molecules. The primary absorption band of advanced glycation end-products (AGEs) lies between 340 nm and 420 nm, with the fluorescence spectrum spanning the 420 nm to 600 nm range [3]. Instruments designed for detecting autofluorescence have been developed and validated through small-scale clinical trials targeting human AGEs. Researchers utilize fiber optic spectrometers for detection, employing an excitation light source ranging from 300 to 400 nm to irradiate the skin and capture signals with a glass fiber optic spectrometer. However, transdermal detection is hindered by the low efficiency of spontaneous fluorescence. In vivo, where the concentration of AGEs is relatively low, the fluorescent peaks of AGEs are significantly diminished by the skin's strong and time-varying spontaneous fluorescence, further compounded by associated shot noise, rendering the AGEs' fluorescence even weaker. This presents substantial challenges in constructing reliable predictive models for AGEs concentration. Furthermore, the propagation characteristics of fluorescence signals are influenced by the skin's color composition, contact pressure, and the optical parameters of biological tissues. Researchers have attempted to circumvent the influence of skin color by analyzing tissues such as the retina and wrist, but the fragile structure of the retina is not conducive to long-term monitoring. Correcting AGEs' fluorescence signals through calibration models requires prior information such as skin type, age, and the subject's habits, among others. The introduction of additional information necessitates preliminary collection, which is not favorable for reducing detection costs. Building on the fluorescence properties of advanced glycation end-products (AGEs), this paper introduces a novel portable skin fluorescence detection device designed for the assessment of AGEs. The device comprises a control module, an LED emission module, and a detection module. The LED emission module emits a 395 nm wavelength laser to excite the AGEs in the skin, causing them to fluoresce. This fluorescence is then captured by a photodiode within the detection module, enabling the measurement of AGEs concentration based on the intensity of the fluorescence signal. To account for the impact of tissue optical parameters on the fluorescence signal, a 520 nm LED is utilized to measure skin reflectance, facilitating the correction of AGEs' fluorescence values. The device is equipped with a miniaturized STM32 microcontroller and a 3.7 v battery, serving as both the control and power modules, which simplifies the process of long-term monitoring of AGEs' fluorescence intensity. To validate the effectiveness of this system, a study was conducted with 14 volunteers to analyze the correlation between AGEs and glycated hemoglobin levels. Moreover, the device was used to perform long-term monitoring of AGEs' fluorescence and blood glucose levels in volunteers, revealing a correlation exceeding 0.95 between the two, indicating that AGEs levels reflect changes in blood glucose levels to a significant extent. Furthermore, by developing a multivariate prediction model, the study also found that AGEs levels are associated with age, BMI, gender, and a physical activity index, providing new insights into predicting AGEs content and blood glucose levels. This integration of portable technology with fluorescence measurement offers a promising tool for the non-invasive and continuous monitoring of biomarkers that are critical for diabetes management, and potentially other age-related conditions.

2. Materials and Methods

2.1. Structural Design of the Portable Autofluorescence Detection System

As depicted in Figure 1a, the device, designed for attachment to the wrist, has a length of 60.00 mm and a width of 50.00 mm, enabling portability and continuous, on-the-go monitoring. It employs a dual-LED configuration with an emission wavelength of 395 nm to elicit the characteristic fluorescence of AGEs. The emitted fluorescence is detected by an S1223 photodetector. To ensure the integrity of the signals and facilitate data processing, the device integrates a three-layered circuit board that controls the system, processes the signals, and manages data handling. The LEDs are encased within a black housing fabricated using 3D printing technology, which serves to prevent light leakage that could potentially interfere with the accuracy of detection. Additionally, filters are embedded within this housing to refine the wavelength specificity, as detailed in the structural representation in Figure 1c. This design exemplifies a seamless amalgamation of functionality and sophistication, crucial for the effective non-invasive monitoring of AGEs in clinical and everyday settings. Table 1 enumerates all of the critical components incorporated within the device, detailing their specifications, parameters, and individual prices. This facilitates the evaluation of the device's operational capabilities and aids in budgetary planning.

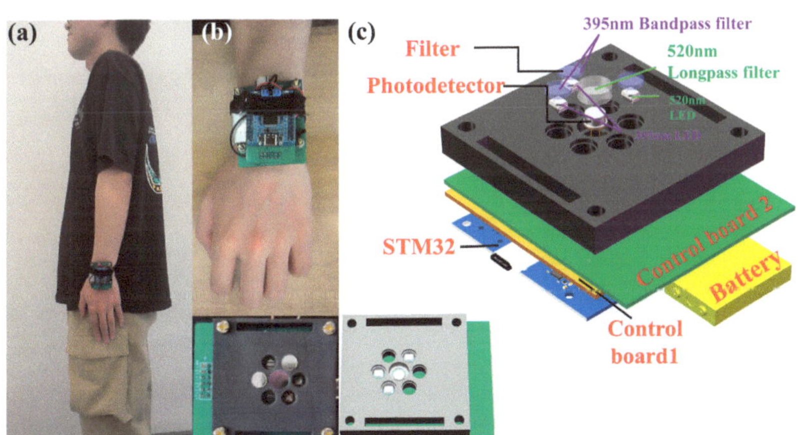

Figure 1. Structural design of the portable autofluorescence detection system. (**a**) Schematic of the device worn on the arm, with dimensions of 60 mm × 50 mm × 20 mm, secured with an elastic band. (**b**) Actual device displaying various modules. (**c**) Three-dimensional structural design illustrating the 395 nm excitation module with a bandpass filter emitting light onto the forearm to stimulate AGEs' fluorescence; the emitted fluorescence is then received by the detector through a 520 nm long pass filter.

Table 1. Key Components of the Portable Autofluorescence Detection System Used in This Study.

Device	Parameters	Cost
Excitation light	395 nm\3 W	0.69$
Calibration light source	520 nm\1 W	0.69$
Batteries	3.7 V\500 mAh	2.00$
Photodetector	0.45 A\W	9.00$
Excitation filter	10 mm\395 ± 10 nm	13.00$
Optical emission filter	10 mm\520 ± 10 nm	4.80$
Total		30.18$

2.2. Enhanced Control Module Design for Precision and Portability

At the heart of the device's control apparatus lies an STM32 circuit board, strategically positioned on the uppermost layer with dimensions of 39.00 mm by 33.40 mm, featuring an STM32 microcontroller chip. The control of LED pulsation is governed by the I/O port voltage fluctuations on the STM32 circuit board, while the onboard analog-to-digital converter (ADC) of the STM32 reads the voltage output from the detection module for analog-to-digital conversion and data processing. To address the issue of insufficient output power from the STM32 microcontroller's I/O ports, a dedicated transistor amplification circuit has been devised. A high-gain NPN transistor with a collector current of 3 A and a collector-emitter breakdown voltage of 30 V forms a common-emitter current amplification circuit with negative feedback, which effectively amplifies the nominal 10 mA current from the control board. By meticulously adjusting the transistor's operating point and the negative feedback network, the 10 mA output from the STM32 chip is amplified to approximately 500 mA, which is sufficient to drive the LED lights, thus ensuring their stable and uniform irradiation. For enhanced portability, a compact charging and discharging module (as shown in Figure 2) is connected to a 3.7 V lithium battery. This module stabilizes the boost output to 5 V, supplying power to the STM32 circuit board and the entire system.

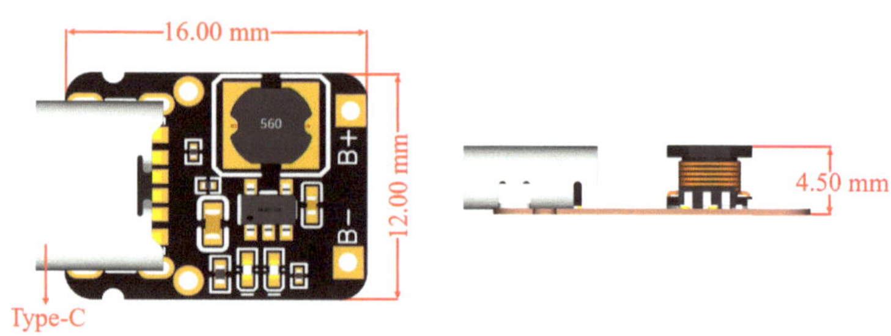

Figure 2. Compact integrated charging and discharging module dimensions: 16.00 mm × 12.00 mm × 2.60 mm.

2.3. Optimized LED Emission Module for Enhanced Autofluorescence Excitation

The study by M. Koetsier et al. suggests that within the 355–405 nm wavelength range, the induced autofluorescence in skin shows no significant variance [25]. Based on these findings, our chosen excitation light source consists of two LED beads with a peak emission wavelength of 395 nm. These LEDs feature a 15 mm spectral half-width, a 60° emission angle, and a 3 W output power, and are soldered onto Control Board 2. Utilizing two 395 nm bandpass filters, extraneous wavelengths from the light source are eliminated, allowing the directed light to excite the autofluorescence of AGEs within the volunteer's wrist skin. Additionally, an LED with a central emission wavelength of 520 nm serves as the calibration light source to mitigate the influence of varying skin types on fluorescence detection. By processing the fluorescence signal intensities under both wavelengths through algorithms, we can more accurately assess the AGEs content. The circuit design includes three unused LED positions, ensuring they do not interfere with the detection outcome.

Control Board 2 measures 60.00 mm in length and 50.00 mm in width, featuring a central circular recess in its design.

2.4. Advanced Detection Module with Precision Filtering and Amplification Circuitry

The intricate design of the detection module is centralized within Control Board 1, a highly specialized spectral analysis circuit board with dimensions of 50.00 mm by 50.00 mm. The photodetector S1223 is expertly interfaced with Control Board 1, penetrating the circular recess of Control Board 2 to sit within a 3D-printed black enclosure. This strategic placement ensures the detector is optimally close to the volunteer's skin for maximum signal capture. A dedicated 520 nm optical filter is adeptly utilized to discriminate against any extraneous light, selectively passing only the fluorescence emanating from AGEs proteins. The S1223 photodetector captures this filtered light, converting it into a measurable voltage signal. To bolster detection precision, Control Board 1 incorporates a bespoke circuit for filtering and signal amplification to ensure that even the most delicate fluorescent signals are effectively discerned and amplified. These processed signals are then routed through an output terminal to the STM32 circuit board, where they undergo meticulous analog-to-digital conversion and further processing for accurate AGEs quantification.

3. Results

3.1. Study on the Correlation between Glycated Hemoglobin and AGEs' Fluorescence Intensity

Glycated hemoglobin (HbA1c) is a product of the binding of glucose in the blood with hemoglobin in red blood cells. It forms a stable compound through the covalent bonding of glucose with the N-terminal valine of the hemoglobin beta chain, and serves as an index of average blood plasma glucose levels over the preceding 8–12 weeks [26,27]. It is a gold standard for assessing glycemic control and gauging the risk of chronic complications in diabetes patients [28]. Notably, individual glucose levels are influenced by dietary habits, health status, and physical exercise load over time [29]. Furthermore, serum AGEs levels correlate with dietary intake, with high-protein or high-fat foods cooked at high temperatures being rich in dietary AGEs [30]. This section explores the correlation between HbA1c levels and AGEs' fluorescence intensity in volunteers. Figure 3a illustrates the process of detecting glycated hemoglobin. To ensure experimental accuracy, blood samples from all volunteers were collected after fasting. The ELISA assay, utilizing a kit designed for blood glucose detection, provided an effective and precise means of quantifying glycated hemoglobin. The ELISA plates were coated with HbA1c-specific antibodies to capture the analyte effectively. The volunteers' blood samples were then incubated in the plate wells at 4 °C for 24 h, allowing for the binding of HbA1c to the coated antibodies. After incubation, the wells were thoroughly washed to remove non-specifically bound proteins and potential interferences. Horseradish peroxidase (HRP)-conjugated secondary antibodies, which have a high affinity for HbA1c, were then added, forming stable immune complexes with the bound HbA1c during a secondary incubation. Additional washing removed any unbound enzyme-labeled secondary antibodies, preparing them for signal detection. TMB, a chromogenic substrate, underwent an oxidation reaction catalyzed by HRP, changing from colorless to blue, indicating the presence of HbA1c. The reaction was initiated by adding TMB to the system and halted at a predetermined time by a stop solution to prevent over-catalysis. This ensured the stability of the color change for reliable optical density (OD) measurements. The OD values were obtained by using a microplate reader, correlating with the HbA1c concentration in the samples. After an ELISA measurement, AGEs' fluorescence intensity was measured on the volunteers' wrists using a fluorescence detector, and the data were statistically fitted with the HbA1c levels determined by ELISA.

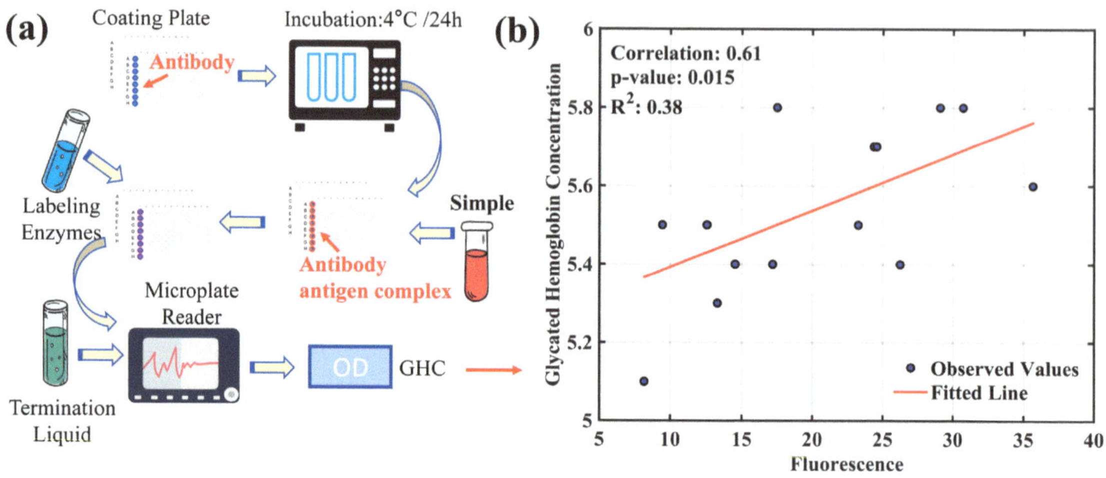

Figure 3. ELISA measurement and regression analysis of glycated hemoglobin and AGEs. (**a**) The ELISA process with coating plates, incubation, antibody-antigen complex formation, addition of labeling enzymes, microplate reading, and application of termination liquid. (**b**) The regression analysis demonstrating the relationship between glycated hemoglobin concentration and AGEs' fluorescence, with a correlation coefficient of 0.61, a *p*-value of 0.015, and a coefficient of determination (R^2) of 0.38.

Figure 3b illustrates the correlation between HbA1c and AGEs' fluorescence values obtained in this experiment, with a coefficient of determination (R^2) of 0.38. This indicates that the changes in HbA1c levels account for 38% of the variation in AGEs' fluorescence intensity, suggesting a moderate correlation between the two. However, the correlation is not strong, implying that other factors, in addition to HbA1c, may influence the formation and accumulation of AGEs.

3.2. Study on the Trends of AGEs' Fluorescence Intensity and Blood Glucose Variation

To investigate the potential connection between the content of AGEs and real-time blood glucose levels, and to analyze their synchrony and correlation at different times of the day, we enlisted three volunteers to perform bi-daily tests of AGEs content at fixed times (8 a.m. and 8 p.m.). This testing continued for 5 days to capture any potential circadian rhythm fluctuations. In addition to measuring the fluorescence intensity of AGEs with a fluorescence detector during each test, the volunteers also used an Abbott blood glucose meter for finger-prick blood sampling to record immediate blood glucose levels.

The experimental data displayed in Figure 4 indicate cosine similarities between the fluorescence intensity of AGEs and blood glucose levels for the three volunteers, which were 0.91, 0.93, and 0.95, respectively. This suggests a significant synchronous fluctuation relationship between AGEs' fluorescence intensity and blood glucose levels during the observed time periods, implying a close association between AGEs formation and daily fluctuations in blood glucose levels, with a certain level of consistency across different individuals. However, while cosine similarity provides a quantifiable method for comparing similarities between the two, it does not directly infer causality. Additionally, we cannot overlook the potential differences among individuals. These differences may be caused by various factors, including but not limited to dietary habits, sleep quality, daily activity levels, and genetic factors. Therefore, future research should consider including more samples and exploring a wider range of potential influencing factors to further investigate how blood glucose levels affect AGEs formation, and how this influence is manifested through circadian rhythm changes.

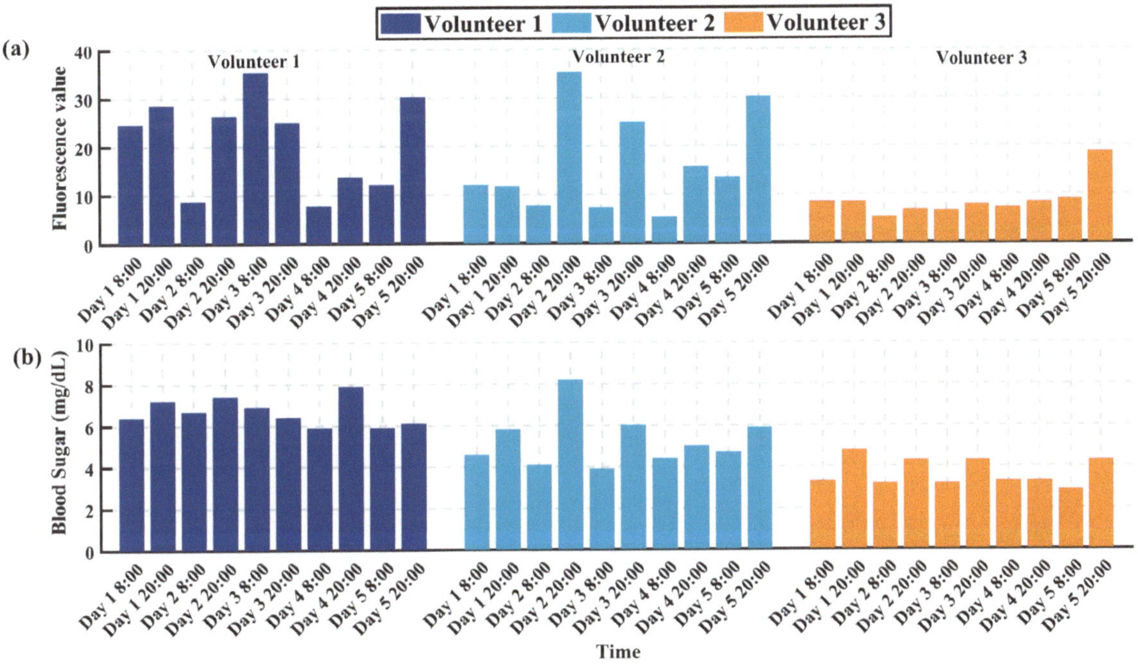

Figure 4. Investigation of AGEs content and real-time blood glucose levels over time. (**a**) Fluorescence intensity of AGEs content for Volunteers 1, 2, and 3 measured at different times. (**b**) Blood glucose levels of Volunteers 1, 2, and 3 measured at different times.

3.3. Validation of a Multivariate Prediction Model Based on Analysis of Factors Influencing Advanced Glycation End-Product Content

A multivariate prediction model validation based on the analysis of factors influencing the content of advanced glycation end-products (AGEs) was conducted. Fluorescence intensity of AGEs in the skin of 14 volunteers was measured using a fluorescence detector, and statistical analysis was performed on potential influencing factors such as age, body mass index (BMI), gender, and physical activity index (Phys Act) of the volunteers. By constructing a correlation heatmap, as shown in Figure 5a, we could visually observe the correlation between AGEs content and these variables.

The statistical analysis revealed a moderate positive correlation between AGEs content and volunteers' age, with a correlation coefficient of 0.3824. This suggests that AGEs accumulation increases gradually with age, consistent with previous findings regarding the correlation between AGEs and the aging process [31,32]. Additionally, a strong positive correlation (correlation coefficient = 0.6627) was observed between AGEs content and BMI, indicating a significant association between obesity and AGEs formation and accumulation. Obese individuals may have higher levels of AGEs, possibly due to chronic low-grade inflammation, increased oxidative stress, and consumption of lipid-rich foods. The correlation between gender (Gender) and AGEs content was weaker (correlation coefficient = 0.1218), suggesting a minor influence of gender on AGEs formation and accumulation. However, this does not entirely rule out the potential role of gender in AGEs-related diseases, which may require larger sample sizes and further investigation.

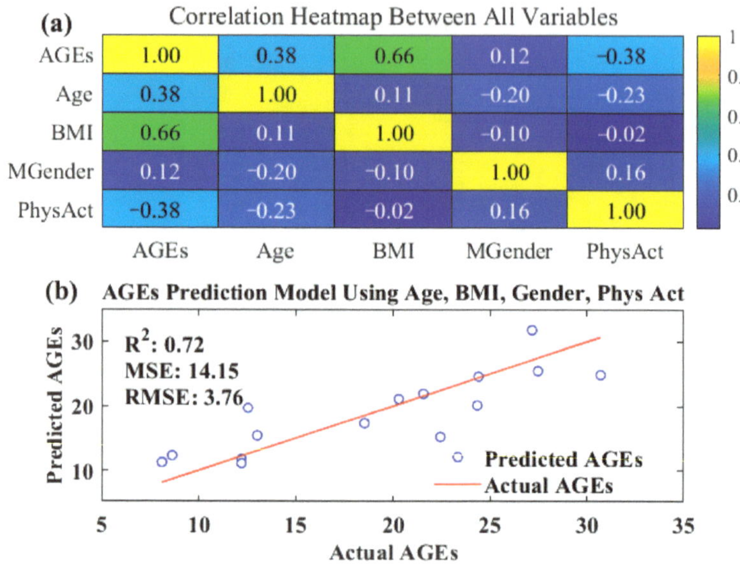

Figure 5. Analysis of influencing factors on AGEs content in human skin. (**a**) Correlation heatmap between AGEs, age, BMI, gender, and physical activity; (**b**) validation of AGEs prediction model using age, BMI, gender, and physical activity.

Notably, AGEs content exhibited a negative correlation with physical activity (PhysAct) (correlation coefficient = −0.3822), implying that higher levels of physical activity may be associated with lower AGEs levels. This finding supports the potential benefits of a proactive lifestyle and exercise in slowing AGEs accumulation and improving health outcomes. Furthermore, a predictive model for individual AGEs content was established using AGEs, gender, and physical activity, which are highly correlated factors. Regression analysis of the predictive model's results against the actual measured AGEs content further validated the association between these factors and AGEs content (see Figure 5b), yielding a coefficient of determination (R^2) of 0.72. This high R^2 indicates a good predictive capability of the model and underscores the significant predictive value of AGEs, gender, and physical activity for AGEs content.

4. Conclusions

This study designs and implements a portable AGEs detection device based on the spontaneous fluorescence of skin AGEs, which effectively measures the fluorescence intensity of skin AGEs. Experimental results indicate a certain degree of correlation between AGEs concentration and blood glucose levels, and through the validation of a multivariate predictive model, reveal the complex influencing factors of AGEs accumulation. The device offers real-time, non-invasive, and portable monitoring, making it suitable for large-scale screening and long-term monitoring. It provides a new tool for the early detection and prevention of chronic diseases such as diabetes, and holds promise for playing a crucial role in the early identification and risk assessment of chronic diseases like diabetes.

5. Discussion

The primary exogenous sources of advanced glycation end-products (AGEs) are derived from dietary intake, specifically from foods high in AGEs. Increased consumption of such foods is closely linked to elevated levels of AGEs in plasma and urine [33–35]. These high-AGEs foods, often processed at high temperatures through methods such as frying and grilling, typically contain high amounts of sugar. Excessive consumption can lead to increased blood glucose levels, establishing a connection between the intake of high-AGEs foods and higher blood glucose levels [36]. The affordability and convenience of the wearable devices used in this study enable us to expand our research into a broader population and implement long-term dietary tracking. This expansion will facilitate the collection of more comprehensive data, thereby enhancing the robustness of our findings.

Moreover, the research has revealed that levels of advanced glycation end-products (AGEs) are correlated with BMI, gender, and physical activity index, offering new perspectives for predicting AGEs content and blood glucose levels. Notably, BMI, which to some extent reflects food intake [37], suggests that excessive intake leads to the accumulation of AGEs. Therefore, a higher BMI may correlate with elevated AGEs levels due to increased consumption of foods rich in AGEs. However, it is crucial to thoroughly consider and account for the various biological and social differences that may underlie the observed variations in AGEs levels and their correlation with blood sugar. For instance, individuals with the same BMI might exhibit markedly different metabolic conditions. Recognizing these factors is essential. To improve the accuracy and reliability of our measurements, we will conduct rigorous dietary control experiments and monitoring. Initially, we will quantify AGEs in the diet and record the sugar content and calories in food to better understand the relationship between AGEs and blood sugar levels. Secondly, we plan to expand our research, which will help reduce variability and provide more robust data.

Additionally, understanding the biochemical pathways through which dietary AGEs influence blood glucose levels could provide further insights into metabolic health and disease prevention. Advanced glycation end-products (AGEs) have been shown to influence various biological processes. For instance, recent studies indicate that AGEs can regulate skin glycation by inhibiting specific transcription activators [38]. Additionally, the use of dual-channel fluorescence probes has enhanced the accuracy of detecting intracellular changes, which is crucial for understanding the relationship between AGEs and blood glucose levels [39]. Integrating these methods and insights into our future research will facilitate a clearer comprehension of the variability in the correlation between AGEs levels and blood glucose. This integration may enhance the predictive capabilities and clinical applications of our system. This line of research could lead to the development of targeted dietary recommendations that mitigate the impact of high-AGEs foods, potentially reducing the risk of diabetes and other related chronic conditions.

In future studies, we also plan to use our wearable devices for continuous monitoring and long-term tracking of AGEs and blood sugar levels. The miniaturization and wearability of our devices will enable us to conduct more extensive experiments and large-scale studies. This could enhance the accuracy and robustness of our findings on the relationship between AGEs and blood sugar levels. By integrating these strategies and methods, we aim to gain a clearer understanding of the variability in the correlation between AGEs levels and blood sugar, thereby enhancing the predictive capabilities and potential clinical applications of our system.

Author Contributions: Conceptualization, J.Z. (JiaXing Zhou) and Y.L.; methodology, J.Z. (JiaXing Zhou); software, Y.L.; validation, J.Z. (JiaXing Zhou); formal analysis, J.Z. (JiaXing Zhou); investigation, Y.L.; resources, Y.L.; data curation, Y.L.; writing—original draft preparation, F.C.; writing—review and editing, J.Z. (JiaXing Zhou); visualization, J.Z. (JinFeng Zhang); supervision, F.C.; project administration, F.C.; funding acquisition, F.C. All authors have read and agreed to the published version of the manuscript.

Funding: This project was funded by the Hainan Province Key Area R&D Program, ZDYF2024SHFZ084, and the Innovational Fund for Scientific and Technological Personnel of Hainan Province, KJRC2023B06.

Institutional Review Board Statement: The study was conducted in accordance with the Declaration of Helsinki, and approved by the Ethics Committee of Hainan Medical College for preliminary experiments on clinical testing of diabetes patients (protocol code 2024-83).

Informed Consent Statement: Written informed consent has been obtained from the patient(s) to publish this paper.

Data Availability Statement: Data available upon request from the authors.

Conflicts of Interest: The authors declare no conflicts of interest.

References

1. Itakura, M.; Yamaguchi, K.; Kitazawa, R.; Lim, S.-Y.; Anan, Y.; Yoshitake, J.; Shibata, T.; Negishi, L.; Sugawa, H.; Nagai, R.; et al. Histone Functions as a Cell-Surface Receptor for AGEs. *Nat. Commun.* **2022**, *13*, 2974. [CrossRef] [PubMed]
2. Twarda-Clapa, A.; Olczak, A.; Białkowska, A.M.; Koziołkiewicz, M. Advanced Glycation End-Products (AGEs): Formation, Chemistry, Classification, Receptors, and Diseases Related to AGEs. *Cells* **2022**, *11*, 1312. [CrossRef] [PubMed]
3. Atzeni, I.M.; Van De Zande, S.C.; Westra, J.; Zwerver, J.; Smit, A.J.; Mulder, D.J. The AGE Reader: A Non-Invasive Method to Assess Long-Term Tissue Damage. *Methods* **2022**, *203*, 533–541. [CrossRef] [PubMed]
4. Hegab, Z.; Gibbons, S.; Neyses, L.; Mamas, M.A. Role of Advanced Glycation End Products in Cardiovascular Disease. *World J. Cardiol.* **2012**, *4*, 90–102. [CrossRef] [PubMed]
5. Vistoli, G.; De Maddis, D.; Cipak, A.; Zarkovic, N.; Carini, M.; Aldini, G. Advanced Glycoxidation and Lipoxidation End Products (AGEs and ALEs): An Overview of Their Mechanisms of Formation. *Free Radic. Res.* **2013**, *47*, 3–27. [CrossRef] [PubMed]
6. Perrone, A.; Giovino, A.; Benny, J.; Martinelli, F. Advanced Glycation End Products (AGEs): Biochemistry, Signaling, Analytical Methods, and Epigenetic Effects. *Oxidative Med. Cell. Longev.* **2020**, *2020*, e3818196. [CrossRef] [PubMed]
7. Zhou, Q.; Cheng, K.-W.; Xiao, J.; Wang, M. The Multifunctional Roles of Flavonoids against the Formation of Advanced Glycation End Products (AGEs) and AGEs-Induced Harmful Effects. *Trends Food Sci. Technol.* **2020**, *103*, 333–347. [CrossRef]
8. Gautieri, A.; Passini, F.S.; Silván, U.; Guizar-Sicairos, M.; Carimati, G.; Volpi, P.; Moretti, M.; Schoenhuber, H.; Redaelli, A.; Berli, M.; et al. Advanced Glycation End-Products: Mechanics of Aged Collagen from Molecule to Tissue. *Matrix Biol.* **2017**, *59*, 95–108. [CrossRef] [PubMed]
9. Loske, C.; Neumann, A.; Cunningham, A.M.; Nichol, K.; Schinzel, R.; Riederer, P.; Münch, G. Cytotoxicity of Advanced Glycation Endproducts Is Mediated by Oxidative Stress. *J. Neural Transm.* **1998**, *105*, 1005–1015. [CrossRef]
10. Meerwaldt, R.; Hartog, J.W.L.; Graaff, R.; Huisman, R.J.; Links, T.P.; den Hollander, N.C.; Thorpe, S.R.; Baynes, J.W.; Navis, G.; Gans, R.O.B.; et al. Skin Autofluorescence, a Measure of Cumulative Metabolic Stress and Advanced Glycation End Products, Predicts Mortality in Hemodialysis Patients. *J. Am. Soc. Nephrol.* **2005**, *16*, 3687. [CrossRef]
11. Wee, B.F.; Sivakumar, S.; Lim, K.H.; Wong, W.K.; Juwono, F.H. Diabetes Detection Based on Machine Learning and Deep Learning Approaches. *Multimed. Tools Appl.* **2024**, *83*, 24153–24185. [CrossRef]
12. Islam, M.M.F.; Ferdousi, R.; Rahman, S.; Bushra, H.Y. Likelihood Prediction of Diabetes at Early Stage Using Data Mining Techniques. In *Computer Vision and Machine Intelligence in Medical Image Analysis*; Gupta, M., Konar, D., Bhattacharyya, S., Biswas, S., Eds.; Springer: Singapore, 2020; pp. 113–125.
13. Evans, M.; Morgan, A.R.; Patel, D.; Dhatariya, K.; Greenwood, S.; Newland-Jones, P.; Hicks, D.; Yousef, Z.; Moore, J.; Kelly, B.; et al. Risk Prediction of the Diabetes Missing Million: Identifying Individuals at High Risk of Diabetes and Related Complications. *Diabetes Ther.* **2021**, *12*, 87–105. [CrossRef] [PubMed]
14. Garagliano, J.M.; Katsurada, A.; Miyata, K.; Derbenev, A.V.; Zsombok, A.; Navar, L.G.; Satou, R. Advanced Glycation End Products Stimulate Angiotensinogen Production in Renal Proximal Tubular Cells. *Am. J. Med. Sci.* **2019**, *357*, 57–66. [CrossRef] [PubMed]
15. Noce, A.; Rovella, V.; Marrone, G.; Cattani, G.; Zingaretti, V.; Limongi, D.; D'Agostini, C.; Sorge, R.; Casasco, M.; Di Daniele, N.; et al. Hemodialysis Biomarkers: Total Advanced Glycation End Products (AGEs) against Oxidized Human Serum Albumin (HSAox). *Acta Diabetol.* **2019**, *56*, 1323–1331. [CrossRef] [PubMed]
16. Maasen, K.; Van Greevenbroek, M.M.; Scheijen, J.L.J.; Van Der Kallen, C.J.; Stehouwer, C.D.; Schalkwijk, C.G. High Dietary Glycemic Load Is Associated with Higher Concentrations of Urinary Advanced Glycation Endproducts: The Cohort on Diabetes and Atherosclerosis Maastricht (CODAM) Study. *Am. J. Clin. Nutr.* **2019**, *110*, 358–366. [CrossRef] [PubMed]
17. Corica, D.; Pepe, G.; Currò, M.; Aversa, T.; Tropeano, A.; Ientile, R.; Wasniewska, M. Methods to Investigate Advanced Glycation End-Product and Their Application in Clinical Practice. *Methods* **2022**, *203*, 90–102. [CrossRef] [PubMed]
18. Li, J.; Cai, F.; Dong, Y.; Zhu, Z.; Sun, X.; Zhang, H.; He, S. A Portable Confocal Hyperspectral Microscope without Any Scan or Tube Lens and Its Application in Fluorescence and Raman Spectral Imaging. *Opt. Commun.* **2017**, *392*, 1–6. [CrossRef]
19. Cai, F.; He, S. Using Graphics Processing Units to Accelerate Perturbation Monte Carlo Simulation in a Turbid Medium. *J. Biomed. Opt.* **2012**, *17*, 040502. [CrossRef] [PubMed]
20. Chen, J.; Cai, F.; He, R.; He, S. Experimental Demonstration of Remote and Compact Imaging Spectrometer Based on Mobile Devices. *Sensors* **2018**, *18*, 1989. [CrossRef]
21. Birukov, A.; Cuadrat, R.; Polemiti, E.; Eichelmann, F.; Schulze, M.B. Advanced Glycation End-Products, Measured as Skin Autofluorescence, Associate with Vascular Stiffness in Diabetic, Pre-Diabetic and Normoglycemic Individuals: A Cross-Sectional Study. *Cardiovasc. Diabetol.* **2021**, *20*, 110. [CrossRef]
22. Li, Y.; Shen, F.; Hu, L.; Lang, Z.; Liu, Q.; Cai, F.; Fu, L. A Stare-Down Video-Rate High-Throughput Hyperspectral Imaging System and Its Applications in Biological Sample Sensing. *IEEE Sens. J.* **2023**, *23*, 23629–23637. [CrossRef]
23. Li, Y.; Shen, F.; Lang, Z.; Cai, M.; Zeng, Q.; Liu, Q.; Cai, F. An Underwater Near-Infrared Spectral Continuum Robot as a Tool for in Situ Detection and Classification. *Measurement* **2023**, *216*, 112913. [CrossRef]

24. Li, Y.; Fu, Y.; Lang, Z.; Cai, F. A High-Frequency and Real-Time Ground Remote Sensing System for Obtaining Water Quality Based on a Micro Hyper-Spectrometer. *Sensors* **2024**, *24*, 1833. [CrossRef] [PubMed]
25. Koetsier, M.; Lutgers, H.L.; Smit, A.J.; Links, T.P.; Vries, R.D.; Gans, R.O.B.; Rakhorst, G.; Graaff, R. Skin Autofluorescence for the Risk Assessment of Chronic Complications in Diabetes: A Broad Excitation Range Is Sufficient. *Opt. Express OE* **2009**, *17*, 509–519. [CrossRef] [PubMed]
26. Pohanka, M. Glycated Hemoglobin and Methods for Its Point of Care Testing. *Biosensors* **2021**, *11*, 70. [CrossRef] [PubMed]
27. Zhan, Z.; Li, Y.; Zhao, Y.; Zhang, H.; Wang, Z.; Fu, B.; Li, W.J. A Review of Electrochemical Sensors for the Detection of Glycated Hemoglobin. *Biosensors* **2022**, *12*, 221. [CrossRef] [PubMed]
28. Selvin, E.; Steffes, M.W.; Zhu, H.; Matsushita, K.; Wagenknecht, L.; Pankow, J.; Coresh, J.; Brancati, F.L. Glycated Hemoglobin, Diabetes, and Cardiovascular Risk in Nondiabetic Adults. *N. Engl. J. Med.* **2010**, *362*, 800–811. [CrossRef] [PubMed]
29. Bazaev, N.A.; Pletenev, A.N.; Pozhar, K.V. Classification of Factors Affecting Blood Glucose Concentration Dynamics. *Biomed. Eng.* **2013**, *47*, 100–103. [CrossRef]
30. Nowotny, K.; Jung, T.; Höhn, A.; Weber, D.; Grune, T. Advanced Glycation End Products and Oxidative Stress in Type 2 Diabetes Mellitus. *Biomolecules* **2015**, *5*, 194–222. [CrossRef]
31. Peng, H.; Gao, Y.; Zeng, C.; Hua, R.; Guo, Y.; Wang, Y.; Wang, Z. Effects of Maillard Reaction and Its Product AGEs on Aging and Age-Related Diseases. *Food Sci. Hum. Wellness* **2024**, *13*, 1118–1134. [CrossRef]
32. Semba, R.D.; Nicklett, E.J.; Ferrucci, L. Does Accumulation of Advanced Glycation End Products Contribute to the Aging Phenotype? *J. Gerontol. Ser. A* **2010**, *65A*, 963–975. [CrossRef] [PubMed]
33. Scheijen, J.L.J.M.; Hanssen, N.M.J.; van Greevenbroek, M.M.; Van der Kallen, C.J.; Feskens, E.J.M.; Stehouwer, C.D.A.; Schalkwijk, C.G. Dietary Intake of Advanced Glycation Endproducts Is Associated with Higher Levels of Advanced Glycation Endproducts in Plasma and Urine: The CODAM Study. *Clin. Nutr.* **2018**, *37*, 919–925. [CrossRef] [PubMed]
34. Mo, Y.; Ma, X.; Li, H.; Ran, X.; Yang, W.; Li, Q.; Peng, Y.; Li, Y.; Gao, X.; Luan, X. Relationship between Glycated Albumin and Glycated Hemoglobin According to Glucose Tolerance Status: A Multicenter Study. *Diabetes Res. Clin. Pract.* **2016**, *115*, 17–23. [CrossRef] [PubMed]
35. Koschinsky, T.; He, C.J.; Mitsuhashi, T.; Bucala, R.; Liu, C.; Buenting, C.; Heitmann, K.; Vlassara, H. Orally Absorbed Reactive Glycation Products (Glycotoxins): An Environmental Risk Factor in Diabetic Nephropathy. *Proc. Natl. Acad. Sci. USA* **1997**, *94*, 6474–6479. [CrossRef] [PubMed]
36. Russell, W.R.; Baka, A.; Björck, I.; Delzenne, N.; Gao, D.; Griffiths, H.R.; Hadjilucas, E.; Juvonen, K.; Lahtinen, S.; Lansink, M.; et al. Impact of Diet Composition on Blood Glucose Regulation. *Crit. Rev. Food Sci. Nutr.* **2016**, *56*, 541–590. [CrossRef] [PubMed]
37. Van Galen, K.A.; Ter Horst, K.W.; Serlie, M.J. Serotonin, Food Intake, and Obesity. *Obes. Rev.* **2021**, *22*, e13210. [CrossRef] [PubMed]
38. Han, J.; Sun, Y.; Wu, T.; Hou, X.; Zheng, S.; Zhang, H.; Lin, T.; Liu, H.; Sun, T. Echinacoside-Zinc Nanomaterial Inhibits Skin Glycation by Suppressing the Transcriptional Activation of the Receptor for Advanced Glycation End-Products. *ACS Nano* **2023**, *17*, 14123–14135. [CrossRef]
39. Li, S.; Wang, P.; Ye, M.; Yang, K.; Cheng, D.; Mao, Z.; He, L.; Liu, Z. Cysteine-Activatable Near-Infrared Fluorescent Probe for Dual-Channel Tracking Lipid Droplets and Mitochondria in Epilepsy. *Anal. Chem.* **2023**, *95*, 5133–5141. [CrossRef]

Disclaimer/Publisher's Note: The statements, opinions and data contained in all publications are solely those of the individual author(s) and contributor(s) and not of MDPI and/or the editor(s). MDPI and/or the editor(s) disclaim responsibility for any injury to people or property resulting from any ideas, methods, instructions or products referred to in the content.

sensors MDPI

Article

Raman Spectroscopy of Disperse Systems with Varying Particle Sizes and Correction of Signal Losses

Erik Spoor [1,*], Viktoria Oerke [1], Matthias Rädle [1] and Jens-Uwe Repke [2]

[1] CeMOS Research and Transfer Center, Mannheim University of Applied Sciences, Paul-Wittsack-Str. 10, 68163 Mannheim, Germany; m.raedle@hs-mannheim.de (M.R.)
[2] Process Dynamics and Operations Group, Technische Universität Berlin, Straße des 17. Juni 135, 10623 Berlin, Germany; jens-uwe.repke@tu-berlin.de
* Correspondence: e.spoor@hs-mannheim.de

Abstract: In this paper, a dispersion of glass beads of different sizes in an ammonium nitrate solution is investigated with the aid of Raman spectroscopy. The signal losses caused by the dispersion are quantified by an additional scattered light measurement and used to correct the measured ammonium nitrate concentration. Each individual glass bead represents an interface at which the excitation laser is deflected from its direction causing distortion in the received Raman signal. It is shown that the scattering losses measured with the scattered light probe correlate with the loss of the Raman signal, which means that the data obtained can be used to correct the measured values. The resulting correction function considers different particle sizes in the range of 2–99 μm as well as ammonium nitrate concentrations of 0–20 wt% and delivers an *RMSEP* of 1.952 wt%. This correction provides easier process access to dispersions that were previously difficult or impossible to measure.

Keywords: disperse phase; continuous phase; optical spectroscopy; particle measurement; process control; Raman spectroscopy; process engineering; UV/VIS spectroscopy; suspension measurement

Citation: Spoor, E.; Oerke, V.; Rädle, M.; Repke, J.-U. Raman Spectroscopy of Disperse Systems with Varying Particle Sizes and Correction of Signal Losses. *Sensors* **2024**, *24*, 3132. https://doi.org/10.3390/s24103132

Academic Editors: Qing Yu, Ran Tu, Ting Liu and Lina Li

Received: 11 April 2024
Revised: 6 May 2024
Accepted: 13 May 2024
Published: 15 May 2024

Copyright: © 2024 by the authors. Licensee MDPI, Basel, Switzerland. This article is an open access article distributed under the terms and conditions of the Creative Commons Attribution (CC BY) license (https://creativecommons.org/licenses/by/4.0/).

1. Introduction

Process analytical technology is a diverse field of measurement techniques for monitoring processes and analyzing the composition of mixtures [1–3]. Optical measurement technology, for example, offers direct access to process parameters such as the concentration of individual educts or products [4–7].

Frequently used optical measurement methods include UV/VIS, infrared, fluorescence and Raman spectroscopy. VIS spectroscopy is often used for the analysis of colors in order to achieve reproducible results [4,6,7]. In the UV and NIR spectrum, on the other hand, a structural analysis of samples is possible. If a substance has active bands in these ranges, changes or shifts can occur with concentration, allowing even the smallest concentrations to be detected [8]. Measurements in the NIR spectrum can require a great deal of calibration and are very sensitive to changing process parameters [9,10]. Another method is fluorescence spectroscopy, which is used in biomedicine, for example [11]. A sample is excited by light irradiation, which then emits a photon of a different wavelength. A major advantage of this measurement technique is its versatility and sensitivity, which means that several fluorophores within the same sample can be examined using different excitation wavelengths. However, this measurement method is limited to substances that emit photons [11,12].

Raman spectroscopy, which is used in this work, is a constantly developing field. The advantage of this technology is that non-contact and non-destructive measurements are possible through a glass pane, which makes process access easy to implement. In addition, Raman spectroscopy is molecule-sensitive and allows for the precise differentiation of different components, with very short measurement times in the range of seconds to milliseconds. In Raman spectroscopy, a target is irradiated by a laser, and the photons interact with the molecules. The valence electrons are thereby raised to an unstable energy level and then drop again, causing photons with altered frequencies, energies and wavelengths to be scattered in all spatial directions. The wavelength shift, also known as the Raman shift, is dependent on the molecular group and can therefore be used for the qualitative analysis of material compositions. In addition, the strength of the resulting signal is linearly dependent on the concentration, which allows for a quantitative evaluation of the target. The use of this measurement technique in homogeneous mixtures is now widespread and much researched [4,6,13–16]. However, the analysis of heterogeneous mixtures represents a special application. Disperse mixtures such as emulsions and suspensions have a high number of interfaces at which both the excitation light and the resulting Raman signal can be scattered. This reduces the power density of the laser at the focal point, leading to a strong attenuation of the measured signal and, therefore, to deviating measurement results or even results that cannot be interpreted [17–21]. A simple variant of signal correction is to compare the characteristic peaks of different components. However, this requires a constant component, such as a solvent, which can be used as a reference. In addition, the reference peak must lie within the measurement range of the spectrometer [21–23]. In the application example of this paper, a Tec5 (Steinbach, Germany) spectrometer is used that has a measuring range up to 3200 cm^{-1} but uses water as the solvent, which has its characteristic peak at 3400 cm^{-1}. More complex calculation models are also used in the literature, which have many parameters to be calibrated but have the associated disadvantage of high computational effort and time-consuming calibration [24,25]. There are also other approaches to solving the problems of measuring dispersed phases, such as refractive index matching. Here, an additional component is added to a phase, which reduces or even completely removes the refractive index difference between the phases. As a result, optical diffraction no longer takes place. However, adding a further component to the system can lead to new problems, such as the overlapping of peaks or the later purification of the product, and is therefore not universally applicable [18,26].

In order to have a simple and more universally applicable correction method, an additional probe is used in this paper that quantifies the losses of the excitation beam. This makes it possible to establish a direct correlation between reduced Raman peaks and increasing light scattering with increasing particle concentration and, finally, to correct the Raman spectra. For this purpose, the concentration of an ammonium nitrate solution in which glass beads are added as a disperse phase is investigated. The glass beads do not have a spectrum of their own that collides with the signal of the ammonium nitrate, which means that the focus of the observation can be placed on the investigation of the scattering effects caused by dispersed particles.

2. Materials and Methods

The basic principle of the measurements is that a Raman probe excites the sample and detects the Stokes signal and, at the same time, a scattering probe detects the losses of the excitation laser at an increased particle concentration. In a fluid without scattering losses due to particles or droplets, the laser passes through the fluid in a straight line, and no excitation photons should reach the scattering probe. Also, no significant amount of the Stokes signal is detected from the Raman measurements. This is partly because the scattering probe is not focused and partly because the detection module used is not designed for Raman measurements and has a much lower sensitivity. As soon as particles are present in the sample, the excitation light from the laser is scattered and detected by the probe. This effect and the correlation with the Raman signal are investigated and used for signal correction.

The measurement setup consists of a macro quartz glass cuvette from Hellma (Müllheim, Germany) with a sample that is placed in the focal point of two probes and positioned on a magnetic stirrer (Figure 1). An InPhotonics (Norwood, MA, USA) RPS785/16-5 probe was used to excite the sample. It has a focus length of 7.5 mm with a spot size of 158 µm and a numerical aperture of 0.27. The probe is connected to a Raman spectrometer MultiSpec

Desk ETH with an LS-LD laser cassette (785 nm excitation wavelength) and an SC-CCD RAMAN spectrometer cassette (detection range 319–3213 cm^{-1} with increments of 1 cm^{-1}) from tec5 (Steinbach, Germany). The laser power can be set between 50 and 500 mW and was adjusted to the maximum output for the measurements. The detector is a scientific thermoelectric-cooled CCD detector with a spectral resolution of 7 cm^{-1}. The position of the probe was adjusted so that the focal point was approx. 2 mm inside the cuvette. Normally, measurements would be taken in the center of a cuvette, but previous studies [27] have shown that a lower penetration depth is beneficial for the signal at higher particle concentrations. The second probe is a self-built scattering probe consisting of seven 200 μm fibers arranged in a circle. It has already been tested in other applications and has proven to be robust and cost-effective in the use of scattered light measurements [28]. This probe does not emit its own light, but it is only used to detect scattered light. For this purpose, it is connected to an MCS 601 UV-NIR C spectrometer from Zeiss (Oberkochen, Germany) (detection range of 190–1015 nm with increments of 0.5 nm) with a spectral resolution of 2.4 nm. The inner components consist of a grating with a diode array.

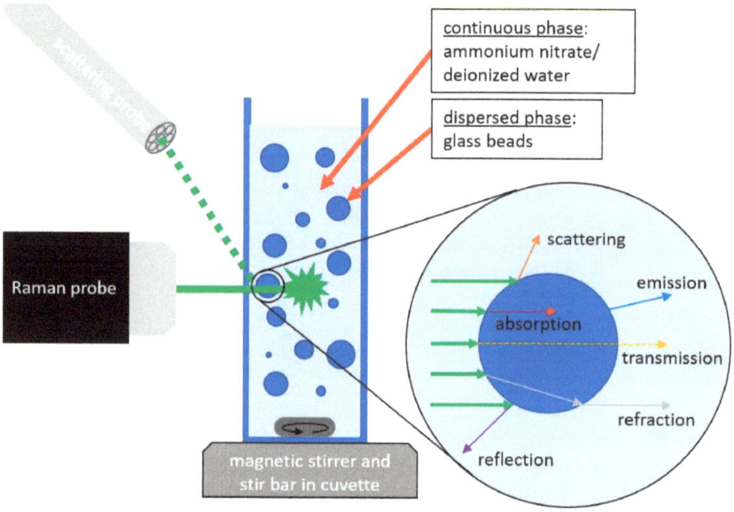

Figure 1. Scheme of the measurement setup with Raman and scattering probes, a cuvette with suspension on a magnetic stirrer and interaction of light.

In order to ensure the consistency of the measurements, a mount was manufactured that allows the Raman probe to be placed vertically to the cuvette at the desired distance and the scattered light probe to be attached aligned with the focal point of the Raman probe (Figure 2).

The basic substances under investigation have already been examined in a previous paper [27]. This involves an ammonium nitrate solution into which glass beads (SiO_2) are dispersed. To improve the understanding of disperse systems and their influence on Raman spectroscopy, four different particle sizes were investigated. The particles used were Omicron NP3 (2.093 μm), Omicron NP5 (4.089 μm), Micropearl (6.604 μm) and Starmixx (99.149 μm). The particle sizes refer to the Sauter diameter, which was measured with a HELOS particle size analyzer from Sympatec (Clausthal-Zellerfeld, Germany) [29]. Five ammonium nitrate solutions with deionized water were prepared for each of the four particle sizes. For each ammonium nitrate solution, 12 glass bead concentrations were then prepared. This corresponds to a total number of 240 samples. The individual concentrations are listed in Table 1 for one particle size at one ammonium nitrate concentration. It is expected that the size of the particles will have a measurable influence on the Raman signal. For example, at the same concentrations, there is a larger number of NP3 particles in the same volume than would be the case with Starmixx. As the number of particles increases, so does the number of interfaces at which light refraction can occur. This means that with increasing particle size, less refraction and therefore signal loss in the Raman spectrum is to be expected. This influence is analyzed in more detail in the following measurements using the selected particle sizes and concentrations shown in Table 1.

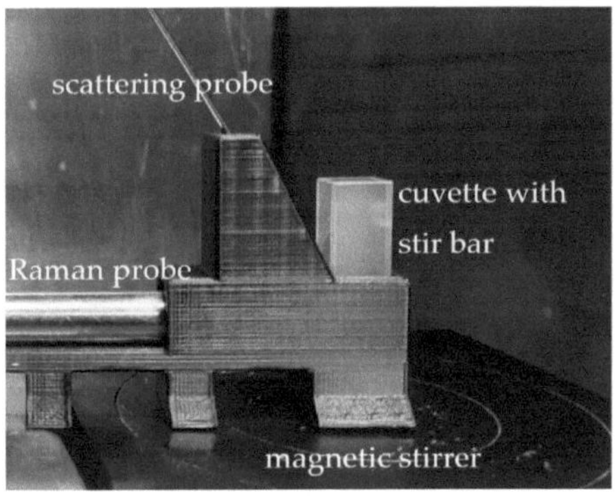

Figure 2. Setup of the Raman and scattering probes and the cuvette on the magnetic stirrer.

Table 1. Overview of the concentrations of ammonium nitrate solutions and particles.

Particle Name (Size)	Ammonium Nitrate Solution/wt%	Particle Concentrations per Ammonium Nitrate Concentration/wt%
NP3 (2.093 µm)	0–20 in 5 wt% steps	0.00–0.25 in 0.05 wt% steps
		0.50–3.00 in 0.50 wt% steps
NP5 (4.089 µm)	0–20 in 5 wt% steps	0.00–0.25 in 0.05 wt% steps
		0.50–3.00 in 0.50 wt% steps
Micropearl (6.604 µm)	0–20 in 5 wt% steps	0.00–0.25 in 0.05 wt% steps
		0.50–3.00 in 0.50 wt% steps
Starmixx (99.149 µm)	0–20 in 5 wt% steps	0.00–0.25 in 0.05 wt% steps
		0.50–3.00 in 0.50 wt% steps

3. Results

Initially, the peak of the Raman spectrum to be examined must be identified. As an example, the spectra of 20 wt% ammonium nitrate with no particles and with 3 wt% particles are shown in Figure 3. Ammonium nitrate (NH_4NO_3) consists of NH_4^+ and NO_3^- ions, which can be identified in the Raman spectrum by their characteristic bands. In the spectrum shown, the NO_3^- ions generate relatively weak signals at 716 cm^{-1}, 1390 cm^{-1} and 1671 cm^{-1} and a strong signal at 1047 cm^{-1}, which is most suitable for identifying the ammonium nitrate concentration. NH_4^+ only produces a weak signal at 1460 cm^{-1}, which overlaps with the signal of the NO_3^- ions [30–32]. The NO_3^- peak at 1047 cm^{-1} is therefore best suited for evaluation. This peak has a FWHM (full width half maximum) of 11 cm^{-1} and is detectable both with and without 3 wt% particles. All intensities of the Raman signal that were measured refer to this peak. The measurement signal of the water content cannot be evaluated in these spectra, as the significant OH peak is at approx. 3400 cm^{-1} and only a small part of the side at 3200 cm^{-1} is recognizable [33,34]. The glass beads themselves do not generate a signal of their own that can be measured in the spectrum.

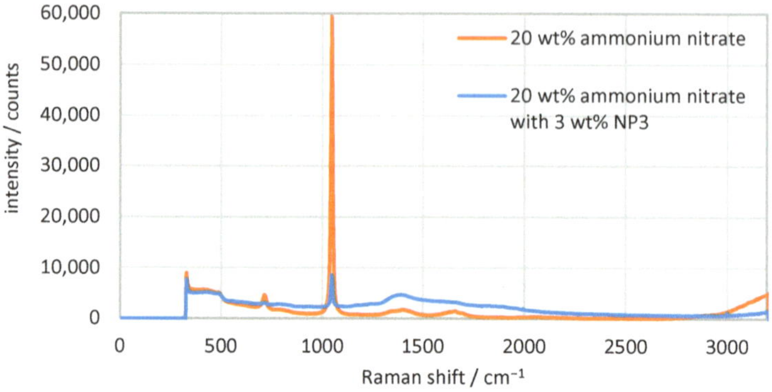

Figure 3. Spectra of the ammonium nitrate solution with 3 wt% NP3 particles and without particles for comparison.

All measured data are processed immediately after acquisition in order to minimize offset errors between the measurement series. In the first step, a baseline correction is performed in Formula (1) by subtracting the mean values of the left edge (1107 cm^{-1}) and right edge (1987 cm^{-1}) from the intensity value of the ammonium nitrate peak.

$$I_{AN} = I_{1047\text{cm}^{-1}} - \frac{I_{1107\text{cm}^{-1}} + I_{1987\text{cm}^{-1}}}{2} \tag{1}$$

After the baseline correction, the database is normalized, as there may be a slight displacement of the probes between the measurements. The measurements at 0 wt% particles can serve as a reference. These measuring points should provide the same values for the same ammonium nitrate concentration regardless of the defined particle type. For correction, all data for each ammonium nitrate concentration are related to the measured values of Micropearl (6.604 µm) at 0 wt% particles (Formula (2)).

$$I_{ANnorm} = \frac{I_{AN} \cdot I_{AN}(6.604\ \mu\text{m};\ 0\ \text{wt\% particles})}{I_{AN}(0\ \text{wt\% particles})} \tag{2}$$

All Raman data of the ammonium nitrate peak are plotted in Figure 4 sorted by particle size over the particle concentration. Because of the previous normalization, the 0 wt% measurement points of all particle sizes and the same ammonium nitrate concentration correspond to each other. In principle, there is a linear correlation between a Raman signal and the concentration of the target. Examining the measurement points at a constant particle concentration, it can be seen that this correlation exists for the ammonium nitrate concentration. However, the observation over the particle concentration provides a steep course up to 0.25 wt% and then a flattening intensity curve. The measurement series of NP3 and NP5 run into saturation towards 3 wt% and then no longer show any significant changes with the particle concentration. The curves of Micropearl and Starmixx, on the other hand, do not yet show any saturation. As the particle size increases, the signal losses of the measurement decrease. While the peak heights for NP3 decrease by a factor of 6 to 10, the peaks for Starmixx only decrease by a factor of 1.6 to 2. The reduced light scattering with increasing particle size can be explained by the fact that, overall, larger particles have a smaller number of boundaries than small particles at the same concentration.

The measurements with the scattering probe are shown as an example in Figure 5, using 20 wt% ammonium nitrate with no particles and with 3 wt% particles.

The produced peak is a combination of Rayleigh scattering, which is scattered in all spatial directions during molecular interactions, and direct excitation radiation, which is deflected by the particles. A differentiation is not possible since both signals have the same wavelength. The peak is located at 785 nm with a FWHM of 2.5 nm, and the spectrum shows no other peaks. With 0 wt% particles, the peak height is approx. 9000 counts and is therefore 333 times smaller than the peak with 3 wt% particles. This shows, on the one hand, that light scattering occurs on the cuvette and in the medium even without particles and, on the other hand, that the light scattering mentioned is not significant compared with the scattering caused by the particles.

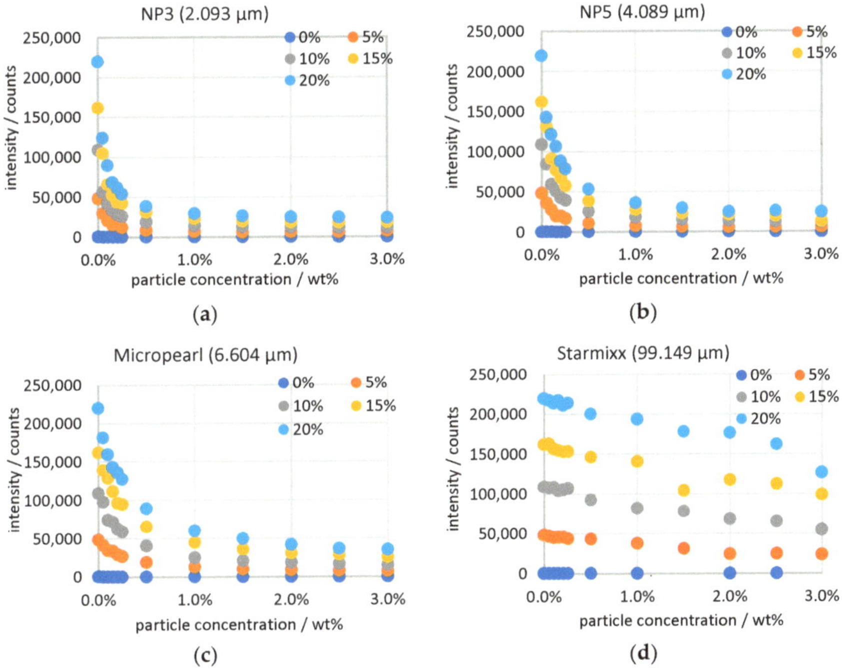

Figure 4. Raman measurements of particle sizes of (**a**) 2.093 µm; (**b**) 4.089 µm; (**c**) 6.604 µm and (**d**) 99.149 µm with increasing particle and ammonium nitrate concentrations.

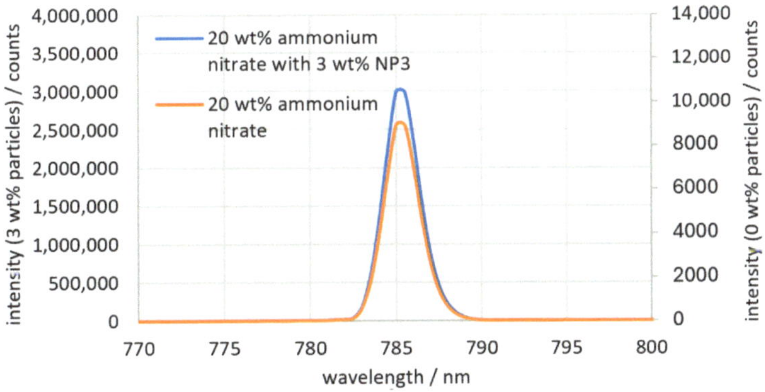

Figure 5. Peak of the excitation laser measured by the scattering probe with 3 wt% NP3 particles and without particles for comparison.

The data are calculated in the same way as in Formula (2) for the Raman data. A baseline correction is not necessary as the baseline has no relevant influence compared with the peak height. The peak height is calculated using Formula (3), which is based on the Beer–Lambert [4] law but uses the reciprocal of the intensity ratio. Typically, the Beer–Lambert law would be used for transmission measurements to calculate the light loss as extinction. For a transmission measurement, the intensity would decrease with the particle concentration, which corresponds to an increase in extinction. However, as the measurements of the scattering probe initially measure the opposite effect, that is, the increase in the measurement signal, as more light is scattered by the particles, the reciprocal value must be used for the purpose of representation in order to continue to depict an increase in light loss or extinction with increasing particle concentration. This is performed in order to have a better representation of the effects occurring and, in addition, to be comparable to the previous paper where an actual transmission setup was used [27].

$$E_\lambda = \log_{10}\left(\frac{I}{I_{0 \text{ wt\% particles}}}\right) \quad (3)$$

Analogous to the Raman data, the scattered light data are sorted by particle size and plotted over the particle concentration (Figure 6). The calculation according to the Beer–Lambert law, which takes the measurement points with 0 wt% particles as a reference, results in an origin in the zero points of the graphs for all measurement series. The curves are comparable to those of the Raman measurements, as here, too, the measurement signal changes very strongly up to 0.25 wt% and then becomes saturated. The correlation can also be logically explained by the fact that the more laser radiation is scattered, the lower the excitation of the sample and the intensity of the Stokes scattering. The correlation can also be seen in the fact that the light scattering decreases with increasing particle size.

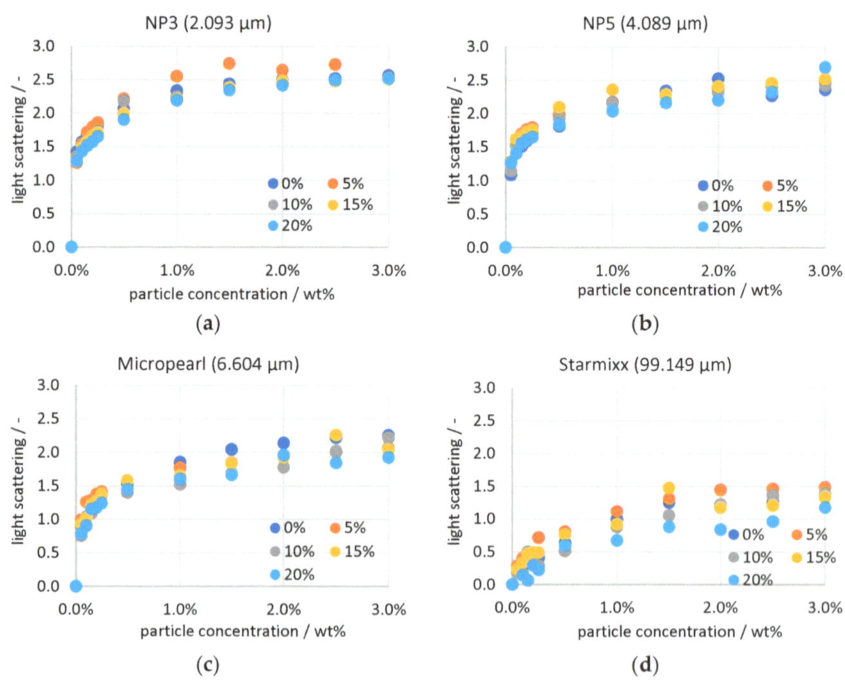

Figure 6. Scattering measurements of particle sizes of (**a**) 2.093 µm; (**b**) 4.089 µm; (**c**) 6.604 µm and (**d**) 99.149 µm with increasing particle and ammonium nitrate concentrations.

One difference to the Raman data is that no dependence of the scattered light on the ammonium nitrate concentration can be measured. This is also to be expected, as the laser radiation should not contain any information about it. The variance in the measured values at a constant particle concentration is on average approx. 0.09 for NP3 and approx. 0.16 for Starmixx. The gradient of the measured value fluctuations with the particle size could be due to the fact that larger particles sink faster and thus have a greater dependence on the stirrer speed.

Although the representation of the signals via the particle concentration allows for an explanation of the scattering effects, it provides an incomplete picture of the relationships. This is due to the fact that the concentration does not contain any information about the size of the particles. In order to obtain a more accurate picture of the relationships, the specific disperse surface of all samples is calculated. This value is defined as the total surface area of all particles per volume. First, the total density of the mixture ρ_t must be calculated from the mass fraction (w) and individual densities (ρ) of ammonium nitrate (AN), distilled water (dw) and particles (p) according to Formula (4):

$$\rho_t = \frac{1}{\frac{w_{AN}}{\rho_{AN}} + \frac{w_{dw}}{\rho_{dw}} + \frac{w_p}{\rho_p}} \tag{4}$$

The total number of particles (N_p) can be determined in Formula (5) from the weighed mass (m_p), the density (ρ_p) and the diameter (D) of the particles:

$$N_p = \frac{m_p}{\rho_p \cdot \frac{\pi \cdot D^3}{6}} \tag{5}$$

The specific disperse surface (S_d) can then be determined from the number of particles (N_p), mixing density (ρ_t), total weight of the solution (m_t) and surface area of a single particle (S) (Formula (6)):

$$S_d = S \cdot N_p \cdot \frac{\rho_t}{m_t} \tag{6}$$

The measured values of all particle sizes at 20 wt% ammonium nitrate are plotted against the specific disperse surface in Figure 7. The new plot shows that all measurement series follow the same course but are distorted when plotted against the concentration. The linearity of the Starmixx measurements can also be explained, as all data fall off in an approximately linear pattern up to a specific disperse surface of approx. 3000 m^2/m^3. The previously described effect that larger particles at the same concentration have fewer interfaces than smaller particles is now reflected in the new plot, whereby at 3 wt%, Starmixx has a specific disperse surface of 810 m^2/m^3 and NP3 of 37,930 m^2/m^3.

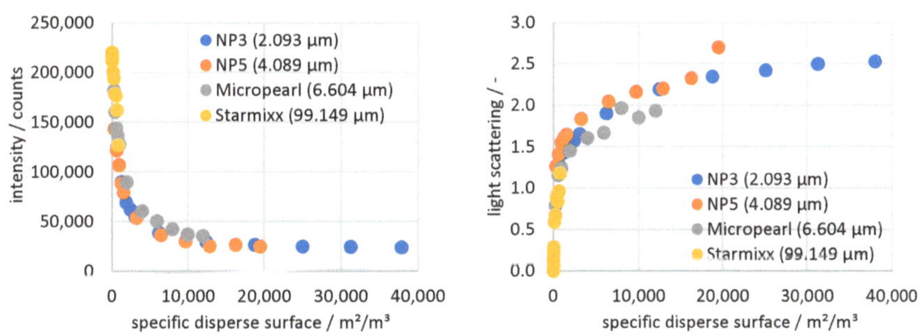

Figure 7. (**a**) Raman and (**b**) scattering data of all four particle sizes with the 20 wt% ammonium nitrate solution plotted over the specific disperse surface of the particles.

To view all the measurement data in a collective context, the Raman data are plotted as a change in peak height (compared to the 0 wt% particle measurement) against the corresponding scattered light value in Figure 8. A trend comparable to the plot against the specific disperse surface can be seen, where the Starmixx data are distributed over lower x-axis values than the remaining data sets. For the NP3, NP5 and Micropearl data points, the spread across the x-axis also correlates with the particle size. The clear difference to

Starmixx is due to the particle size being up to 47 times larger. Nevertheless, the overall picture shows a clear trend between the peak height change and light scattering. Especially in the range below a scattering of 1, the deviations of 0.02 from the trend are quite small. Above 1, the curve becomes more of a scatter plot and shows a deviation of 0.57 from the trend. The diagram demonstrates the direct correlation between Raman peak and scattering losses, making the measurement data independent of any knowledge of the particle size or specific dispersed surface area. All factors influencing the measurement signal are reflected in the form of the scattering loss.

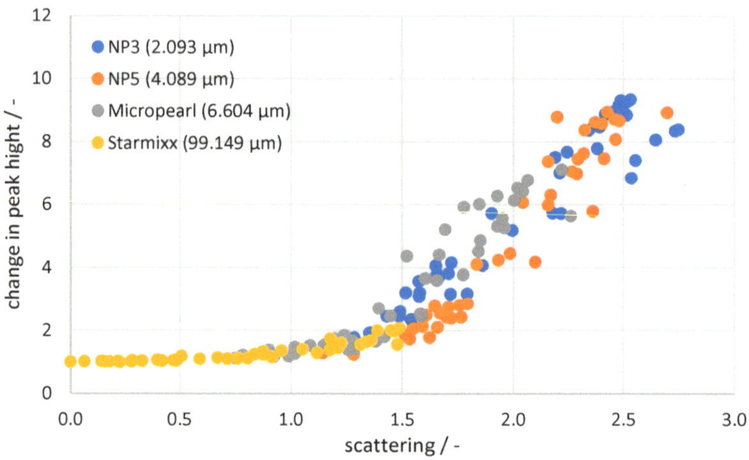

Figure 8. Change in the peak height of all samples plotted over the scattering of light, sorted by particle size.

To obtain a regression that can be used to correct the Raman values, the amount of data is first reduced by taking the mean values from the same particle concentration (Figure 9a). Starting from this, different trend lines are adjusted iteratively until the highest possible regression is achieved. In order to obtain a good adaptation, it is necessary to split the data at a scatter value of 0.9 and to describe the data below 0.9 by a second-degree polynomial function and the data above 0.9 by a third-degree polynomial function. The trend lines and formulas obtained are shown in Figure 9b.

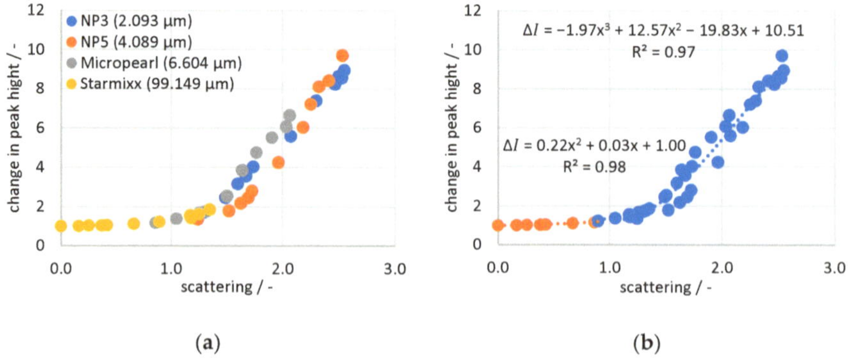

Figure 9. Average change in the peak height per particle concentration plotted over the average scattering of light per particle concentration: (**a**) sorted by particle size and (**b**) sorted by best regression with regression formulas.

From the formulas shown in Figure 9b, the change in peak height can be calculated as the correction factor ΔI. If this is multiplied by the measured intensity I_{AN} in Formula (7), the result is the corrected intensity $I_{AN-corr}$, which should match a measured value without particles:

$$I_{AN-corr} = I_{AN} \cdot \Delta I \tag{7}$$

For a clear display, the values are converted into mass fractions of the predicted ammonium nitrate concentration (ω_{AN}), which can be performed using the proximity Equation (8). The relationship originates from the regression of the measured Raman data without particles:

$$\omega_{AN} = 9 \cdot 10^{-7} \cdot I_{AN-corr} + 0.0023 \tag{8}$$

The mean corrected ammonium nitrate concentrations of the measurement series are shown in Figure 10. A theoretically optimal correction should lie on the corresponding horizontal grid line at 0, 5, 10, 15 and 20 wt%, respectively. All deviations from this are errors in the prediction. As a measure of the error, the *RMSEP* (Root Mean Squared Error of Prediction) is calculated according to Formula (9) [35], plotted as an error indicator in Figure 10 and listed individually in Table 2.

$$RMSEP = \sqrt{\frac{1}{N} \cdot \sum_{i=1}^{N} (y_i - \overline{y}_1)^2} \tag{9}$$

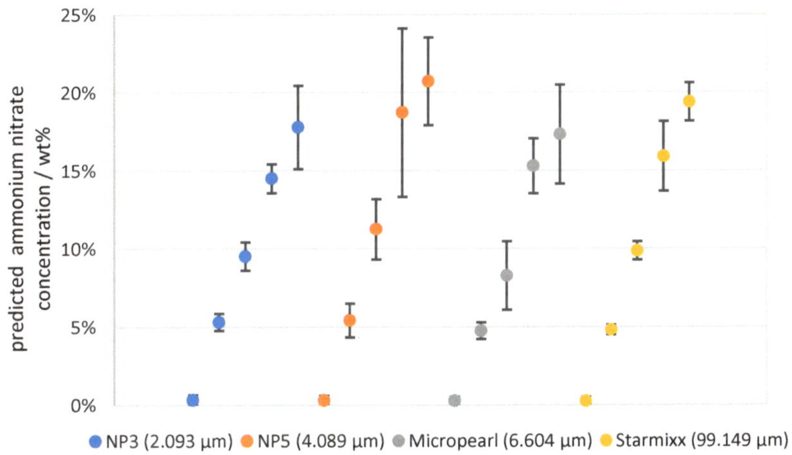

Figure 10. Calculated prediction for the ammonium nitrate concentration based on the measured scattering of light with increasing particle concentration and plotted *RMSEP* as an indicator of the prediction error.

The highest deviation with an *RMSEP* of 5.386 wt% occurs with NP5 and 15 wt% ammonium nitrate. The region at 0.1–0.2 wt% particles is significant for this, as the deviation there ranges from 7.2 to 10.2 wt%. The Raman data and scattered light data initially show no clear anomalies. Figure 8 shows that the NP5 data tend to be lower than other data points and are therefore below the regression curve. This results in an over-correction, which can be seen in the data points in question. The reverse is true for Micropearl at 10 wt% ammonium nitrate, for example, where the data points are above the regression curve and under-correct the prediction by 3.2–4.1 wt%. The best predictions occur in the range of the 0 and 5 wt% ammonium nitrate concentrations, where the deviation is as low as 0.25 wt% for Starmixx at 0 wt% ammonium nitrate and 0.311 wt% at 5 wt% ammonium nitrate. Overall, this results in an *RMSEP* of 1.176–2.896 wt% for the average observation per particle size and an *RMSEP* of 1.952 wt% when considering all measurement data. As mentioned before, the lower concentrations have a much lower *RMSEP* compared with the average value and some outliers have a higher value. But considering that the signal approaches zero without the correction, an overall averaged *RMSEP* of 1.952 wt% of the prediction gives a good indication of the actual value of the data. While this accuracy is quite high for precise concentration determinations and compared with homogeneous mixtures, the accuracy is well suited for reaction tracking or trend measurements.

Table 2. Calculated *RMSEP* for all predicted ammonium nitrate concentrations.

Particle Name (Size)	AN-Conc./wt%	*RMSEP* per AN-Conc./wt%	*RMSEP* per Particle Size/wt%	Overall Averaged *RMSEP*/wt%
NP3 (2093 μm)	0	0.325	1.356	1.952
	5	0.544		
	10	0.906		
	15	0.925		
	20	2.668		
NP5 (4089 μm)	0	0.296	2.896	
	5	1.072		
	10	1.942		
	15	5.386		
	20	2.815		
Micropearl (6604 μm)	0	0.260	1.906	
	5	0.538		
	10	2.183		
	15	1.761		
	20	3.154		
Starmixx (99,149 μm)	0	0.250	1.176	
	5	0.311		
	10	0.578		
	15	2.219		
	20	1.225		

4. Discussion

The scope of this work was to investigate the relationship between Raman measurements and dispersive systems based on a simple system of ammonium nitrate solution and glass beads and to find a way to correct the signal losses. In the previous work [27], this was performed with only one constant particle size by taking a separate transmission measurement with a separate light source. This measurement was a reference for how much light is lost due to dispersion. This has now been optimized by first replacing the transmission measurement with a scattered light probe, which measures the laser radiation, and then performing both measurements simultaneously. In addition, the substance system was expanded to include more particle sizes to obtain a more complex assessment.

The measurements of 2–99 μm particles and 0–3 wt% particle concentrations in 0–20 wt% ammonium nitrate solutions show the relationship between increasing particle concentration, decreasing Raman signal and increasing scattered light signal. The Raman data of NP3 (2.093 μm), NP5 (4.089 μm) and Micropearl (6.604 μm) provide a steep, almost linear decrease up to 0.25 wt% and then change into a flattening curve that runs almost parallel to the x-axis at 3 wt%. The Starmixx (99.149 μm) data, on the other hand, show a deviating curve over the particle concentration, as the data points fall almost linearly and show significantly lower losses. This can be explained by plotting the measured data over the specific disperse surface area.

It can be concluded that the 99.149 μm particles at the same concentration as the 2.093 μm particles have a significantly lower specific disperse surface area, which means that there are fewer boundary surfaces and, therefore, less signal interference can occur. This plot also shows that the same course of signal loss occurs for all particle sizes. For this reason, a representation via the concentration only provides a distorted image and delivers an incomplete picture. The Raman data were then plotted over the scattered light data as a change in peak height to obtain a correction function. This plot also illustrates the relationship between scattered light data and specific dispersive surface area, as the plots show the same distribution of data across the x-axis. Starmixx provides significantly lower scattered light values because of its lower specific dispersive surface area (approx. 1.5 at approx. 800 m^2/m^3) and NP3 provides the highest scattered light values as it has the largest specific dispersive surface area (approx. 2.7 at approx. 38,000 m^2/m^3). This correlation between scattered light and specific dispersive surface area shows that the scattered light measurement is a good measure and indirectly contains information on particle size and concentration.

The resulting correction was then applied to all measurement data, resulting in an average *RMSEP* for all measurement data of 1.952 wt%. However, deviations of up to 10 wt% can occur when considering individual data points. This is due to the fact that a generally applicable model was used, which can be applied to all measurement data, but is therefore also more susceptible to deviating measurement series. An average *RMSEP* of 1.952 wt% appears to be quite high for low concentrations, since in the example of 5 wt% measurements, this would represent a deviation of 40% of the measured value. On closer inspection, however, the deviations in this range are significantly lower, with an *RMSEP* of 0.311–1.072 wt% depending on the particle size under consideration. Even though for some data points, such as 15 wt% ammonium nitrate with NP5 particles, the prediction shows significant deviations, it still reflects values in the correct order of magnitude. The mean *RMSEP* is therefore not representative of every prediction of the ammonium nitrate concentration but provides a rough overview for a broad application of the model. Compared with the measurement of homogeneous mixtures, an average *RMSEP* of 1.952 wt% can be quite high, but the accuracy is still sufficient to carry out reaction tracing or investigations of concentration trends. A more specific regression and a lower *RMSEP* for individual ammonium nitrate concentrations or particle sizes would be conceivable but would have the disadvantage of greater calibration effort and would require knowledge of the ammonium nitrate concentration or particle size. However, despite deviations in individual measurement data, it can be shown that a correction is possible using this measurement technique. In further work, the findings of this study will be transferred from suspensions to emulsions and applied to more complex systems.

5. Conclusions

It can be concluded that the performed correction is a good and simple option to correct measurement data of the continuous phase that are influenced by disperse systems. The scattered light database contains the required information on particle size and concentration and achieves an *RMSEP* of 1.952 wt% when considering all measurement data. This means that no further specific knowledge of the size or number of particles is required.

Author Contributions: Conceptualization, E.S.; methodology, E.S.; validation, E.S.; formal analysis, E.S.; investigation, E.S. and V.O.; resources, M.R.; data curation, E.S. and V.O.; writing—original draft preparation, E.S.; writing—review and editing, M.R. and J.-U.R.; visualization, E.S.; supervision, M.R. and J.-U.R.; project administration, E.S. All authors have read and agreed to the published version of the manuscript.

Funding: This research received no external funding.

Institutional Review Board Statement: Not applicable.

Informed Consent Statement: Not applicable.

Data Availability Statement: The data presented in this study are available on request from the corresponding author.

Conflicts of Interest: The authors declare no conflicts of interest.

References

1. Kessler, R.W.; Kessler, W.; Zikulnig-Rusch, E. A Critical Summary of Spectroscopic Techniques and their Robustness in Industrial PAT Applications. *Chem. Ing. Tech.* **2016**, *88*, 710–721. [CrossRef]
2. Simon, L.L.; Pataki, H.; Marosi, G.; Meemken, F.; Hungerbühler, K.; Baiker, A.; Tummala, S.; Glennon, B.; Kuentz, M.; Steele, G.; et al. Assessment of Recent Process Analytical Technology (PAT) Trends: A Multiauthor Review. *Org. Process Res. Dev.* **2015**, *19*, 3–62. [CrossRef]
3. Bakeev, K.A. (Ed.) *Process Analytical Technology: Spectroscopic Tools and Implementation Strategies for the Chemical and Pharmaceutical Industries*, 2nd ed.; Wiley: Chichester, UK, 2010; ISBN 978-0-470-72207-7.
4. Parson, W.W.; Burda, C. *Modern Optical Spectoscopy: From Fundamentals to Applications in Chemistry, Biochemistry and Biophysics*; Springer International PU: Berlin/Heidelberg, Germany, 2023; ISBN 3-031-17221-3.
5. Ritgen, U. *Analytische Chemie I*; Springer Berlin Heidelberg: Berlin/Heidelberg, Germany, 2020; ISBN 3662604949.
6. Günzler, H. *IR-Spektroskopie: Eine Einführung, 4, Vollständig Überarbeitete und Aktualisierte Auflage (Online-Ausg.)*; Wiley-VCH GmbH & Co. KGaA: Weinheim, Germany, 2003; ISBN 978-3-527-30801-9.
7. Vahur, S.; Knuutinen, U.; Leito, I. ATR-FT-IR spectroscopy in the region of 500–230 cm^{-1} for identification of inorganic red pigments. *Spectrochim. Acta A Mol. Biomol. Spectrosc.* **2009**, *73*, 764–771. [CrossRef] [PubMed]
8. Shinde, G.; Godage, R.K.; Jadhav, R.S.; Manoj, B.; Aniket, B. A Review on Advances in UV Spectroscopy. *Res. J. Sci. Technol.* **2020**, *12*, 47. [CrossRef]
9. Li, W.; Luo, Y.; Wang, X.; Gong, X.; Huang, W.; Wang, G.; Qu, H. Development and Validation of a Near-Infrared Spectroscopy Method for Multicomponent Quantification during the Second Alcohol Precipitation Process of Astragali radix. *Separations* **2022**, *9*, 310. [CrossRef]
10. Wang, H.-P.; Chen, P.; Dai, J.-W.; Liu, D.; Li, J.-Y.; Xu, Y.-P.; Chu, X.-L. Recent advances of chemometric calibration methods in modern spectroscopy: Algorithms, strategy, and related issues. *TrAC Trends Anal. Chem.* **2022**, *153*, 116648. [CrossRef]
11. Mycek, M.-A. (Ed.) *Handbook of Biomedical Fluorescence*; Dekker: New York, NY, USA, 2003; ISBN 978-0-8247-0955-6.
12. Lakowicz, J.R. (Ed.) *Principles of Fluorescence Spectroscopy*; Springer US: Boston, MA, USA, 2006; ISBN 978-0-387-31278-1.
13. Krishnan, R.S.; Shankar, R.K. Raman effect: History of the discovery. *J. Raman Spectrosc.* **1981**, *10*, 1–8. [CrossRef]
14. Vaskova, H. A powerful tool for material identification: Raman spectroscopy. *Int. J. Math. Models Methods Appl. Sci.* **2011**, *5*, 1205–1212.
15. Martínez, R.I.; Ramírez, A.O.; González, R.L.; León, R.D.d.; Cárdenas, L.V.; Martínez, E.T.; Elizondo, A.D.; Vielma, B.R. Polymerization Reactor Monitoring by In-line Raman Spectrometry. *J. Mater. Sci. Eng. A* **2017**, *7*, 303–310. [CrossRef]
16. Frauendorfer, E.; Hergeth, W.-D. Industrial application of Raman spectroscopy for control and optimization of vinyl acetate resin polymerization. *Anal. Bioanal. Chem.* **2017**, *409*, 631–636. [CrossRef]
17. Wiederseiner, S.; Andreini, N.; Epely-Chauvin, G.; Ancey, C. Refractive-index and density matching in concentrated particle suspensions: A review. *Exp. Fluids* **2011**, *50*, 1183–1206. [CrossRef]
18. Kollhoff, R.T.; Kelemen, K.; Schuchmann, H.P. Local Multiphase Flow Characterization with Micro Particle Image Velocimetry Using Refractive Index Matching. *Chem. Eng. Technol.* **2015**, *38*, 1774–1782. [CrossRef]
19. Kortüm, G. *Reflectance Spectroscopy: Principles, Methods, Applications*; Softcover repr. of the hardcover 1. ed. 1969; Springer: New York, NY, USA, 2014; ISBN 978-3-642-88073-5.
20. Paul, A.; Meyer, K.; Ruiken, J.-P.; Illner, M.; Müller, D.-N.; Esche, E.; Wozny, G.; Westad, F.; Maiwald, M. Process spectroscopy in microemulsions—Raman spectroscopy for online monitoring of a homogeneous hydroformylation process. *Meas. Sci. Technol.* **2017**, *28*, 35502. [CrossRef]
21. Meyer, K.; Ruiken, J.-P.; Illner, M.; Paul, A.; Müller, D.; Esche, E.; Wozny, G.; Maiwald, M. Process spectroscopy in microemulsions—Setup and multi-spectral approach for reaction monitoring of a homogeneous hydroformylation process. *Meas. Sci. Technol.* **2017**, *28*, 35501. [CrossRef]
22. Schalk, R.; Braun, F.; Frank, R.; Rädle, M.; Gretz, N.; Methner, F.-J.; Beuermann, T. Non-contact Raman spectroscopy for in-line monitoring of glucose and ethanol during yeast fermentations. *Bioprocess Biosyst. Eng.* **2017**, *40*, 1519–1527. [CrossRef] [PubMed]
23. van den Brink, M.; Pepers, M.; van Herk, A.M. Raman spectroscopy of polymer latexes. *J. Raman Spectrosc.* **2002**, *33*, 264–272. [CrossRef]
24. Chen, X.; Laughlin, K.; Sparks, J.R.; Linder, L.; Farozic, V.; Masser, H.; Petr, M. In Situ Monitoring of Emulsion Polymerization by Raman Spectroscopy: A Robust and Versatile Chemometric Analysis Method. *Org. Process Res. Dev.* **2015**, *19*, 995–1003. [CrossRef]
25. Chang, C.; Feng, L.-F.; Gu, X.-P.; Zhang, C.-L.; Dai, L.-K.; Chen, X.; Hu, G.-H. In Situ Raman Spectroscopy Real-Time Monitoring of a Polyester Polymerization Process for Subsequent Process Optimization and Control. *Ind. Eng. Chem. Res.* **2022**, *61*, 17993–18003. [CrossRef]

26. Hufnagel, T.; Rädle, M.; Karbstein, H.P. Influence of Refractive Index Differences on the Signal Strength for Raman-Spectroscopic Measurements of Double Emulsion Droplets. *Appl. Sci.* **2022**, *12*, 9056. [CrossRef]
27. Spoor, E.; Rädle, M.; Repke, J.-U. Raman Spectroscopy of Glass Beads in Ammonium Nitrate Solution and Compensation of Signal Losses. *Sensors* **2024**, *24*, 314. [CrossRef]
28. Spoor, E.; Welzel, S.; Nieken, U.; Rädle, M. UV/VIS-Spectroscopic Inline Measurement for the Detection of Fouling Processes during the Polymerization of N-Vinylpyrrolidone. *Reactions* **2023**, *4*, 176–188. [CrossRef]
29. Schmitt, L.; Meyer, C.; Schorz, S.; Manser, S.; Scholl, S.; Rädle, A.M. Use of a Scattered Light Sensor for Monitoring the Dispersed Surface in Crystallization. *Chem. Ing. Tech.* **2022**, *94*, 1177–1184. [CrossRef]
30. Gillen, G.; Najarro, M.; Wight, S.; Walker, M.; Verkouteren, J.; Windsor, E.; Barr, T.; Staymates, M.; Urbas, A. Particle Fabrication Using Inkjet Printing onto Hydrophobic Surfaces for Optimization and Calibration of Trace Contraband Detection Sensors. *Sensors* **2015**, *15*, 29618–29634. [CrossRef] [PubMed]
31. Diaz, D.; Hahn, D.W. Raman spectroscopy for detection of ammonium nitrate as an explosive precursor used in improvised explosive devices. *Spectrochim. Acta A Mol. Biomol. Spectrosc.* **2020**, *233*, 118204. [CrossRef] [PubMed]
32. Lundeen, J.W.; Tobias, R.S. Evidence for ion pairing and solvation from Raman spectra of solutions of ammonium, silver, potassium, sodium, and calcium nitrates in liquid ammonia. *J. Chem. Phys.* **1975**, *63*, 924–934. [CrossRef]
33. Walrafen, G.E.; Blatz, L.A. Weak Raman bands from water. *J. Chem. Phys.* **1973**, *59*, 2646–2650. [CrossRef]
34. Sun, Q.; Qin, C. Raman OH stretching band of water as an internal standard to determine carbonate concentrations. *Chem. Geol.* **2011**, *283*, 274–278. [CrossRef]
35. Faber, N.M. Estimating the uncertainty in estimates of root mean square error of prediction: Application to determining the size of an adequate test set in multivariate calibration. *Chemom. Intell. Lab. Syst.* **1999**, *49*, 79–89. [CrossRef]

Disclaimer/Publisher's Note: The statements, opinions and data contained in all publications are solely those of the individual author(s) and contributor(s) and not of MDPI and/or the editor(s). MDPI and/or the editor(s) disclaim responsibility for any injury to people or property resulting from any ideas, methods, instructions or products referred to in the content.

sensors MDPI

Article

Impact of Reducing Statistically Small Population Sampling on Threshold Detection in FBG Optical Sensing

Gabriel Cibira [1,*], Ivan Glesk [2], Jozef Dubovan [2] and Daniel Benedikovič [2]

[1] Institute of Aurel Stodola, Faculty of Electrical Engineering and Information Technology, University of Zilina, Komenskeho 843, 03101 Liptovsky Mikulas, Slovakia
[2] Department of Multimedia and Information-Communication Technologies, Faculty of Electrical Engineering and Information Technology, University of Zilina, Univerzitna 1, 01026 Zilina, Slovakia; ivan.glesk@uniza.sk (I.G.); jozef.dubovan@uniza.sk (J.D.); daniel.benedikovic@uniza.sk (D.B.)
* Correspondence: gabriel.cibira@uniza.sk

Abstract: Many techniques have been studied for recovering information from shared media such as optical fiber that carries different types of communication, sensing, and data streaming. This article focuses on a simple method for retrieving the targeted information with the least necessary number of significant samples when using statistical population sampling. Here, the focus is on the statistical denoising and detection of the fiber Bragg grating (FBG) power spectra. The impact of the two-sided and one-sided sliding window technique is investigated. The size of the window is varied up to one-half of the symmetrical FBG power spectra bandwidth. Both, two- and one-sided small population sampling techniques were experimentally investigated. We found that the shorter sliding window delivered less processing latency, which would benefit real-time applications. The calculated detection thresholds were used for in-depth analysis of the data we obtained. It was found that the normality three-sigma rule does not need to be followed when a small population sampling is used. Experimental demonstrations and analyses also showed that novel denoising and statistical threshold detection do not depend on prior knowledge of the probability distribution functions that describe the FBG power spectra peaks and background noise. We have demonstrated that the detection thresholds' adaptability strongly depends on the mean and standard deviation values of the small population sampling.

Keywords: statistically small population sampling; two-sided sampling; one-sided sampling; threshold detection; fiber Bragg gratings; FBG sensing; optical spectrum analysis

Citation: Cibira, G.; Glesk, I.; Dubovan, J.; Benedikovič, D. Impact of Reducing Statistically Small Population Sampling on Threshold Detection in FBG Optical Sensing. *Sensors* **2024**, *24*, 2285. https://doi.org/10.3390/s24072285

Academic Editors: Qing Yu, Ran Tu, Ting Liu and Lina Li

Received: 1 March 2024
Revised: 27 March 2024
Accepted: 31 March 2024
Published: 3 April 2024

Copyright: © 2024 by the authors. Licensee MDPI, Basel, Switzerland. This article is an open access article distributed under the terms and conditions of the Creative Commons Attribution (CC BY) license (https://creativecommons.org/licenses/by/4.0/).

1. Introduction

Measuring the resonant wavelengths of fiber Bragg grating (FBG) sensors, finding their fingerprints, or classifying FBGs themselves in optical sensing systems require denoising the acquired FBG spectral peaks [1–7]. Meanwhile, all important properties of FBG spectral peaks have to be preserved.

To mitigate the impact of the background noise on the quality of the optical signals in telecommunications and sensing systems, hardware pre-processing using real-time wavelength filtering has been typically used [8–10]. Software approaches based on verification or post-detection algorithms can be leveraged to detect signals in a complex or fluctuating background noise environment [11,12]. In the latter case, a variety of statistical detection techniques are usually applied. Such methods utilize a preset level of an allowable detection threshold τ, relying on the signal-to-noise ratio (SNR) evaluation [12,13].

The reviewing article [14] deals with Brillouin scattering methods of determining the frequency shift of signals and approximation methods in fiber optic metrology and distributed optical fiber sensing. Based on available filtering, fitting, approximation, correlation, and control techniques, they aim to improve denoising and detection of non-FBG sensing via artificial intelligence approaches. This may be applied to improve qualitative parameters (such as resolution and accuracy) of the measurement equipment. Similarly, Brillouin optical time domain reflectometry [15,16] and optical frequency domain reflectometry [17,18] methods allow for improved signal clarity, better measurement accuracy, and enhanced signal resolution. The essence of these methods consists of processing the specified number of averaged data points in time through the data set while taking the set number of data points in the time window. This is used in various measurement techniques and is gaining attraction in distributed sensing applications.

Our focus has been on a quasi-distributed system based on FBG sensing in optical fiber, where methods based on similar (but not the same) principles can be used [19]. The goal has been to avoid complex mathematical procedures, fingerprinting databases, or artificial intelligence-based techniques [19–21]. In [22,23], we reported on a denoising technique based on a digital sliding window. A statistical detector was introduced to detect the spectral power of FBGs in an additive mixture of the signal and background noise. The statistical detector controls the power level depending on the given threshold level τ, as shown in Figure 1, indicating that the power level above the threshold belongs to the FBG signal. As is typical for FBG sensing, the statistical probability density function (pdf) achieves higher values for the FBG power spectra compared to the pdf of the background noise. This is because most of the measured values of the reflected FBG power signal are higher than the background noise values.

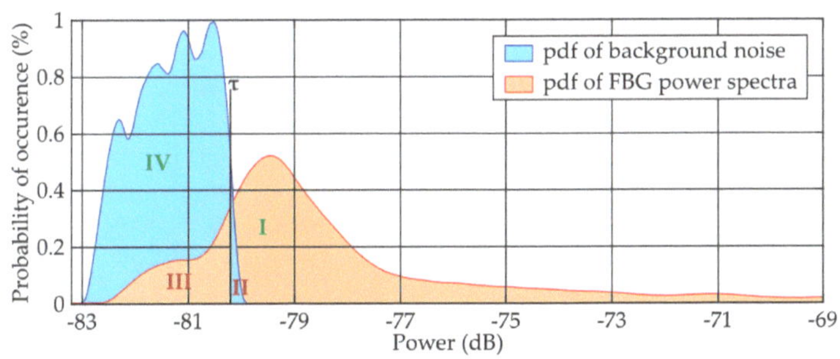

Figure 1. Principle of statistical detection of the FBG power spectra peaks in the additive mixture of the signal and background noise, where pdf means the respected probability density function, τ is the threshold in dB and I–IV are zones of correct or incorrect detection determined by the τ and pdfs.

Detection of FBG power spectra using threshold τ in the overlapping power zone (where the additive mixture contains both FBG signals and background noise) brings some risk of either loss (Zone III) of FBG detection or false alarms (Zone II) due to the noise detection. Signals detected above the threshold τ and simultaneously above the background noise (Zone I) indicate the correct decision about the presence of the FBG. Finally, the presence of FBG below the threshold (Zone IV) is also evaluated as the correct decision if it originated from background noise, called the "rejecting detection hypothesis". This is depicted in Figure 2. It is noted that the calculation of the detection threshold τ is based on Bayesian decision theory [24–26]. Typically, practical applications often aim to either maximize correct detection in Zone I or minimize false alarms in Zone II. Zone III in Figure 2 represents the detection loss and also shows the "significance level of the hypothesis test". Zone IV of the correct rejection of the hypothesis also shows the "power of the test". For example, if the desired $p_{FA} = 10^{-3}$, the power of the hypothesis test is sufficiently high at $1 - p_{FA} = 0.999$. It is understood that numerical values from Zones III and IV indicate whether or not statistical sampling is representative.

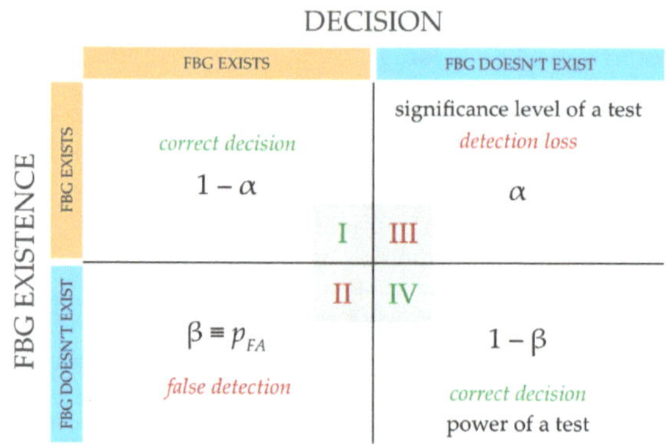

Figure 2. Basic principle of the decision-making process, where I–IV are zones of correct or incorrect detection determined by the threshold τ and pdfs.

Neyman and Pearson showed [27] that the likelihood ratio test will maximize the power of the test for a fixed population sampling for a given false alarm probability. Therefore, the likelihood ratio test is statistically the strongest hypothesis test in the sampling signal detection theory. Then, any monotonic likelihood ratio function (i.e., pdf of FBG power spectra vs. pdf of background noise) can be used as the decision variable based on comparing mean values μ (of the raw observations across the entire sample) against the threshold τ. As a consequence, optimal decisions without prior knowledge of likelihood functions can be constructed regardless of the prior knowledge of the Gaussian process regression method (or similar) [28].

In a previous study [23], a fixed K-number of the discrete power spectral samples was processed using the sliding window technique where the K-number corresponded to a number of discrete wavelength steps within the FBG bandwidth. As was shown, K can be smaller. This depends on the requirement to either increase the threshold stability or to reduce the computational complexity. Statistical tests of reliability and validity showed the limits on the smallest K in the population sampling [24–29].

In this article, we focus on determining the least necessary but sufficient number of significant samples in statistically small population sampling while minimizing the impact on statistical numerical characteristics.

1. First, we investigate the impact of two-sided sliding window sampling around the cell under the test;
2. Second, we investigate the impact of a one-sided sliding window.

In both these methods, K discrete steps related to the bandwidth of FBGs are applied.

Next, K will be gradually reduced to the smallest population sampling. This population sampling reduction will be conducted with respect to minimizing the impact on statistical detection of the FBG power spectra.

In experimental demonstrations, we will investigate how reducing K in the sliding window will impact the SNR and detection.

2. Statistical Thresholding Using Two-Sided Small Population Sampling

In this section, we study and analyze the symmetrical K-size two-sided population sampling, composed of left and right sub-windows, see Figure 3.

Figure 3 illustrates the main principle of statistical detection based on a sliding window. As we have already explained, the comparator in the statistical detector decides whether the power level in the cell under the test (CUT) is above the threshold level τ. After the presence of FBG power spectra peaks was evaluated, the window was shifted by one wavelength step, and the adjacent cell became the CUT. This is why the window is called a sliding window along the waveband. The waveband comprises N number of cells each containing different power levels of the signal with noise.

Based on Bayesian decision theory and the Neyman–Pearson approach, as well as the minimum required false detection, p_{FA}, the statistical detector is not allowed to exceed the preset value of p_{FA}. In other words, at $p_{FA} = 10^{-3}$, a maximum of 1 false threshold detection is allowed from 1000 CUTs investigated.

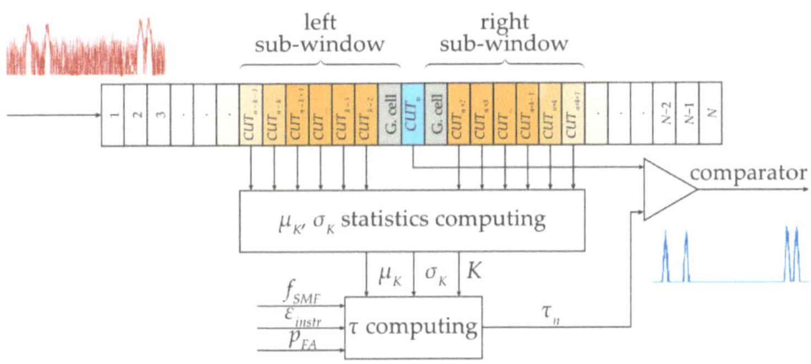

Figure 3. Concept of the statistical threshold detector of FBG power spectra peaks level with a symmetric two-sided K-size sliding window.

2.1. Statistical Threshold Calculation

The calculation of the statistical threshold τ uses statistical characteristics of the additive mixture of the signal and background noise. This comprises the mean μ_K and the standard deviation σ_K. Both characteristics are calculated from the fixed number of K cells from the left and the right sub-windows. This can be called K-sized population sampling around the CUT. By default, a symmetric K-size window is chosen due to the typical symmetric Gaussian shape of the reflected FBG power spectra peaks [8–11]. The size of K depends on the properties of FBGs, including the FBG bandwidth B_{FBG}, sensing interrogator resolution δ_{sens}, and the effects of attenuation. In the following example, let's assume B_{FBG} = 0.8 nm and δ_{sens} = 0.008 nm. Thus, the above-threshold power can be approximated from M discrete wavelength steps as $M \cong B_{FBG}/\delta_{sens} = 100$. As a rule, it is recommended to keep $K \cong M$.

First, the calculated threshold τ has to contain the noise function f_{SMF} describing the attenuation approximation of the single-mode fiber (SMF-28). Second, an instrumental error function ε_{instr} should be included. The ε_{instr} is a sum of all instrumentation errors and includes fluctuations in the wavelength discretization, quantization, deviations, or offsets due to internal or external environmental changes. Procedures for calculating total instrumentation error are explained in [30] using qualitative parameters of involved devices in the given experimental setup. In this case, the FBG sensing instrumentation error is influenced by wavelength measurements with an accuracy of ±10 pm and power measurements of ±2%, regardless of whether a slow or fast scanning mode was used. Generally, both f_{SMF} and ε_{instr} functions are stabilized during long-term use, assuming stable operating conditions for the optical fiber and FBG sensing interrogator.

Next, the calculation of τ has to include the required p_{FA} (for example, $p_{FA} = \times10^{-3} \ldots \times10^{-6}$). The smaller the p_{FA}, the higher values of τ can be achieved. However, the threshold values should range from the minimum value slightly above the background noise energy E_{min} up to the maximum expected value of the additive mixture E_{max} (the FBG power spectra peak value with background noise). Due to the above, the p_{FA} is parameterized in the range from E_{min} to E_{max}. For large sampling populations, a full parametrization is typically required and is equal to 1. However, for decreasing K-size, the p_{FA} is parameterized with lower weights. The condition $K < M$ allows for the adequate weakening of the p_{FA} parametrization.

Finally, the calculated threshold τ has to include the additive mixture $(N_0 + E_S)_k$ obtained in the given k^{th} CUT. This value should be weighted by both the mean μ_K and the standard deviation σ_K. Both are obtained from the K-size population sampling within the sliding window. If the K-size is reduced, the accuracy deteriorates and the parametrization of the $(N_0 + E_S)_k$ value is weakened accordingly. On the contrary, the increased standard deviation increases the contribution of the additive mixture in the calculation of τ.

To conclude, the parameterized p_{FA} and $(N_0 + E_S)_k$, will affect the fast dynamic adaptation of the threshold τ. The calculation of the threshold τ is given by Equation (1):

$$\tau = \varepsilon_{instr} + f_{SMF} + \frac{-E_{max} - E_{min}}{ln\frac{1}{p_{FA}}}\frac{K-G}{K} + (N_0 + E_S)_k \mu_K \sigma_K \ (dB), \quad (1)$$

where G is the number of guard cells in the neighborhood of the CUT that do not participate in the threshold calculation. In general, the more guard cells, the smaller the weight of the parameterized p_{FA}.

2.2. Experimental Demonstration and Results

The experimental setup is shown in Figure 4. Figure 4a shows the investigation of FBG power spectra in a reflection mode by an optical interrogator, and Figure 4b shows a transmission mode by a stand-alone detector/optical spectrum analyzer.

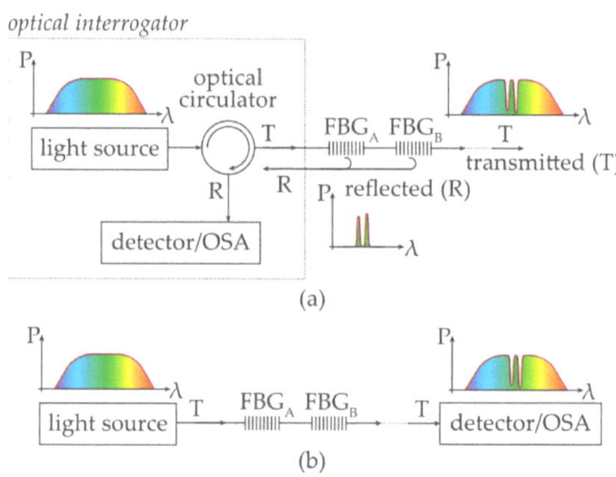

Figure 4. FBG sensing experimental setup using (**a**) an interrogator; (**b**) a stand-alone light source and a detector/optical spectrum analyzer.

The non-linear attenuation of the used optical fiber (G.652. D SMF) and the creation of the approximate broadband f_{SMF} attenuation function is described in [31]. In the experimental demonstration, various optical fiber lengths are considered, representing the range of attenuation between $-1 \ldots -45$ dB. Several FBG optical sensors with a bandwidth of $B_{FBG} \cong 0.8$ nm and maximum attenuation of -20 dB at the resonant wavelength λ_{FBG} are connected to optical fiber.

The digitized additive mixture of the signal and background noise (in the spectral domain) is continuously processed in the predefined wavelength sliding window of different K sizes. This sliding window systematically shifts and μ_K, σ_K, and τ are dynamically calculated (Equation (1)) for each of the CUT_k, see Figure 3. The G neighboring guard cells are excluded from μ_K, σ_K, and τ computing.

In Sections 2.2.1 and 2.2.2, different threshold values of τ will be determined and investigated for different values of p_{FA} for different FBG power spectra peaks and the presence of the background noise.

The interrogator processes discrete power values for each discrete wavelength in the presence of the quantization noise. To compare different effects of instrumental distortion, a commercial interrogator and a table-top analyzer with a different wavelength resolution are used in experimental investigations.

2.2.1. Experimental Investigation of Two-Sided Small Population Sampling Using Interrogator

The commercial interrogator Sylex S-line S-400 [32] was used in this study, having the wavelength resolution of δ_{sens} = 0.08 nm. Because of slightly changing BFBG under the influence of fluctuating noise, the sampling with $M = 9 \ldots 13$ discrete values and B_{FBG} $\cong$ 0.8 nm was selected. Four FBG sensors (A, B, C, and D) were deployed within the C-band. The experimental results for various detection thresholds are shown in Figure 5. In addition, interfering spectra with ten times narrower FBG bandwidths (I to X) were implemented to demonstrate the advantage of dynamic threshold adaptation. The variety of detection conditions due to partial overlap of FBG power spectra, and their varying density distribution were investigated. The calculation of threshold τ follows the procedure as described in Section 2.1 and includes all the listed components. It respects the schematic diagram shown in Figure 3. Sliding windows of size K = {8, 10, 12, 16, 24, 32, 40, 60} scan individual cells (symmetrically on left and right). This enables a step-by-step calculation of variables μ_K and σ_K within the C-band. Obtained results are shown in Figure 5.

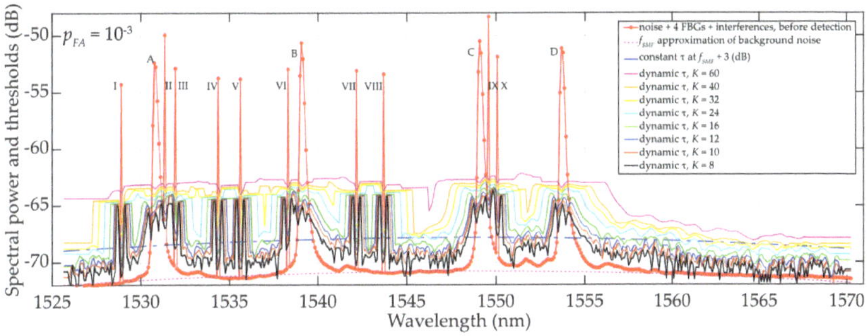

Figure 5. Results of dynamic statistical threshold detection of the wideband (A, . . ., D) and narrow-band (I, . . ., X) FBGs with different power levels, $p_{FA} \approx 10^{-3}$ and $K = 8 \ldots 60$.

Due to a rapidly rising or descending σ_K at FBG power spectra peaks edges, the dynamic threshold "shakes", especially when K = 8. In this case, the method is not appropriate for denoising threshold detection. However, the results for K = 10 or K = 12 indicate already adapted threshold τ to the additive mixture of signals and fluctuated background noise (see Equation (1)). Despite "shaky" thresholds also being seen here, they are ~0.5 to 1.5 dB, respectively, above the background noise, and therefore, the statistical detection of FBG power spectra peak levels becomes more reliable. A further increase in K over M results in "shakeless" and increased τ values, thus yielding a safer detection of FBG power spectra. Therefore, the recommended setting is $K \cong M$.

2.2.2. Experimental Investigation of Two-Sided Small Population Sampling Using Table-Top Analyzer

The analyzer AQ6370C [33] is used to process the transmitted power spectra with an oversampled wavelength resolution of δ_{sens} = 0.0035 nm. Here, the sampling is conducted with discrete values of $M = 90 \ldots 120$ and $B_{FBG} \cong$ 0.8 nm. Within the C-band, four FBG sensors (A, B, C, and D) in the C-band are used. The results of various detection threshold scenarios are shown in Figure 6. Here, the attenuation of the optical fiber was assumed -35 dB with the highly fluctuated background noise ($\sigma_{K\,N0} \cong$ 4.3 dB) and input power of $SNR_{in} \cong$ 8.5 dB. Despite these unfavorable conditions, the detectability and adaptability of τ are improved, compared to the previous case study.

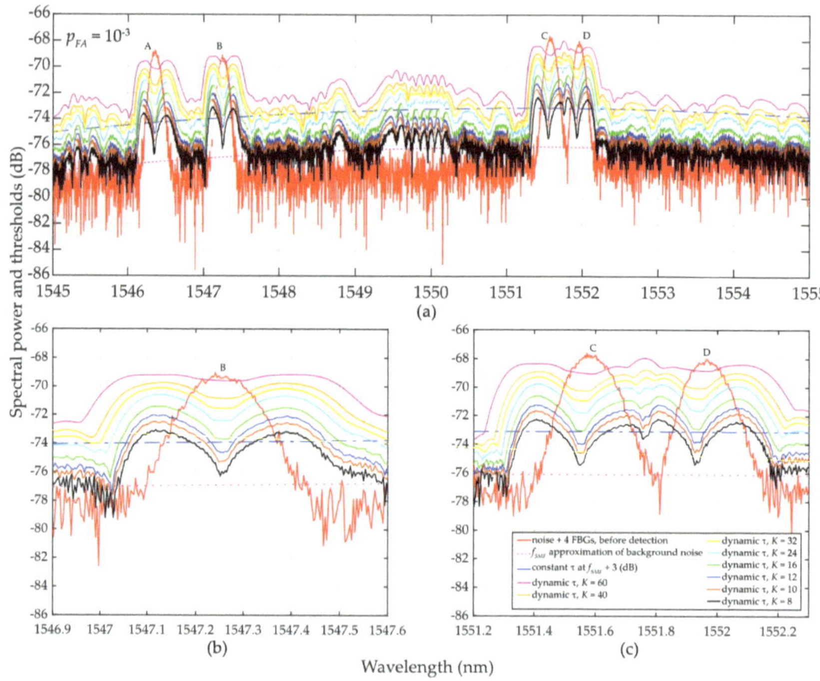

Figure 6. Results of dynamic statistical threshold detection of the wideband FBGs with different power levels, $p_{FA} \approx 10^{-3}$ and $K = 8 \ldots 60$: (**a**) for (A, …, D) FBGs; (**b**) detailed view for B FBG; (**c**) detailed view for C and D FBGs.

As in the previous case study described in Section 2.2.1, thresholds for τ are also "shaky" for the same reasons. However, for $M = 90 \ldots 120$ and $K = 8 \ldots 60$, the results obtained are significantly better despite those unfavorable detection conditions. Surprisingly, even for $K = 12$ or $K = 16$, the threshold detection results are acceptable and are comparable to the previous results in Section 2.2.1 for $K \cong M$.

In Figure 6, a sudden/significant drop in the threshold values τ in the close proximity of the FBG power spectra peaks can be noted. The deeper the drop of threshold values (especially when K is much less than M), the higher the difference, which helps to improve the SNR. These value differences are illustrated in Figure 7.

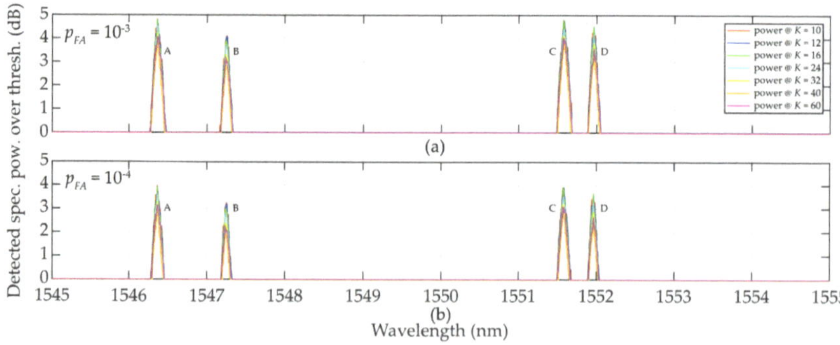

Figure 7. Illustration of value differences of the wideband (A, …, D) FBGs when $K = 10 \ldots 60$: (**a**) $p_{FA} \approx 10^{-3}$; (**b**) $p_{FA} \approx 10^{-4}$.

In Figure 7a, $p_{FA} \approx 10^{-3}$ and $K = 10 \ldots 32$, a random low-level false detection occurred for power levels below 1.5 dB and the detected FBG power level spectra approaching 5 dB.

When $p_{FA} \approx 10^{-4}$ (case Figure 7b), all thresholds rise to their higher level. As a consequence, no false detections were observed for any K-size. This maintained the reliable detection of the FBG power spectra peak levels without false detections. However, the highest level of $K = 60$ (the strictest threshold) causes a decrease in those values above the threshold.

2.3. Threshold Behavior Analysis and Discussion

In this subsection, a mathematical analysis of the threshold calculation is presented and implications for the detection of the FBG power spectra peaks are derived.

Let us first analyze the parameterization of the 3rd component of Equation (1). We assume a typical FBG power attenuation ranging in the interval ($E_{max} \ldots E_{min}$) = 20 dB and a typical maximum value of $p_{FA} \leq 10^{-3}$. As a result, the 3rd component in Equation (1) ranges from −1.45 to −2.8 dB for the population sampling $K = 4 \ldots 60$, and assuming 1 guard cell adjacent to the CUT in each of the sub-windows:

$$\frac{-E_{max}-E_{min}}{ln\frac{1}{p_{FA}}}\frac{K-G}{K} = \frac{|E_{max}-E_{min}|}{ln\frac{1}{p_{FA}}}\frac{K-G}{K} = \frac{20}{ln\frac{1}{0.001}}\frac{K-G}{K} =$$

$$= \frac{20}{-6.90776}\frac{K-G}{K} = -2.895\frac{K-G}{K} = -2.895\begin{cases}\frac{4-2}{4} = 0.5\\ \vdots\\ \frac{60-2}{60} = 0.9\overline{6}\end{cases} = \begin{cases}-1.4476\\ \vdots\\ -2.7985\end{cases}(dB). \quad (2)$$

If $K = 60$, the 3rd component parametrization is equal to −2.8 and will cause a reduced threshold τ. On the contrary, for $K = 4$, the 3rd component parametrization is equal to −1.45 and will cause an increased τ, see Equation (1). This property can be used to set the value of τ which will be used later.

Next, the 4th component in Equation (1) will be analyzed in the presence of background noise only ($(N_0 + E_S)_k = N_{0k}$). Here, the K value affects the threshold calculation through changes of statistical characteristics of μ_K and σ_K as follows:

$$\mu_K = \frac{1}{K}\sum_{i=1}^{K} x_i = N_0 = \sum_{i=1}^{K} x_i\begin{cases}\frac{1}{30-2} = \frac{1}{28} = 0.0357\\ \vdots\\ \frac{1}{60-2} = \frac{1}{58} = 0.0172\end{cases}(dB), \quad (3)$$

$$\sigma_K = \sqrt{\frac{\sum_{i=1}^{K}(x_i-\mu_K)^2}{K-1}} = \frac{\sqrt{\sum_{i=1}^{K}(x_i-\mu_K)^2}}{\sqrt{K-1}} = \sqrt{\sum_{i=1}^{K}(x_i-\mu_K)^2}\begin{cases}\frac{1}{\sqrt{30-2}} = 0.189\\ \vdots\\ \frac{1}{\sqrt{60-2}} = 0.131\end{cases}(dB). \quad (4)$$

Finally, the total contribution of the 4th component to the threshold calculation of Equation (1) without the occurrence of FBG power spectra peak is:

$$\mu_K\sigma_K = \sum_{i=1}^{K} x_i\sqrt{\sum_{i=1}^{K}(x_i-\mu_K)^2}\begin{cases}0.0357{\cdot}0.189 = 0.006747\\ \vdots\\ 0.0172{\cdot}0.131 = 0.002253\end{cases}(dB). \quad (5)$$

From the above, it can be shown that an approximately 3-fold increase in the contribution of the 4th component (from 0.002253 in the case of $K = 60$ to 0.006747 in the case of $K = 30$, respectively) can be achieved, thus contributing to the threshold level increases.

Let's now analyze the μ_K-parameterization of $(N_0 + E_S)_k$ in Equation (1) when the approximately Gaussian-shaped FBG power spectra peak overlaps with the sliding window. Using the example from Section 2.2.1, and using $K = M/2 \cong 60$, the mean values keep increasing from the lowest to the highest values just in 30 wavelength steps. At the 60th step, where CUT_k contains the FBG power spectra peak maxima, the two-sided sampling reaches the value $\mu_{K=60} \cong 2 \cdot 0.25 \cdot (E_{max} \ldots E_{min})$ above the N_0 noise level, see Figure 8a. When the sliding window touches the falling edge of the FBG power spectra peak and starts to leave it, the mean value μ_K starts decreasing. However, for $M/2 \cong 60$ and $K = 30$, the behavior of the mean values will remain mostly unchanged. To be noted, for the 60th step, the shorter the sliding window, the higher the mean value μ_K. When $K = 10$, $\mu_{K=10} \cong 0.707 \cdot (E_{max} \ldots E_{min})$.

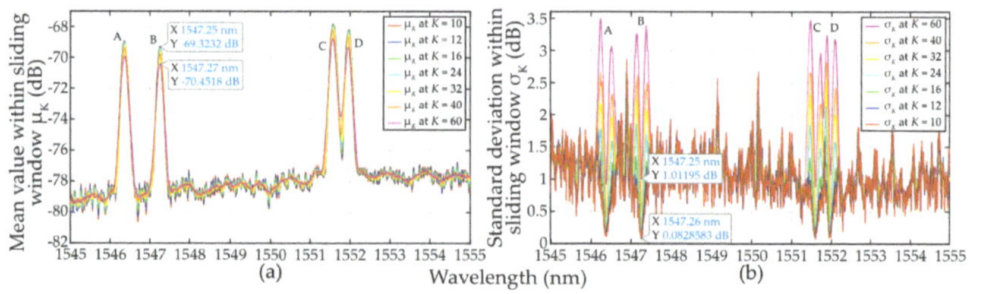

Figure 8. Statistical characteristics of the additive mixture of the (A, ..., D) FBGs signal and background noise within sliding windows when $K = 10 \ldots 60$: (**a**) mean values μ_K; (**b**) standard deviation values σ_K.

Now we analyze the σ_K-parameterization of $(N_0 + E_S)_k$ in Equation (1) when the approximately Gaussian shaped FBG power spectra peak overlaps with the sliding window. This is shown in Figure 8b. For $K = M/2 \cong 60$, the standard deviation values achieve the highest values in the ~30th and ~90th steps. Here, the square root multiplier in Equation (4) achieves the widest span of input values. It is worth noting that the σ_K-parametrization on the leading and falling edges can reach similar effects as the μ_K-parametrization. This depends on the steepness of the edges. When the sampling window slides from the 30th to the 90th step, the σ_K value drops. For $M \cong 60$ and $K = 30$, the behavior of the standard deviation values will be similar to the case of $K = M/2 \cong 60$. To be noted, the longer the sliding window, the larger the standard deviation σ_K. This is the origin of threshold adaptability.

Finally, a comparison of the magnitudes μ_K and σ_K in Figure 8 shows the difficulty of meeting the three-sigma rule (known also as the 68-95-99.7 rule) that is used to verify the normality of population sampling. This rule is also used for percentage quantification of reliability of population sampling from selected values (here, the selection of cells in the two-sided sliding window) in the following way. If ~68% of these values are from the interval $(\mu \pm 1\sigma)$, ~95% from the interval $(\mu \pm 2\sigma)$, and 99.7% from the interval $(\mu \pm 3\sigma)$, respectively, a randomly selected sample can be considered the Gaussian normal distribution. As an example, we analyze one of the measured FBG spectral peaks measured by the table-top analyzer (see Section 2.2.2). Here, the FBG B maxima of μ_K values spanning from -70.3659 to -69.3485 dB (depending on K-size) is reached for $\lambda_{FBG} = 1547.25$ nm (see Figure 7a). This corresponds to values of σ_K spanning from 0.08286 to 1.01195 dB (see Figure 7b). Based on the three-sigma rule, 99.7% of the values should have been within the interval $(-69.3485 \pm 3 \cdot 0.08286) = (-69.5971 \ldots -69.0999)$ dB for two-sided $K = 10$ but is not. As shown in Figure 6, the values inside the sliding window are from the interval $(-69.308 \ldots -68.734)$ dB. Similarly, the three-sigma rule is not fulfilled for two-sided $K = 60$ because the values span in interval $(-78.688 \ldots -68.734)$ dB, which is out of the required interval $(-70.3959 \pm 3 \cdot 1.01195) = (-73.43175 \ldots -67.36)$ dB, see Figure 6. The same applies to the other K-sizes and μ and σ values of other wavelengths. However, it needs to be noted that in cases for $K = 10 \ldots 16$, the three-sigma rule is less broken compared to cases for $K = 40 \ldots 60$. In spite of this, the use of two-sided small population sampling is reliable for successful statistical threshold detection. This is illustrated by results in Figures 5–7.

3. Statistical Thresholding Using One-Sided Small Population Sampling

In this section, we study and analyze the impact of the population sampling using one (left) sided window having an asymmetrical $K/2$-size, see Figure 9.

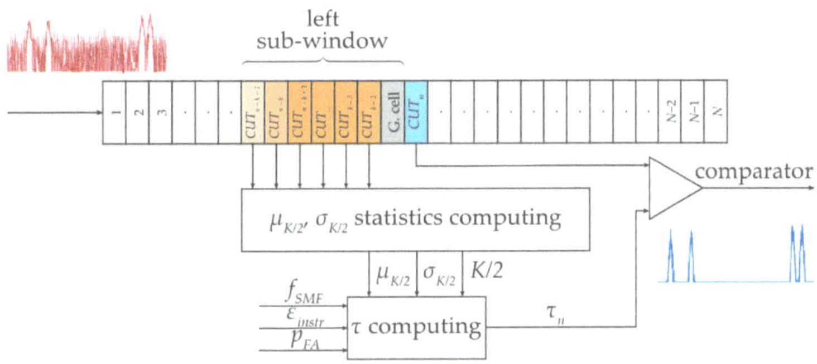

Figure 9. Concept of the statistical threshold detector of FBG power spectra peaks level with an asymmetric one-sided $K/2$-size sliding window.

3.1. Statistical Threshold Calculation

The calculation of the statistical threshold τ uses the mean $\mu_{K/2}$ values and the standard deviation $\sigma_{K/2}$ values obtained from the fixed number of $K/2$-cells contained in the left sub-window. Please note that the sliding window is asymmetrically located, in this case sitting on the left side, see Figure 9. This reduces the computation complexity by excluding the right-side sub-window from population sampling. In the next step, we will investigate the impact of this approach on the quality of the threshold detection results. Since the shape of the reflected FBG power spectra is typically a symmetric Gaussian function, we need to learn if in this approach the $\mu_{K/2}$, $\sigma_{K/2}$, and τ would differ from the two-sided μ_K and $\sigma_{K/2}$, respectively. Similar to Section 2.1, the calculation of the threshold τ will use Equation (1).

3.2. Experimental Demonstration and Results

To compare the effect of halving the population sampling, the same considerations, instrumentation, and conditions are applied in the experimental investigation as in the previous Section 2.2.

3.2.1. Experimental Investigation of One-Sided Small Population Sampling Using Interrogator

Here, as described in Section 2.2.1, the same commercial interrogator and the same deployment scenario of four FBG sensors A, B, C, and D along the C-band with $M = 9 \ldots 13$, interfered by narrowband FBGs I . . . X, was used.

Results are shown in Figure 10 indicating various threshold detection. As the sliding window approaches individual FBG power spectra levels (Note: sliding window shifts from lower to higher wavelengths), the thresholds keep increasing very slowly. This is due to a weak mean value of $\mu_{K/2}$. Please compare the results in Figure 5 and the discussed effect of μ_K in Section 2.3. After passing through the FBG power spectra peaks, the threshold behavior stabilizes. This is similar to the behavior shown in Figure 5. However, the cell reduction from K to $K/2$ noticeably causes higher fluctuations of $\sigma_{K/2}$ thus calculated τ, especially when $K/2 < M$. Therefore, this scenario is not recommended for system operation. On the other hand, thresholds τ for $K/2 = 8 \ldots 16 \cong M$ adapt very well to the signal/noise behavior. For further increases in $K/2$, when $K/2 > M$, threshold τ rises accordingly, which may lead to the power loss related to the right side of the FBG power spectra. In summary of the above, it is recommended to use $K/2 \cong M$.

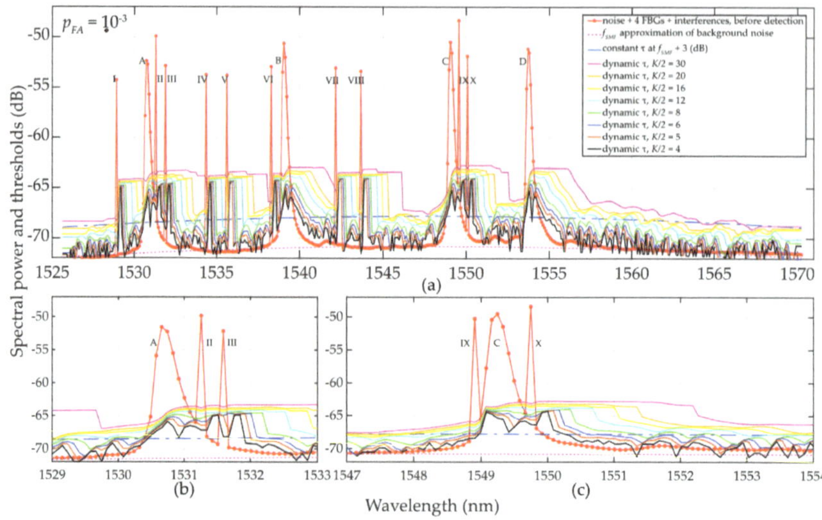

Figure 10. Results of dynamic statistical threshold detection (**a**) of the wideband (A, . . ., D) and narrowband (I, . . ., X) FBGs with different power levels, $p_{FA} \approx 10^{-3}$ and *K*/2 = 4 . . . 30; (**b**) detailed view for A, I and II FBGs; (**c**) detailed view for C, IX and X FBGs.

3.2.2. Experimental Investigation of One-Sided Small Population Sampling Using Table-Top Analyzer

The same table-top analyzer was used to process the transmitted power spectra of four deployed FBG sensors within the optical fiber C-band using values *M* = 90 . . . 120. The experimental results of various detection thresholds τ are shown in Figure 11.

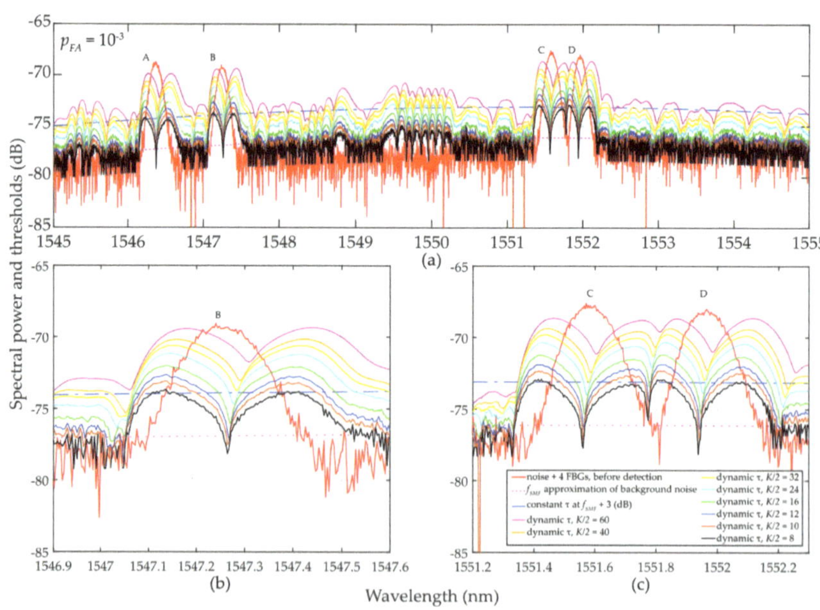

Figure 11. Results of dynamic statistical threshold detection of the wideband FBGs with different power levels, $p_{FA} \approx 10^{-3}$ and *K*/2 = 4 . . . 30 (**a**) for (A, . . ., D) FBGs; (**b**) detailed view for B FBG; (**c**) detailed view for C and D FBGs.

It can be noted again that the thresholds "shake", here slightly more than in the case illustrated in Section 2.2.2. This is due to the smaller mean $\mu_{K/2}$ values compared to "shaky" μ_K values. As the sliding window approaches FBG power spectra peaks, the threshold values increase slowly and better adapt to FBG power spectra levels with background noise. After passing the FBG power spectra peak maxima, the behavior of the threshold levels stabilizes similarly to those in Figure 6. A greater "shaking" was observed for $K/2 = 4 \ldots 6 << M$. This causes rising in false detection. For $K/2 = 8 \ldots 16$ (smaller than M), the detection results are acceptable and comparable to the investigation in Section 2.2.2 with $K = 12$ or $K = 16$ when $K = M$. The previously observed phenomenon of dropping τ values in the vicinity of FBG power spectra peaks appeared also here and again helped to improve the SNR, see Figure 12.

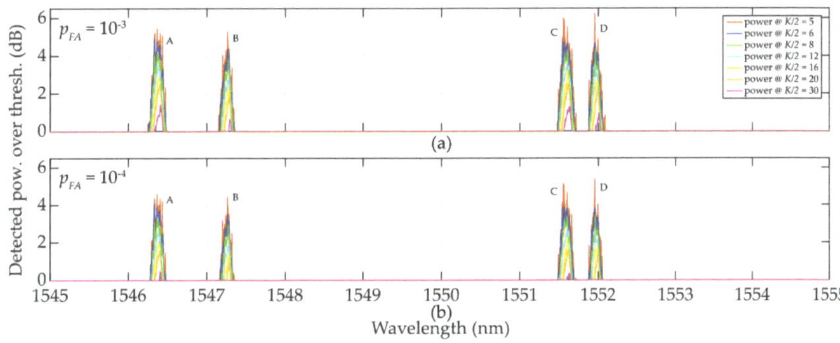

Figure 12. Illustration of value differences of the wideband (A, …, D) FBGs when $K/2 = 8 \ldots 30$: (**a**) $p_{FA} \approx 10^{-3}$; (**b**) $p_{FA} \approx 10^{-4}$.

The detected power levels of the 4 FBGs by using one-sided sliding windows for $K/2 = 5 \ldots 30$ are shown in Figure 12. In cases when $K/2 = 5$ or $K/2 = 6$, thresholds are "shaky" thus the threshold detection is not reliable, leading to increased false detections. In contrast to Figure 7, all FBG power spectra are tilted and sharpened. This is an artifact caused by one-sided population sampling. When $p_{FA} \approx 10^{-3}$ (see Figure 12a), false detections are noted at levels below 1 dB when $K/2 = 8$. No false detections occur for $K/2 > 8$. Contrary to the results shown in Figure 7, here the detected power levels of the FBG power spectra are slightly higher, thus the one-sided population sampling performs better than using two-sided population sampling in Section 2.2.2.

When $p_{FA} \approx 10^{-4}$, shown in Figure 12b, all threshold levels are increased. Therefore, no false detections are noted when $K/2 = 8 \ldots 30$. Here, the large fluctuation of the background noise resulted in the "shaky" threshold behavior. Similar to the situation in Figure 12a, the $K/2 = 12 \ldots 30$ values allow maintaining the FBG power spectra levels at the reliable detection level thus without false detections. However, in contrast to Figure 7, there is no loss in the FBG power spectra detection when $K/2 = 60$.

3.3. Threshold Behavior Analysis and Discussion

In this subsection, a brief analysis of the threshold calculation and behavior is given.

Let's analyze the parameterization of the 3rd component of Equation (1). Considering the same conditions as in Section 2.3, but half K to $K/2 = 2 \ldots 30$, Equation (2) will be as follows:

$$\frac{-E_{max} - E_{min}}{ln\frac{1}{p_{FA}}} \frac{K/2 - G}{K/2} = \frac{20}{ln\frac{1}{0.001}} \frac{K/2 - G}{K/2} = 2.895 \begin{cases} \frac{2-1}{2} = 0.5 \\ \vdots \\ \frac{30-1}{30} = 0.9\bar{6} \end{cases} = \begin{cases} -1.4476 \\ \vdots \\ -2.7985 \end{cases} (dB). \quad (6)$$

The solutions of Equations (2) and (6) in terms of K is a value equal to $K/2$ that is identical for both one-sided and two-sided population sampling.

Let us now analyze the parameterization of the last component of Equation (1) with the presence of the background noise only, $(N_0 + E_S)_k = N_{0k}$. Here, the $K/2$ value affects the threshold calculation through changes of statistical characteristics of the $\mu_{K/2}$ and $\sigma_{K/2}$:

$$\mu_{K/2} = \frac{1}{K/2}\sum_{1=1}^{K/2} x_i = N_0 = \sum_{1=1}^{K/2} x_i \begin{cases} \frac{1}{15-1} = \frac{1}{14} = 0.0714 \\ \vdots \\ \frac{1}{30-1} = \frac{1}{29} = 0.0344 \end{cases} (dB), \tag{7}$$

$$\sigma_{K/2} = \sqrt{\frac{\sum_{i=1}^{K/2}(x_i - \mu_{K/2})^2}{K/2-1}} = \sqrt{\sum_{i=1}^{K/2}(x_i - \mu_K)^2} \begin{cases} \frac{1}{\sqrt{15-1}} = 0.267 \\ \vdots \\ \frac{1}{\sqrt{30-1}} = 0.186 \end{cases} (dB) \tag{8}$$

Here, the numerical values of $\mu_{K/2}$ found in Equation (7) are half of the μ_K values found in Equation (3). Because the number of the selected values x_i is also halved, then $\mu_{K/2} \cong \mu_K$ in cases of uniform statistical distribution. The numerical values of $\sigma_{K/2}$ found from Equation (7) differ from σ_K given by Equation (4). In summary, due to computational demands, the selection of $K/2$ over K is preferable but is governed by the availability of a number of cells with required properties.

Based on Equation (7), we have also analyzed the $\mu_{K/2}$-parameterization of $(N_0 + E_S)_k$ in the case of when the leading edge of FBG power spectra overlaps with the sliding window. This is shown in Figure 13a in the case of small population sampling when $K/2 = 4 \ldots 30$. The increase of the mean $\mu_{K/2}$ values is similar to the case of μ_K when $K = 8 \ldots 32$ (see Figure 8a). Overall, the behavior of the mean of $\mu_{K/2}$ values related to any FBG power spectra in the presence of noise is nearly identical, except when $K = 32 \ldots 60$ where μ_K is slightly lower, see Figure 8, and σ_K fluctuates massively.

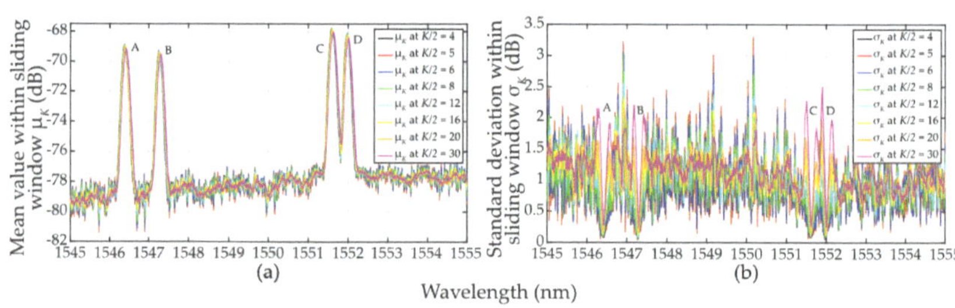

Figure 13. Calculated statistical characteristics of the (A, …, D) FBG power spectra levels plus background noise within sliding windows when $K/2 = 4 \ldots 30$: (**a**) mean values μ_K; (**b**) standard deviation values σ_K.

Based on Equation (8), we then analyzed the $\sigma_{K/2}$-parameterization of $(N_0 + E_S)_k$ in the case when the leading and falling edges of FBG power spectra are part of the sliding window. For $K/2 << M$ or $K/2 < M$, the standard deviation exhibits two local maxima in close proximity to the FBG power spectra peak, see Figure 13b. It can be seen that the $\sigma_{K/2}$ value drops in between the two $\sigma_{K/2}$ maxima. The smaller the population sampling, the deeper the drop of $\sigma_{K/2}$. When comparing $\sigma_{K/2}$ in Figure 13b to σ_K in Figure 8b, due to smaller population sampling, the $\sigma_{K/2}$ maxima values are smaller. The left one-sided population sampling causes a higher rise of the left $\sigma_{K/2}$ maxima compared to the one on its right. This obscures the threshold detection levels. From the above and Figures 10 and 11, it can be concluded that the thresholds on the left side of the FBG power spectra peaks are better adapted to the signal plus background noise levels. Therefore, the outcomes of the "$K/2$ approach" described in Section 3 are superior to those "K approach" described in Section 2.

Finally, a comparison of the magnitudes of μ_K and σ_K in Figure 13 indicates that the three-sigma rule has not been met. Figure 14 reflects an in-depth analysis of the non-fulfillment of the three-sigma rule. It is based on 8750 measurement points taken within the optical fiber C-band. The results obtained for different sizes of K from 4 to 60 are shown for two-sided (blue line) and one-sided (magenta line) sliding windows, respectively. It can be seen that the three-sigma rule is better fulfilled for sliding window $K = 11, \ldots, 16$ (when compared to the rest of the cases). Near plateau response has been observed for both plotted dependencies when K was between 22 and 60 leading to 45 to 56 cases of the three-sigma rule violations. The local minima of both dependencies can be seen around $K = 12$ (11 to 16 is acceptable). In a given additive mixture of the FBG power spectra signal and background noise, those values are considered as optimal and, therefore, recommended for denoising and threshold detection. It can also be concluded that, despite the fact that the three-sigma rule is *not* fulfilled, the detection using one-sided small population sampling would also be reliable. This can be confirmed by examining Figures 5–7 and 10–12.

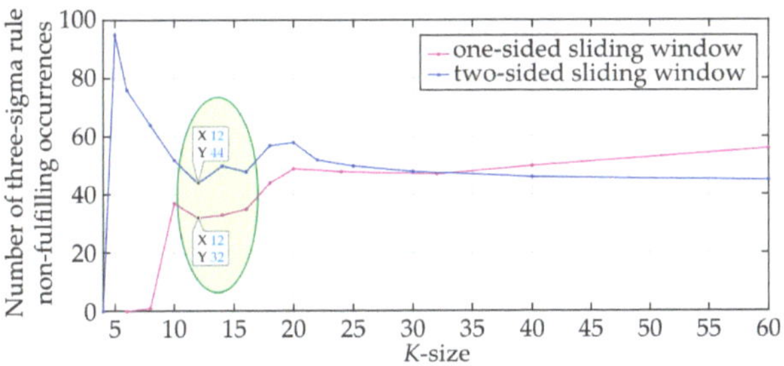

Figure 14. Number of three-sigma rule non-fulfilling occurrences for the mixture of the FBG power spectra and background noise for $K = 4, \ldots, 60$ when using a two-sided (blue line)/one-sided (magenta line) sliding window, respectively.

4. Conclusions

In our recent [23] digital sliding window denoising technique, a number of discrete power spectral population samples were processed. In order to increase computational efficiency, it is possible in some cases to reduce population sampling while maintaining the success of statistical detection. In this article, we focused on determining the small population sampling for sufficient detection of fiber Bragg gratings power spectra in an optical fiber sensing system. For such statistical threshold detection, the highest allowed number of false detections is set, which is based on the Bayesian principle.

In this article, the two-sided and one-sided statistical detectors have been introduced with reduced population sampling. In addition to the explanation of the method and the introduction of the main algorithms, a mathematical assessment of the impact of statistical characteristics, mean, and standard deviation, are presented for various population sampling reductions. Next, reduced population sampling is applied using two common instrumentations for fiber optic sensing: a commercial interrogator with standard wavelength resolution and a laboratory analyzer with improved wavelength resolution. We thereby confirmed the success of the statistical threshold detection under various conditions of fluctuating background noise, signal-to-noise ratio, approaching the adjacent fiber Bragg grating power spectra, and interferences by other signals. As from the demonstrated examples, statistical characteristics' impact on statistical threshold detection was deeply analyzed for different false detection requirements. We have also shown that for the two-sided $K = 11 \ldots 16$ and for the one-sided $K/2 = 5 \ldots 8$ population sampling the majority of cases obey the three-sigma rule. As a result, in the case of a reduced number of samples (11 to 16), the denoising and detection will benefit from implementing the two-sided sliding window. Similarly, in the case of implementing the one-sided sliding window, using 5 to 12 samples is recommended. For higher K values, where the three-sigma rule is only loosely fulfilled, some decrease in the detected FBG power spectra will be observed.

Author Contributions: Conceptualization, G.C.; methodology, G.C.; software, G.C.; validation, G.C., D.B. and J.D.; formal analysis, G.C.; investigation, G.C. and J.D.; resources, G.C.; data curation, G.C. and J.D.; writing—original draft preparation, G.C.; writing—review and editing, G.C., D.B. and I.G.; visualization, G.C. and J.D.; supervision, G.C.; project administration, G.C. and D.B.; funding acquisition, D.B. All authors have read and agreed to the published version of the manuscript.

Funding: This research was funded by the Slovak Grant Agency (VEGA) under Grant 1/0113/22 and in part by the Slovak Research and Development Agency under Grant APVV-21-0217.

Institutional Review Board Statement: Not applicable.

Informed Consent Statement: Not applicable.

Data Availability Statement: Data files or software solution can be made available after meaningful agreement with the authors.

Conflicts of Interest: The authors declare no conflict of interest. The funders had no role in the design of the study; in the collection, analyses, or interpretation of data; in the writing of the manuscript; or in the decision to publish the results.

References

1. Li, H.; Li, K.; Li, H.; Meng, F.; Lou, X.; Zhu, L. Recognition and classification of FBG reflection spectrum under non-uniform field based on support vector machine. *Opt. Fiber Technol.* **2020**, *60*, 102371. [CrossRef]
2. Mustapha, S.; Kassir, A.; Hassoun, K.; Dawy, Z.; Abi-Rached, H. Estimation of crowd flow and load on pedestrian bridges using machine learning with sensor fusion. *Autom. Constr.* **2020**, *112*, 103092. [CrossRef]
3. Lv, Z.; Wu, Y.; Zhuang, W.; Zhang, X.; Zhu, L. A multi-peak detection algorithm for FBG based on WPD-HT. *Opt. Fiber Technol.* **2022**, *68*, 102805. [CrossRef]
4. Yan, Q.; Che, X.; Li, S.; Wang, G.; Liu, X. π-FBG fiber optic acoustic emission sensor for the crack detection of wind turbine blades. *Sensors* **2023**, *23*, 7821. [CrossRef] [PubMed]
5. Zhichao, L.; Xi, Z.; Taoping, S.; Jiahe, M. Heartbeat and respiration monitoring based on FBG sensor network. *Opt. Fiber Technol.* **2023**, *81*, 103561. [CrossRef]
6. Liu, Q.; Yu, Y.; Han, B.S.; Zhou, W. An improved spectral subtraction method for eliminating additive noise in condition monitoring system using fiber Bragg grating sensors. *Sensors* **2024**, *24*, 443. [CrossRef] [PubMed]
7. Zhuang, Y.; Han, T.; Yang, Q.; O'Malley, R.; Kumar, A.; Gerald, R.E., II; Huang, J. A Fiber-optic sensor-embedded and machine learning assisted smart helmet for multi-variable blunt force impact sensing in real time. *Biosensors* **2022**, *12*, 1159. [CrossRef] [PubMed]
8. Madsen, C.K.; Lenz, G. Optical all-pass filters for phase response design with applications for dispersion compensation. *IEEE Phot. Technol. Lett.* **1998**, *10*, 994–996. [CrossRef]
9. Kumar, S.; Sengupta, S. Efficient detection of multiple FBG wavelength peaks using matched filtering technique. *Opt. Quantum Electron.* **2022**, *54*, 89. [CrossRef]
10. Tosi, D. Review and analysis of peak tracking techniques for fiber Bragg grating sensors. *Sensors* **2017**, *17*, 2368. [CrossRef]
11. Guo, Y.; Yu, C.; Ni, Y.; Wu, H. Accurate demodulation algorithm for multi-peak FBG sensor based on invariant moments retrieval. *Opt. Fiber Technol.* **2020**, *54*, 9. [CrossRef]
12. Meshcheryakov, R.; Iskhakov, A.; Mamchenko, M.; Romanova, M.; Uvaysov, S.; Amirgaliyev, Y.; Gromaszek, K. A Probabilistic approach to estimating allowed SNR values for automotive LiDARs in 'smart cities' under various external influences. *Sensors* **2022**, *22*, 609. [CrossRef] [PubMed]
13. Chen, Y.; Liu, Z.; Liu, H. A method of fiber Bragg grating sensing signal de-noise based on compressive sensing. *IEEE Access* **2018**, *6*, 28318–28327. [CrossRef]
14. Krivosheev, A.I.; Barkov, F.L.; Konstantinov, Y.A.; Belokrylov, M.E. State-of-the-art methods for determining the frequency shift of Brillouin scattering in fiber-optic metrology and sensing. *Instrum. Exp. Tech.* **2022**, *65*, 687–710. [CrossRef]
15. Lu, Y.; Zhu, T.; Chen, L.; Bao, X. Distributed vibration sensor based on coherent detection of Phase-OTDR. *J. Light. Technol.* **2010**, *28*, 3243–3249. [CrossRef]
16. Peled, Y.; Motil, A.; Tur, M. Fast Brillouin optical time domain analysis for dynamic sensing. *Opt. Express* **2012**, *20*, 8584–8591. [CrossRef]
17. Liu, Q.; Fan, X.; He, Z. Time-gated digital optical frequency domain reflectometry with 1.6-m spatial resolution over entire 110-km range. *Opt. Express* **2015**, *23*, 25988–25995. [CrossRef] [PubMed]
18. Bai, Q.; Zhang, K.; Liang, C.; Wang, Y.; Gao, Y.; Zhang, H.; Jin, B. Fast strain measurement in OFDR with the joint algorithm of wavelength domain differential accumulation and local cross-correlation. *J. Light. Technol.* **2023**, *41*, 6599–6607. [CrossRef]
19. Anfinogentov, V.; Karimov, K.; Kuznetsov, A.; Morozov, O.G.; Nureev, I.; Sakhabutdinov, A.; Lipatnikov, K.; Hussein, S.M.R.H.; Ali, M.H. Algorithm of FBG spectrum distortion correction for optical spectra analyzers with CCD elements. *Sensors* **2021**, *21*, 2817. [CrossRef]

20. Kahandawa, G.C.; Epaarachchi, J.; Wang, H.; Canning, J.; Lau, K.T. Extraction and processing of real time of embedded FBG sensors using a fixed filter FBG circuit and an artificial neural network. *Measurement* **2013**, *46*, 4045–4051. [CrossRef]
21. Encinas, L.S.; Zimmermann, A.C.; Veiga, C.L.N. Fiber Bragg grating signal processing using artificial neural networks, an extended measuring range analysis. In Proceedings of the 2007 SBMO/IEEE MTT-S International Microwave and Optoelectronics Conference, Salvador, Brazil, 29 October–1 November 2007; pp. 671–674. [CrossRef]
22. Cibira, G. Simplified statistical thresholding techniques for dynamic bandwidth allocation in shared Super-PON. In Proceedings of the 2022 ELEKTRO, Krakow, Poland, 23–26 May 2022; p. 5. [CrossRef]
23. Cibira, G.; Glesk, I.; Dubovan, J. SNR-based denoising dynamic statistical threshold detection of FBG spectral peaks. *J. Light. Technol.* **2023**, *41*, 2526–2539. [CrossRef]
24. Spiegelhalter, D.J.; Best, N.G.; Carlin, B.P.; van der Linde, A. Bayesian measures of model complexity and fit. *J. Roy. Stat. Soc. Ser. B* **2002**, *64*, 583–639. [CrossRef]
25. Johnson, N.L.; Kotz, S.; Balakrishnan, N. *Continuous Univariate Distributions*, 2nd ed.; John Wiley & Sons: New York, NY, USA, 1994; Volume 1, pp. 573–627.
26. Jacod, J.; Protter, P. *Probability Essentials*, 2nd ed.; Springer: Berlin/Heidelberg, Germany, 2004.
27. Neyman, J.; Pearson, E.S. On the problem of the most efficient tests of statistical hypotheses. *Philos. Trans. Roy. Soc. Lond. A* **1933**, *231*, 694–706. Available online: https://royalsocietypublishing.org/doi/epdf/10.1098/rsta.1933.0009 (accessed on 23 February 2024).
28. Griffith, T.; Baker, S.-A.; Lepora, N.F. The statistics of optimal decision making: Exploring the relationship between signal detection theory and sequential analysis. *J. Math. Psychol.* **2021**, *103*, 102544. [CrossRef]
29. Jones, A.R. *Probability, Statistics and Other Frightening Stuff*, 1st ed.; Routledge—Taylor & Francis Group: New York, NY, USA, 2018; Volume II, pp. 1–439.
30. Cibira, G. PV cells electrical parameters measurement. *J. Electr. Eng.* **2017**, *7*, 74–77. [CrossRef]
31. ITU-T, G. 652: Characteristics of a Single-Mode Optical Fibre and Cable. Available online: https://www.itu.int/rec/T-REC-G.652-201611-I/en (accessed on 23 February 2024).
32. Sensing Systems. Available online: https://www.sylex.sk/products/sensing-systems/interrogators/ (accessed on 23 February 2024).
33. Optical Spectrum Analyzer. Available online: https://cdn.tmi.yokogawa.com/files/uploaded/BUAQ6370C_01EN.pdf (accessed on 23 February 2024).

Disclaimer/Publisher's Note: The statements, opinions and data contained in all publications are solely those of the individual author(s) and contributor(s) and not of MDPI and/or the editor(s). MDPI and/or the editor(s) disclaim responsibility for any injury to people or property resulting from any ideas, methods, instructions or products referred to in the content.

sensors

MDPI

Article

Raw Spectral Filter Array Imaging for Scene Recognition

Hassan Askary [1], Jon Yngve Hardeberg [1,2,*] and Jean-Baptiste Thomas [1,2,3]

1 Department of Computer Science, NTNU—Norwegian University of Science and Technology, 2815 Gjøvik, Norway; hassan.askary@ntnu.no (H.A.); jean@spektralion.com (J.-B.T.)
2 Spektralion AS, 2815 Gjøvik, Norway
3 Imagerie et Vision Artificielle (ImVIA) Laboratory, Department Informatique, Electronique, Mécanique (IEM), Université de Bourgogne, 21000 Dijon, France
* Correspondence: jon.hardeberg@ntnu.no

Abstract: Scene recognition is the task of identifying the environment shown in an image. Spectral filter array cameras allow for fast capture of multispectral images. Scene recognition in multispectral images is usually performed after demosaicing the raw image. Along with adding latency, this makes the classification algorithm limited by the artifacts produced by the demosaicing process. This work explores scene recognition performed on raw spectral filter array images using convolutional neural networks. For this purpose, a new raw image dataset is collected for scene recognition with a spectral filter array camera. The classification is performed using a model constructed based on the pretrained Places-CNN. This model utilizes all nine channels of spectral information in the images. A label mapping scheme is also applied to classify the new dataset. Experiments are conducted with different pre-processing steps applied on the raw images and the results are compared. Higher-resolution images are found to perform better even if they contain mosaic patterns.

Keywords: spectral filter array; scene recognition; convolutional neural networks

1. Introduction

Scene recognition is a challenging computer vision task that entails classifying an image into various scene categories based on the present visual information [1]. In contrast to object recognition, it requires modeling of the entire context in the image, including object presence, spatial location, illumination condition, viewing angle, distance, and scale [1,2]. It has applications in autonomous driving, robotics [3,4], video surveillance [5–7], augmented reality [8], and image retrieval [9,10]. It is a difficult task for the machine due to the large interclass similarities and intraclass variations present in different scene categories such as *book store*, *library*, and *archive*, all having similar objects present in the image and having similar layouts and ambient conditions [2].

In this work, the problem of scene recognition in raw spectral filter array (SFA) images is investigated using convolutional neural networks (CNN). The goal is to assess the effectiveness of using raw SFA images for this task. Usual spectral imaging acquisition setups consist of either capturing images in different spectral bands by cycling through multiple optical filters or by capturing the whole multispectral range using diffraction gratings, but one line at a time. Both of these approaches have a limitation of high acquisition time depending on the number of spectral bands or image size. They are also prone to artifacts due to movement during acquisition. Spectral filter array (SFA) technology [11] solves both of these problems by capturing the multispectral image in a single exposure at the expense of spatial resolution. It is similar to the color filter array (CFA) in RGB cameras. It is based on a single sensor overlayed with a Bayer-like pattern of different spectral filters with different spectral sensitivities over each pixel. The number of spectral bands used depends on the design of the SFA pattern. Demosaicing must be performed to reconstruct the full-resolution multispectral image, lowering the spatial resolution compared to other

Citation: Askary, H.; Hardeberg, J.Y.; Thomas, J.-B. Raw Spectral Filter Array Imaging for Scene Recognition. *Sensors* **2024**, *24*, 1961. https://doi.org/10.3390/s24061961

Academic Editors: Qing Yu, Ran Tu, Ting Liu and Lina Li

Received: 5 February 2024
Revised: 6 March 2024
Accepted: 14 March 2024
Published: 19 March 2024

Copyright: © 2024 by the authors. Licensee MDPI, Basel, Switzerland. This article is an open access article distributed under the terms and conditions of the Creative Commons Attribution (CC BY) license (https://creativecommons.org/licenses/by/4.0/).

spectral imaging methods. Demosaicing is an ill-posed problem, where interpolation is required to reconstruct the missing intensity values for each pixel. It also introduces estimated values that might be incorrect, which requires extra processing to rectify. This rectification process is scene-specific and requires identification of the targeted scene beforehand. To avoid these problems, in this work, scene recognition is performed on SFA images without demosaicing them. This speeds up the acquisition time even further because no pre-processing step is applied, and this also enables exploitation of spectral bands for scene classification. Furthermore, a large and diverse raw SFA dataset for scene recognition is introduced, and finally CNN models are investigated to perform scene recognition in raw SFA images.

One of the earliest works in scene recognition is by Szummer and Picard [12]. They classified scene images into *indoor* or *outdoor* categories based on low-level image features. They used the Ohta color space and multi-resolution simultaneous autoregressive model [13] to represent color and texture information. They computed these features on sub-blocks of the input image and then classified them; finally, they combined the classification result from each sub-block to obtain a final prediction using the K-nearest neighbor model. The approach was tested on a fairly small dataset of 1300 images and only for binary classification. Oliva and Torralba [14] proposed the Spatial Envelope representation for general scene classification. It is a global feature representation of the scene image. It describes a scene using five perceptual properties: naturalness, openness, roughness, ruggedness, and expansion. The classification prediction is performed using K-nearest neighbors. The authors also assembled a large dataset consisting of 8100 images over 4000 categories of natural scenes and 3500 categories of urban scenes. The Spatial Envelope representation does not consider local object information, making it sensitive to occlusions and spatial variations [1]. To overcome this, the Bag-of-Visual-Words (BoVW) framework was introduced in which local feature descriptors are extracted from the image. Then, the feature descriptors are quantified in terms of visual words. The image can now be classified on the basis of the frequency of occurrence of these visual words. Fei-Fei and Perona [15] proposed an approach where the scene image is first represented as a bag of codewords, then a probabilistic Bayesian hierarchical model is learned for each class. The model can learn to categorize the local regions of the image in an unsupervised way. It requires only the ground truth categories of the images for training. The model showed limitations in classifying complex indoor scenes because the BoVW approach does not take into account the spatial relationship of local features. To improve on this, Lazebnik et al. [16] proposed Spatial Pyramids. They repeatedly subdivided the image and computed the histogram of the local image features over the subregions. This hierarchical multiscale representation is a generalized form of the BoVW framework capturing spatial information. However, it is not invariant to geometric variation.

The recognition of outdoor scenes is easier than the recognition of indoor scenes. Indoor scene recognition is more difficult because of high inter-class variability present in the images, such as images of library, archive, and book store look similar. Quattoni and Torralba [17] tackled improving performance in indoor scene classification tasks. They devised a prototype-based model that combines global and local discriminative features. The model is based on the idea that images containing similar objects must have similar labels and that the presence of some objects in a scene is more important than that of others for determining the scene label. The authors created prototype images by annotating discriminative regions of interest in those images. Then, spatial pyramids were used extract features from query image, and the features were compared with the prototype regions of interest for similarity.

Until recently, approaches to recognizing scenes have relied on handcrafted features and classical machine learning models such as support vector machines [18] and K-nearest neighbors. Krizhevsky et al. [19] demonstrated the feasibility and superior performance of using deep convolutional neural networks (CNNs) in the large-scale image classification task, ushering in a new era in computer vision. It allows for end-to-end learning of the classification task. The CNN model is composed of a set of convolution layers and then another set of fully connected layers. The convolution layers extract features from the dataset, while the fully connected layers perform the classification. The entire network is tasked with minimizing the loss function using gradient descent, enabling it to automatically learn to extract useful discriminative features and perform classification. Deep learning models outperform classical methods by a large margin; however, they require large datasets and more time for training.

Zhou et al. [20] used the CNN model for scene recognition and also introduced a new large-scale scene recognition dataset called Places [21] with 10 million images. The Places-CNN model achieved state-of-the-art performance on existing benchmark datasets and on the new Places dataset. After this, many variants of deep learning models have been used for scene recognition tasks, improving performance, and pushing the state of the art forward. Some notable works include DAG-CNN [22], which uses a hierarchical CNN model to improve the extraction of local feature and gradient flow, and GAP-CNN [23], which replaces fully connected layers with global average pooling layers, biasing the model to attend to class-specific regions of the scene and reduce the number of learnable parameters.

Most of the work in scene recognition uses RGB images. The performance of scene recognition algorithms can be improved by exploiting additional spectral bands. Brown and Süsstrunk [24] proposed an extension of the Scale-Invariant Feature Transform [25] descriptor for multispectral images for scene recognition. Xiao et al. [26] extended the CENTRIST [27] descriptor to use multispectral images for scene recognition by capturing joint channel information from the RGB and NIR channels. Recently, Sevo and Avramović [28] used the convolutional neural network (CNN) on multispectral images of scenes to predict the scene label. However, in all of these works, one point to note is that the dataset consists of images with only four channels, RGB+NIR. Additionally, Elezabi et al. [29] collected a dataset of raw SFA images of textures to perform texture classification using CNNs and also investigated the impact of different illumination and exposure variations on performance.

To the best of our knowledge, there is no dataset of raw spectral filter array images of indoor and outdoor scenes. Also, to the best of our knowledge, there has been no prior work solving task of scene recognition in raw SFA images using CNNs.

This paper is organized as follows. Section 2 covers the details of the novel raw SFA dataset. Section 3 introduces our architecture to solve scene recognition in raw SFA images based on CNNs. The results are presented in Section 4, and finally the conclusions are presented in Section 5.

2. Dataset

A novel dataset consisting of raw SFA images of indoor and outdoor scenes was collected, entitled CID:Places. The dataset was collected using the SILIOS CMS-C SFA camera [30]. It captures nine bands ranging from 430 nm to 700 nm with a resolution of 1280×1024. Figure 1 shows the arrangement of the SFA pattern along with the spectral bands of the sensor. The dataset is comprised of various indoor and outdoor scenes. All images are 8-bit raw and mosaiced. Each image has a label indicating whether it is an indoor or an outdoor scene, as well as the specific scene category. In total, it has 402 raw SFA images, of which 201 are indoor scenes and the other 201 are outdoor scenes. It consists of 24 specific scene categories that are shown together with the number of images in Figure 2. Figure 3a shows a random sample of outdoor images, and Figure 3b shows a random sample of indoor images.

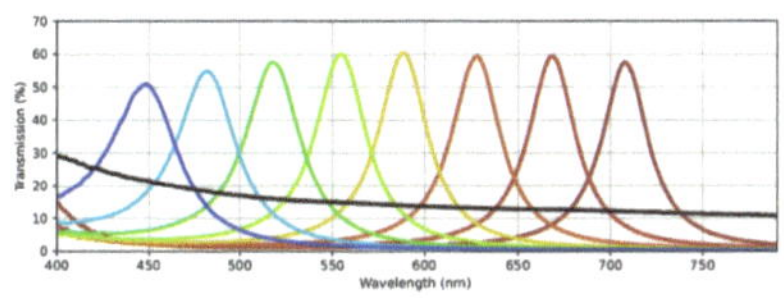

699nm	656nm	614nm
425nm	PAN	573nm
465nm	505nm	545nm

Figure 1. Arrangement of spectral bands in SFA pattern of SILIOS CMS-C sensor as well as transmission and wavelengths of spectral bands of each filter. Reproduced from [29–31].

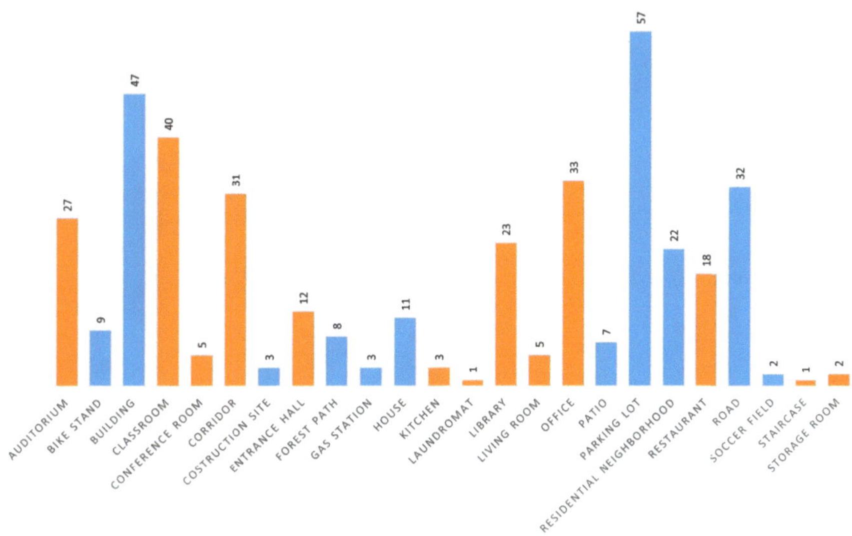

Figure 2. All scene categories and their sizes in our raw SFA scene recognition dataset. Blue color means outdoor scene and orange color means indoor scene.

(a) (b)

Figure 3. Random sample of indoor and outdoor datasets. (**a**) Outdoor raw SFA images. (**b**) Indoor raw SFA images.

The SILIOS CMS-C camera was mounted on a Joby GorillaPod 5K tripod. To capture the scenes, a 12.5 mm lens with a widest aperture of f/1.3 was used. The camera was connected to a Windows laptop with the IDS uEye Cockpit [32] program running. Two people were needed to carry out the captures. One person framed the picture, monitored the histogram, modified the parameters, triggered the capture on the laptop, and the other person held the camera setup. All images were captured in 8-bit sensor raw using the uEye Cockpit 2023 software. It allows for live view of what the camera is seeing along with the image histogram. It also performs live auto-exposure to properly expose the images, although in some extreme lighting situations the lens aperture and focus were manually adjusted.

In the dataset, the *classroom* category is the largest indoor class, and the smallest are *laundromat* and *staircase*. On the other hand, *parking lot* is the largest outdoor class and *soccer field* is the smallest. Very few images are found in the *construction site, soccer field, laundromat, staircase,* and *storage room* classes due to the limited encounters with these scenes during acquisition trips. The dataset was collected on and around a university campus.

Images of *library, office,* and *restaurant* classes were captured under varying lighting conditions. These classes have high dynamic range conditions with daylight entering through the windows, while the camera is exposed to the indoor light level. The dataset also contains images captured at night in artificial lighting. Examples of these images are shown in Figure 4.

Figure 4. Examples of similar scenes taken during day time and night time. Top row corresponds to images taken at night under artificial lighting and bottom row corresponds to images taken during the day time.

The category naming scheme of the Places dataset [21] was followed, with the *bike stand* class being an exception, as it is not present in the Places dataset. This scheme was chosen for its convenience in training Places-CNN with this dataset, given that Places is a widely recognized large-scale scene recognition dataset.

3. Methodology

This section covers the details of the proposed method for classifying scenes in raw SFA images. The proposed model is based on Places-CNN [20]. The model is not trained on the raw SFA dataset; instead, the pretrained weights of the Places-CNN are used. For details of the Places-CNN training methodology, we refer to [20]. Places-CNN is trained on RGB images of the Places dataset [21] so it cannot be used readily with the raw SFA images, which consist of one channel. We introduce a three-pathway network which accepts three pseudo-RGB images, performs inference on each image independently, and finally combines the Softmax probability scores. The three RGB images are obtained from the 9-band raw SFA image. This scheme enables full utilization of the spectral information.

Considering the raw SFA to be a grayscale image reformulates the problem and shifts the multispectral aspect to be implicit in the model. It also makes the model applicable to any nine-channel multispectral camera. We selected three bands from the 3×3 filter array to create a 3-channel pixel in the pseudo-RGB image. Figure 5 shows the selected bands that form the pseudo-RGB pixels in each of the three pseudo-RGB images. These bands were selected based on their wavelengths that correspond to the red, green, and blue colors in the visible wavelength range. One exception is that the panchromatic band is assigned to the B channel in the pseudo-RGB 3 image. It was assigned because it was left over after all other bands were selected. Figure 6 shows an example of these three pseudo-RGB images. These pseudo-RGB images have a resolution of 427×342, while the original raw SFA is 1280×1024.

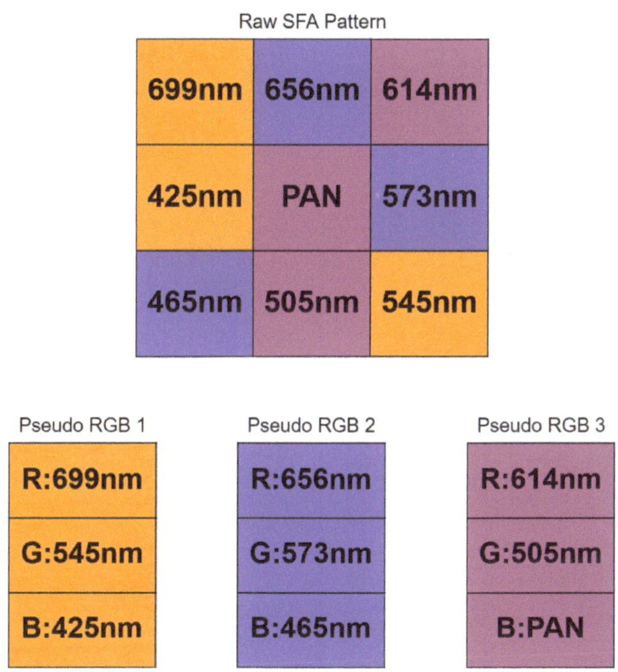

Figure 5. Selected bands that form pseudo-RGB pixel in each pseudo-RGB image.

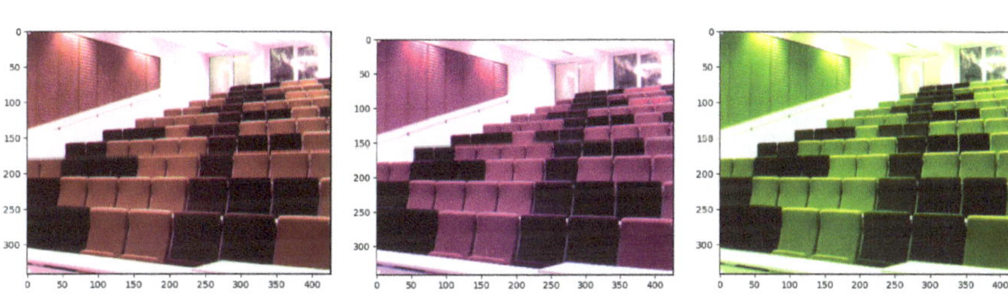

(**a**) (**b**) (**c**)

Figure 6. Example Pseudo-RGB images. (**a**): Pseudo-RGB 1. (**b**): Pseudo-RGB 2. (**c**): Pseudo-RGB 3.

The proposed model illustrated in Figure 7 takes three input images with three channels, the inference on each image is performed independently by a pretrained 11 million parameter Places-CNN network, and finally the prediction is calculated by combining the softmax probabilities of all three networks and selecting the class with the highest score. The Places-CNN architecture is a residual network with skip connections [33] consisting of

18 residual layers. All three networks have shared weights and return a 365 length vector of Softmax probabilities corresponding to each class of the Places dataset. The three resulting probability vectors are summed element-wise and then divided by three to normalize back to the 0 to 1 range. Then, this normalized vector is sorted in descending order, and the highest scoring class is picked.

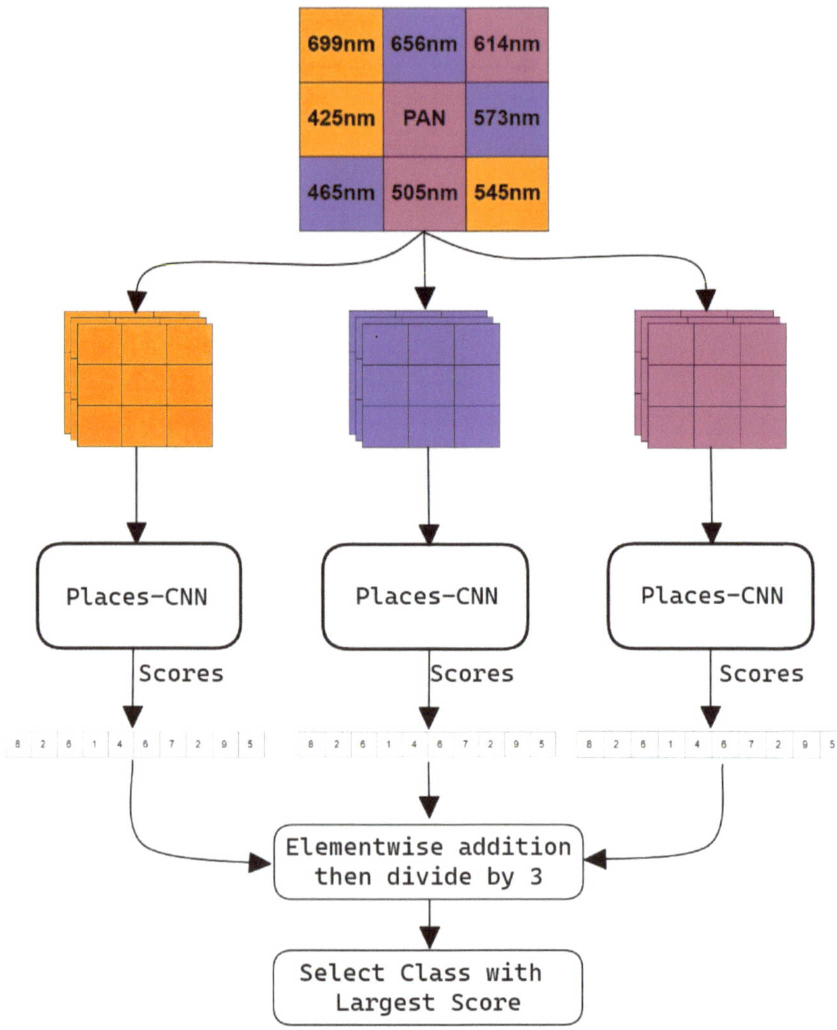

Figure 7. Proposed methodology with Three-Pathway Network.

The Places dataset on which Places-CNN is trained contains 365 fine-grained classes. It includes specific classes such as *apartment building*, *office building*, *hospital*, etc. Our dataset has 24 general classes, including those that do not exist in Places (*bike stand*). Therefore, it encapsulates all buildings in the *building* class that does not exist in the Places dataset. To solve this mismatch, a label mapping is performed before combining the scores. So, all specific classes are replaced with general classes that exist inside our dataset, and their scores are summed. All Places dataset labels are analyzed, and the visually and semantically similar classes are mapped to the general class label in our dataset. Figure 8 shows all the label mappings from the Places dataset labels to the labels of our dataset.

Indoor vs. outdoor binary classification is also performed. The Places dataset assigns an additional indoor/outdoor label to the scene class label. After inference, to predict whether the image is of an indoor scene or an outdoor scene, the first 10 largest scores and their corresponding classes are taken and a majority vote of indoor/outdoor labels determines the resulting category.

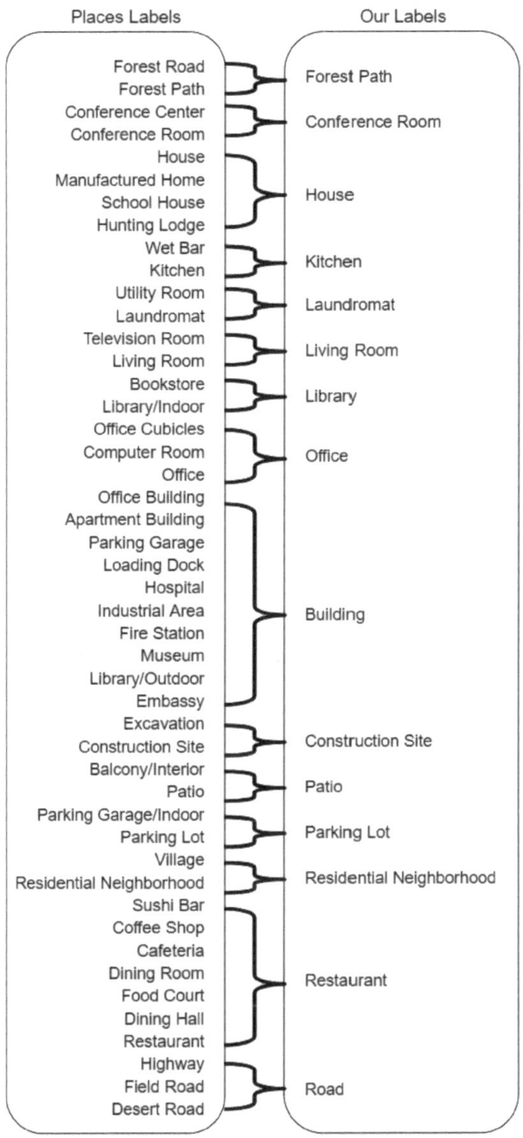

Figure 8. Mapping of Places dataset labels to our raw SFA dataset (CID:Places) labels.

4. Results

In this section, experiments are performed to assess the effectiveness of using pre-trained Places-CNN and the proposed three-pathway network for scene recognition in raw SFA images. Accuracy and F1 scores are considered for both indoor vs. outdoor classification and scene classification. Six models with different configurations are compared. Class activation maps returned by Places-CNN are also examined to explain the judgments.

The details of the configurations compared are as follows:

Config 1: **Raw SFA image as input to Places-CNN.** The raw SFA image is a single channel image with the mosaic patterns indicating the 9 bands. It is treated as a grayscale image. The single channel is duplicated along the *z*-axis to obtain a three-channel image. It is sent to the unmodified pretrained Places-CNN for inference.

Config 2: **Pseudo-RGB 1 image as Input to Places-CNN.** The first pseudo-RGB image constructed by selecting band 699 nm as R, 545 nm as G, and 425 nm as B as shown in Figure 5 is sent as input to Places-CNN for inference and metrics are computed.

Config 3: **Pseudo-RGB 2 image as Input to Places-CNN.** The second pseudo-RGB image is used as input to the unmodified Places-CNN.

Config 4: **Pseudo-RGB 3 image as Input to Places-CNN.** The third pseudo-RGB image is used as input.

Config 5: **Grayscale image as Input to Places-CNN.** The middle panchromatic channel is taken and a three-channel grayscale image is produced by duplicating the value three times along the *z*-axis. The size is similar to that of the pseudo-RGB images, and the mosaic pattern seen in Configuration 1 is absent. Figure 9 shows an example grayscale image.

Config 6: **Three Pseudo-RGB images as Input to Three-pathway Network.** The proposed method is as follows: three pseudo-RGBs are constructed and sent to the three inputs of the three-pathway network to perform inference on each image independently, and then the results are combined.

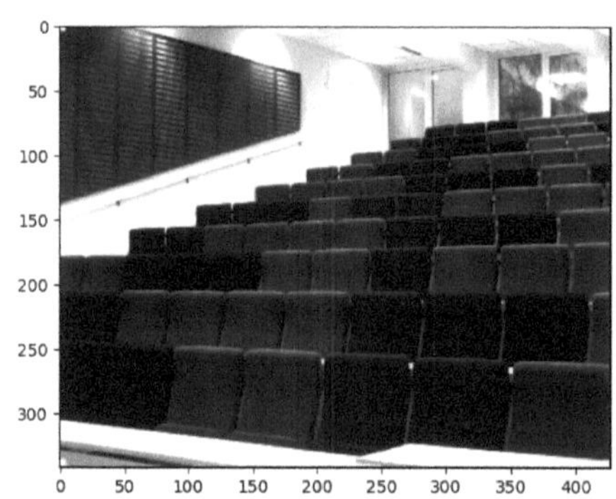

Figure 9. Example grayscale image.

Indoor vs. outdoor accuracies and F1 scores are presented in Table 1. All configurations performed very well, achieving almost perfect accuracy. Configuration 1 where we input the raw SFA image performed the best; we can see from Figure 10a that it made only 3 errors. Configuration 4 performed the worst; in Figure 10b, we can see that it incorrectly predicted 14 outdoor scenes as indoor. Finally, Configuration 6, the proposed method, misclassified 5 images as outdoor, as seen in Figure 10c. Overall, the performance of all approaches is very similar.

The performance metrics for the scene recognition task are shown in Table 2. Configuration 1 has the best accuracy, while Configuration 6 has the highest F1 score. Since the dataset for scene recognition is imbalanced, unlike for indoor vs. outdoor classification, the F1 score is the more useful metric here. Figure 11 shows the confusion matrix for Configuration 6 which is the proposed method. Confusion matrices for other configurations are available in Appendix A. Overall, the performance is not good, with the best F1 score of 0.63 and an accuracy of 0.59, and there is a big difference compared to indoor vs. outdoor classification performance.

Table 1. Accuracy and F1 scores on the indoor vs. outdoor task. The red text indicates the highest values.

Configuration	Accuracy	F1 Score
1: Raw SFA	0.99	0.9901
2: Pseudo-RGB 1	0.9826	0.9829
3: Pseudo-RGB 2	0.9876	0.9877
4: Pseudo-RGB 3	0.9652	0.9663
5: Grayscale	0.9801	0.9804
6: Three-pathway	0.9876	0.9874

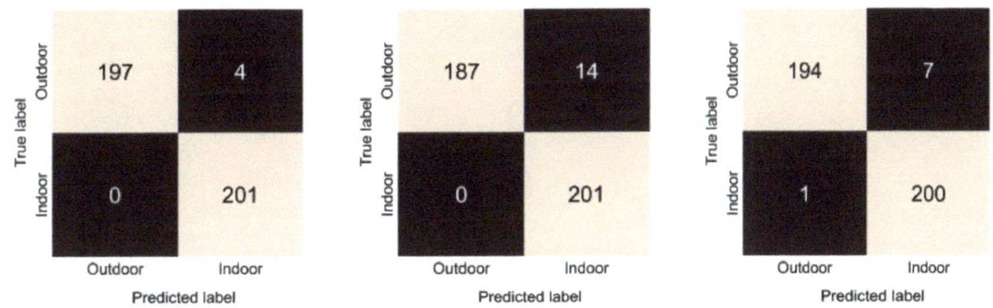

Figure 10. Confusion matrices of the indoor vs outdoor classification task. (**a**): Confusion matrix of Configuration 1. (**b**): Confusion matrix of Configuration 4. (**c**): Confusion matrix of Configuration 6.

Table 2. Accuracy and F1 scores on the scene recognition task. The red text indicates the highest values.

Configuration	Accuracy	F1 Score
1: Raw SFA	0.5995	0.6313
2: Pseudo-RGB 1	0.5547	0.6202
3: Pseudo-RGB 2	0.5572	0.6193
4: Pseudo-RGB 3	0.5224	0.569
5: Grayscale	0.5697	0.6064
6: Three-pathway	0.5771	0.6354

The model struggles with classes that are related to each other. The parking lot is the most misclassified category. It is confused with the building category. In the dataset, there are many parking lots next to or in front of buildings. The parking lot is also confused with the junkyard. Both categories contain images of cars parked in a line. Similarly, the office is confused with the conference room, the restaurant with the classroom because both have arranged tables and chairs, and the residential neighborhood with the building. The model struggles to distinguish subtle details; for example, the junkyard has cars that are not in good condition, or the restaurants usually have tablecloths and other decorations on the tables while classrooms do not.

Two main reasons for the disparity in performance of both tasks is that the indoor vs. outdoor classification decision is taken with a majority vote of the top 10 scores, while for scene recognition only the top 1 score is considered. Configuration 6 performs better on the more difficult scene recognition task, demonstrating a better bias–variance trade-off based on the F1 score. This is because it combines the decision of three networks. Another reason is that the model is not trained on our dataset, and thus the input is out of distribution for it. Configuration 4 has the worst performance due to the panchromatic channel set to the blue channel, resulting in the most color-incorrect image compared to the other pseudo-RGB images.

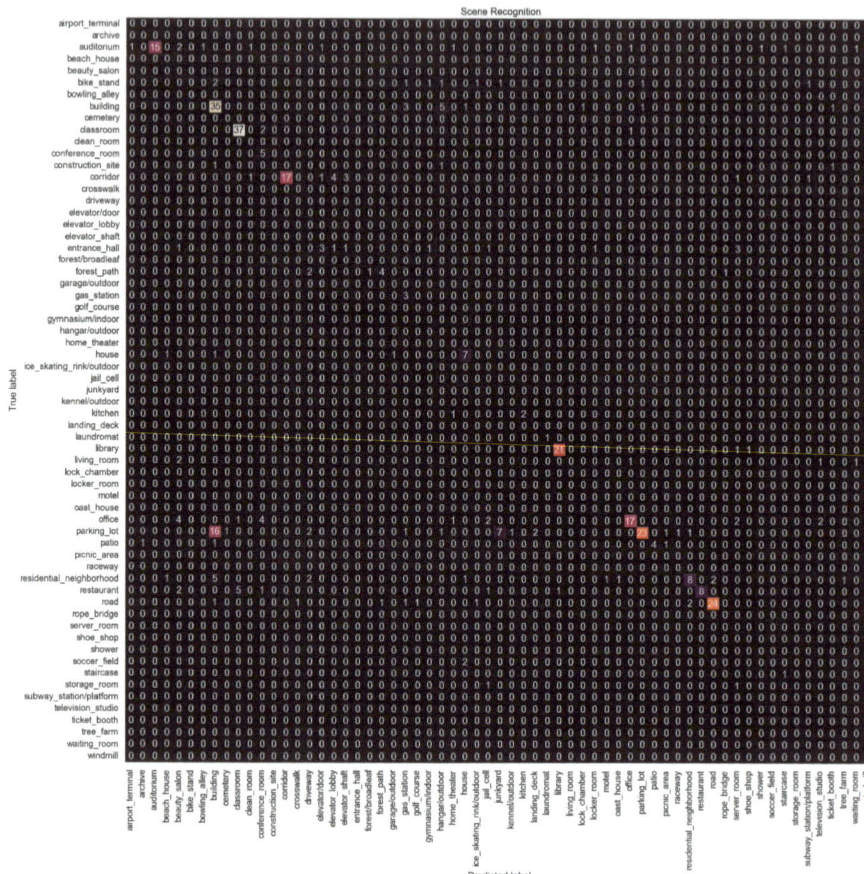

Figure 11. Confusion matrix of Configuration 6 on the scene recognition task.

Further analyzing the configurations, we generate class activation maps from the model. The class activation map is a heat map that indicates which area of the image the model found to be the most discriminating or helpful in its classification. We analyzed class activation maps for two correct classifications and two incorrect classifications, one for the indoor case and one for the outdoor case. Figure 12 shows the class activation maps for an image of a building that was correctly classified by all configurations. All configurations focus on the different parts of the building in the image, which explains their correct predictions. A similar pattern is seen in Figure 13 where the models focus on the display, the cubicles, and the bottom of the revolving chairs to correctly predict the image belonging to the *office* class. Then, we considered misclassification cases. In Figure 14, the image of the parking lot is misclassified as a building. The class activation maps indicate that the models paid attention to the building in the background rather than the cars parked in front. Finally, Figure 15 shows class activation heat maps of an image of an auditorium incorrectly classified. The models focused on the top right of the image, where the staircase and its railing are along with some tables. The misclassifications for this image were varied. Configurations 1, 4 and 6 were classified as *jail cell*, Configurations 2 and 3 as *bowling alley*, and Configuration 5 as *staircase*. The class activation maps did not explain the reason for these predictions. Class activation maps provided some insight into the behavior of neural networks but not the entire explanation. Neural networks remain difficult to explain.

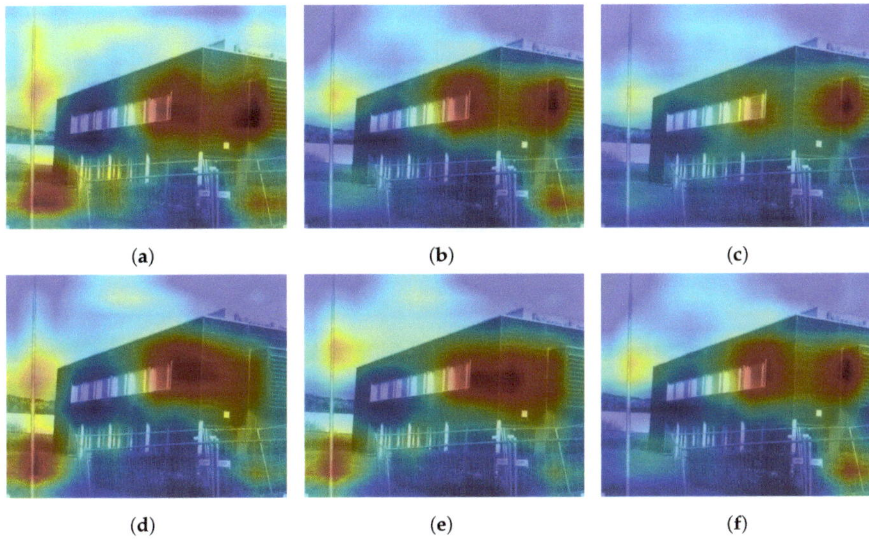

Figure 12. Class activation maps of a *building* image correctly classified by all configurations. (**a**) Configuration 1. (**b**) Configuration 2. (**c**) Configuration 3. (**d**) Configuration 4. (**e**) Configuration 5. (**f**) Configuration 6.

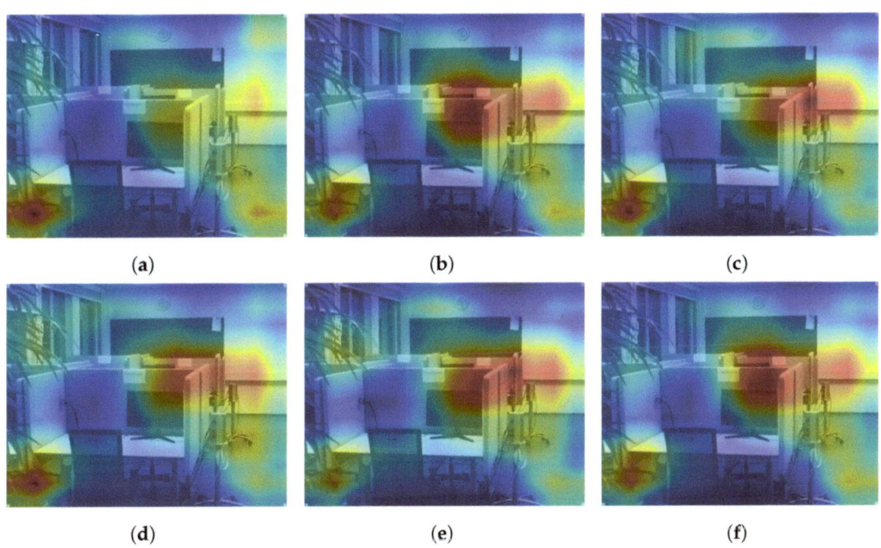

Figure 13. Class activation maps of an *office* image correctly classified by all configurations. (**a**) Configuration 1. (**b**) Configuration 2. (**c**) Configuration 3. (**d**) Configuration 4. (**e**) Configuration 5. (**f**) Configuration 6.

As mentioned earlier, the scene recognition prediction is based on the class with the largest softmax score, while the indoor vs. outdoor classification considers the majority class in the top 10 largest scoring labels. Increasing the top k scores used for the decision improves performance. We considered the example of misclassification shown in Figure 14 where the model predicted the *building* class instead of *parking lot*. The image has the building in the background, while the parking lot is in the foreground. This image can be correctly classified as both *building* and *parking lot*.

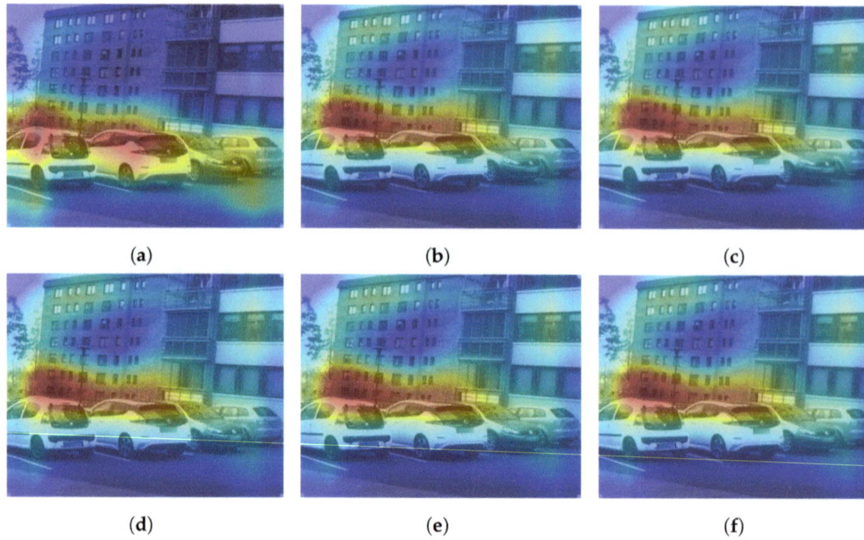

Figure 14. Class activation maps of a *parking lot* image incorrectly classified as *building* by all configurations. (**a**) Configuration 1. (**b**) Configuration 2. (**c**) Configuration 3. (**d**) Configuration 4. (**e**) Configuration 5. (**f**) Configuration 6.

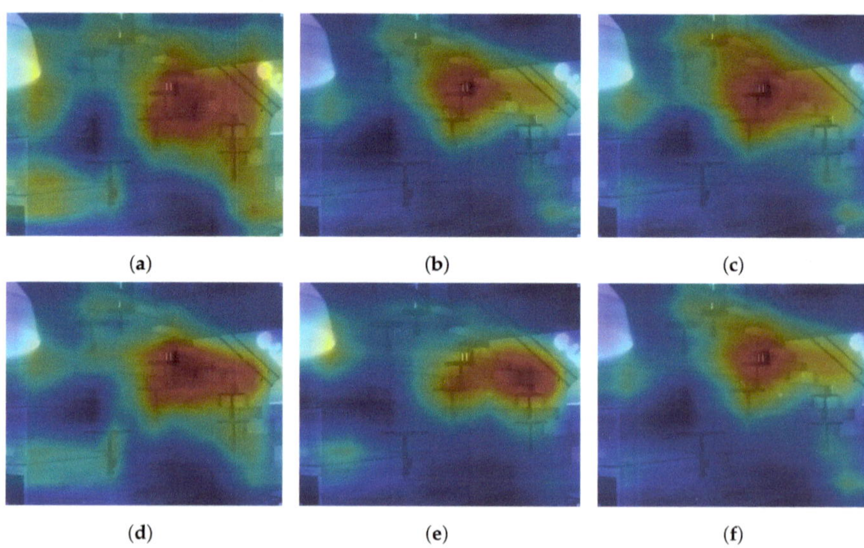

Figure 15. Class activation maps of an *auditorium* image incorrectly classified as *jail cell* by Configurations 1, 4 and 6, as *bowling alley* by Configurations 2 and 3, as *staircase* by Configuration 5. (**a**) Configuration 1. (**b**) Configuration 2. (**c**) Configuration 3. (**d**) Configuration 4. (**e**) Configuration 5. (**f**) Configuration 6.

Table 3 presents the results in which the top k = 1, 2, 3, 5, and 10 scores were considered and if the correct label was present, the image was marked as correctly classified. As the considered top k scores increase, performance also increases. At the top k = 10, the same level of performance is reached as the indoor vs. outdoor classification. For the top k = 10, Configuration 1 is the best performing configuration, while Configuration 6 is the third best. Figure 16 compares the improvements in accuracy and the F1 score as K increases. There is an improvement of approximately 10% when increasing k by one. The improvement slows to approximately 5% after the top k = 3 and higher. Converting the objective to multi-label classification improves performance. However, it is important to emphasize that this is not needed if the model is trained on the dataset as the highest scoring category is most likely to be the correct one. Reasons for not retraining the model are discussed in Section 5.

Table 3. Top K scene recognition accuracy and F1 score. If the label is present in the top k predictions, then the classification is correct. The red text indicates the highest values.

Configuration	Top K	Accuracy	F1 Score
1: Raw SFA	1	0.5995	0.6313
	3	0.7761	0.8043
	5	0.8408	0.8602
	10	0.8955	0.909
2: Pseudo-RGB 1	1	0.5547	0.6202
	3	0.7438	0.7897
	5	0.8109	0.842
	10	0.8731	0.8946
3: Pseudo-RGB 2	1	0.5572	0.6193
	3	0.7562	0.7976
	5	0.8159	0.8391
	10	0.8607	0.8847
4: Pseudo-RGB 3	1	0.5224	0.569
	3	0.7463	0.7785
	5	0.8085	0.8328
	10	0.8582	0.8763
5: Grayscale	1	0.5697	0.6064
	3	0.7463	0.7735
	5	0.8408	0.8614
	10	0.8806	0.8958
6: Three-pathway	1	0.5771	0.6354
	3	0.7711	0.8058
	5	0.8433	0.8674
	10	0.8706	0.8914

For scene recognition, Configuration 1 performs best overall. In Configuration 1, the raw SFA image is duplicated along the z-axis to convert to three channels and input to a pretrained Places-CNN model. The image has a resolution of 1280 × 1024 while the image in all other configurations is smaller at 427 × 342. However, the image in Configuration 1 has mosaic artifacts, whereas the images in other configurations do not. Comparing Configuration 1 and Configuration 5, Configuration 1 still performs better. In Configuration 5, the image is a grayscale image constructed from the panchromatic channel duplicated along the z-axis three times. Both images are grayscale (Configuration 1 raw SFA is treated as grayscale), and the difference is in resolution and mosaic artifacts. Table 4 shows the results when the resolution of the grayscale image (Configuration 5) is increased from 427 × 342 to 1280 × 1024 and is compared with Configuration 1. It also shows the result when the resolution of the images in Configuration 1 is decreased to match the images in Configuration 5 (427 × 342). Increasing the resolution of Configuration 5 improves the results slightly, but does not match what is achieved by Configuration 1. Decreasing the resolution of Configuration 1 decreases the results slightly, but not enough, to match the metrics obtained by Configuration 5. More experimentation is required to know why Configuration 1 which has mosaic artifacts works best. Increasing the resolution of images in Configuration 5 improves performance, and decresing the resolution of images in Configuration 1 degrades performance. Resizing images to a bigger size results in blurry images as the process interpolates more pixels. So, if the images in Configuration 5 have a native resolution of 1280 × 1024, they will be sharper and might match the better results of simply using a raw SFA image. The model benefits from the higher resolution of the raw

SFA image enough that the noise of mosaic pattern does not cause the performance to be worse than the configurations where the images are smaller.

Further comparisons were made with the selection of other channels to construct a grayscale image. Only one channel was selected from the nine bands and duplicated on the z-axis to form a three-channel pixel. The results were similar and can be found in Appendix B.

Configuration 6, which is the proposed model, surpassed Configuration 1 with raw SFA at K = 1, 3, and 5. It was the best performing model at these values of K. The model utilizes the raw SFA image by constructing three pseudo-RGB images and performing inference independently. The results of the three forward passes were combined, and the prediction was chosen. This introduced robustness and reduced noise in the predictions, leading to better results.

Table 4. Comparison of scene recognition accuracy and F1 score of Configurations 1, 5, 1 resized to 427 × 342, and 5 resized to 1280 × 1024. The red text indicates the highest values.

Configuration	K	Acccuracy	F1 Score
1: Raw SFA	1	0.5995	0.6313
	3	0.7761	0.8043
	5	0.8408	0.8602
	10	0.8955	0.909
1: Raw SFA (resized to 427 × 342)	1	0.5945	0.6257
	3	0.7711	0.7958
	5	0.8408	0.8602
	10	0.8955	0.909
5: Grayscale	1	0.5697	0.6064
	3	0.7463	0.7735
	5	0.8408	0.8614
	10	0.8806	0.8958
5: Grayscale (resized to 1280 × 1024)	1	0.5697	0.6068
	3	0.7488	0.7763
	5	0.8458	0.8671
	10	0.8781	0.8933

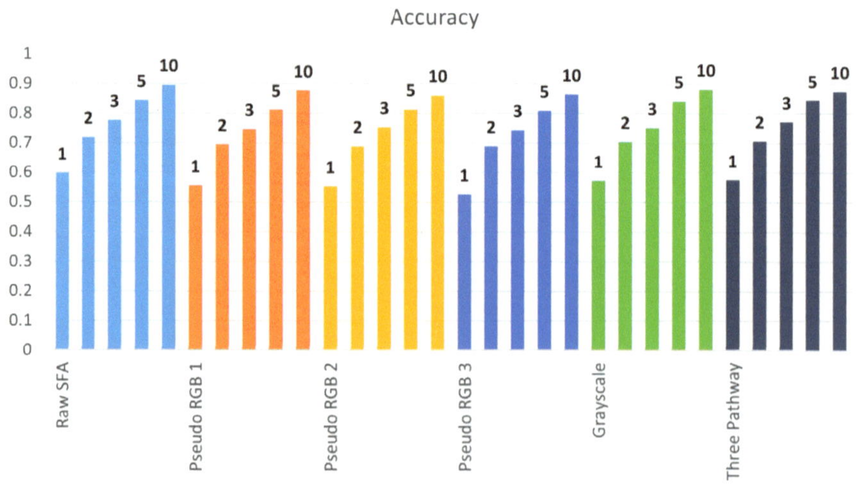

Figure 16. *Cont.*

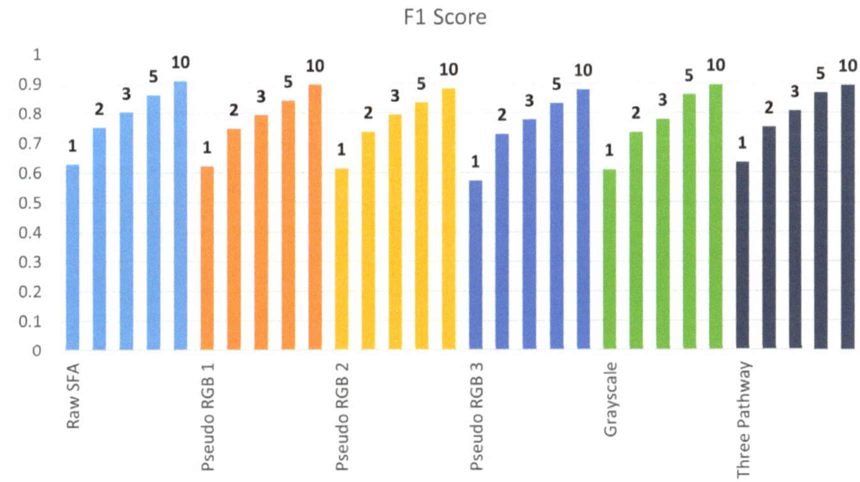

Figure 16. Top K accuracies and F1 scores for each configuration.

5. Conclusions

The aim of this work was to assess the effectiveness of using raw spectral filter array imaging for scene recognition. To achieve this, a raw SFA scene dataset was acquired using the SILIOS CMS-C spectral camera and labeled with indoor/outdoor class, as well as scene class following the labels in the Places dataset.

The pretrained Places-CNN was used as the convolution neural network model for scene recognition and indoor vs. outdoor classification. It was trained on the Places dataset with 10 million images and 365 classes. Six configurations (type of input; Configuration 6 also has a different architecture) and variations were evaluated, one of which was a novel architecture that utilized the individual bands of the spectral filter array by separating them into individual images. All models achieved F1 scores above 90% on the indoor vs. outdoor classification task. F1 scores were not good on the multi-class scene recognition task with the proposed model achieving the best score of 63%. Further experiments were carried out to improve performance on the scene recognition task by considering the top K prediction scores for the decision. When K = 10, the scene recognition F1 scores reached 90% for all models.

Experiments were conducted to explain the good performance of Configuration 1. In Configuration 1, the raw SFA image is treated as a grayscale image. The pixels are duplicated along the *z*-axis to form a three-channel image because Places-CNN requires a three-channel image as input. It retains all the spectral information in the image, albeit with redundancy. The raw SFA image contains the mosaic pattern; however, it has the highest resolution of all the other configurations. In Configuration 5, the middle panchromatic channel is selected and duplicated over the *z*-axis to form a grayscale image. These two Configurations are compared because there are visual similarities to explain the affect of presence of mosaic pattern and resolution. It is found that higher resolution leads to better predictions.

Places-CNN was used pretrained on the Places dataset. It was not trained on our custom raw SFA dataset, that is why scene recognition performance was limited when considering only the highest scoring label in the prediction. However, it was not below 50% accuracy, indicating that due to its large-scale training it has the ability to extract relevant features and discriminate them. Places-CNN was not fine-tuned on our dataset because our dataset is highly imbalanced with some classes, such as the laundromat that contains only one image. Fine-tuning on it results in high accuracies and overfitting. The dataset contains 402 images; more images need to be collected to make training a neural network viable.

The pseudo-RGB images were constructed from the selection of the spectral bands from the raw SFA image. More experimentation can be performed to optimize the selection of the bands. Another comparison which was not conducted was with a demosaiced RGB image of the same scenes.

In this work, the role of illuminations was not explored. Further investigation can be carried out to determine whether correcting the illumination in the raw SFA captures has an impact. Higher resolution was found to have a positive impact on performance regardless of mosaic patterns. Further experiments can be conducted to explain this behavior. The Places-CNN model was not trained on the raw SFA dataset. A logical next step is to collect more data and fine-tune the model on it. Additionally, smaller architectures can be explored, such as Mobilenet [34], to make deployment on edge devices possible for real-time applications.

Author Contributions: Conceptualization, J.-B.T. and J.Y.H.; methodology, H.A. and J.Y.H.; software, H.A.; validation, H.A., J.Y.H. and J.-B.T.; formal analysis, H.A.; investigation, H.A.; resources, J.Y.H.; data curation, H.A.; writing—original draft preparation, H.A.; writing—review and editing, J.-B.T., J.Y.H. and H.A.; visualization, H.A.; supervision, J.Y.H. and J.-B.T.; project administration, J.Y.H. and J.-B.T. All authors have read and agreed to the published version of the manuscript.

Funding: This research received no external funding.

Institutional Review Board Statement: Not applicable.

Informed Consent Statement: Not applicable.

Data Availability Statement: The collected raw SFA dataset CID:Places will be made available on the NTNU Colourlab website, https://colourlab.no/cid (accessed on 4 February 2024).

Acknowledgments: We acknowledge our colleagues Sonain Jamil, Riestiya Zain Fadillah, and Rafique Ahmed for their help in capturing the dataset.

Conflicts of Interest: The authors declare no conflicts of interest.

Abbreviations

The following abbreviations are used in this manuscript:

MDPI	Multidisciplinary Digital Publishing Institute
SFA	Spectral Filter Array
CNN	Convolutional Neural Network
CFA	Color Filter Array
RGB	Red Green Blue
BoVW	Bag of Visual Words
NIR	Near Infrared

Appendix A. Scene Recognition Confusion Matrices for Other Configs

Scene recognition task confusion matrices for Configurations 1, 2, 3, 4, and 5 are presented in Figures A1–A5.

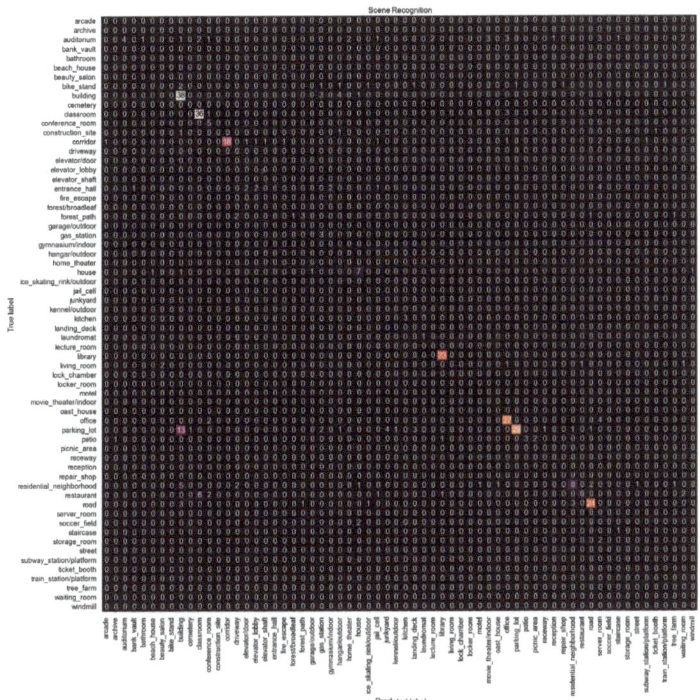

Figure A1. Confusion matrix of Configuration 1 on the scene recognition task.

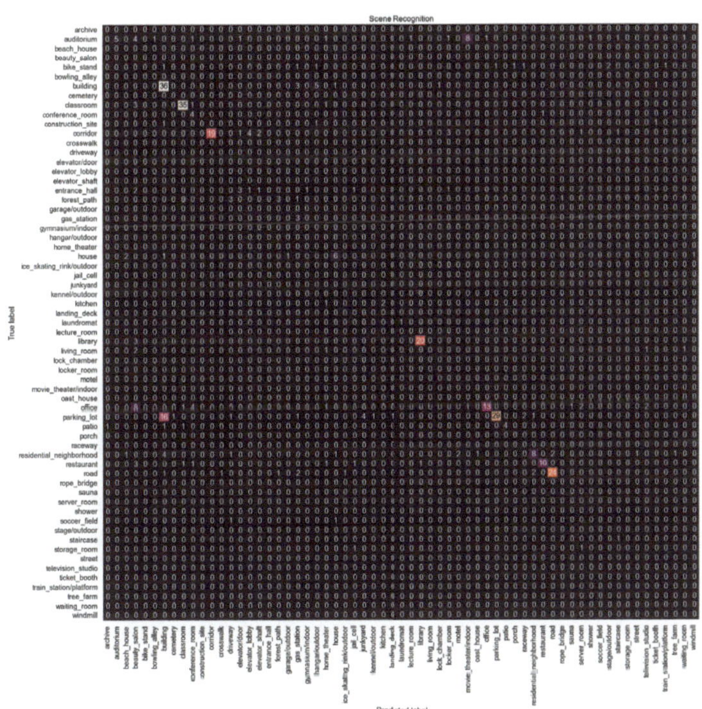

Figure A2. Confusion matrix of Configuration 2 on the scene recognition task.

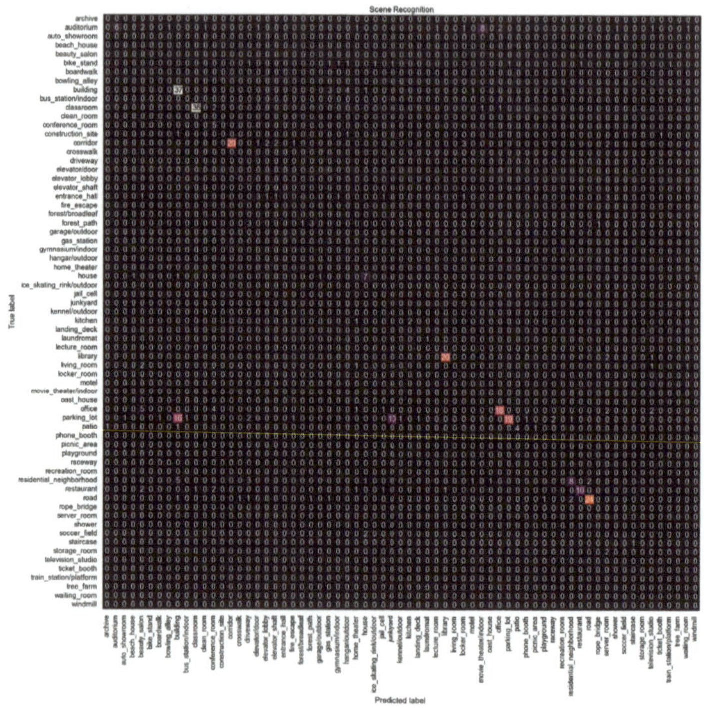

Figure A3. Confusion matrix of Configuration 3 on the scene recognition task.

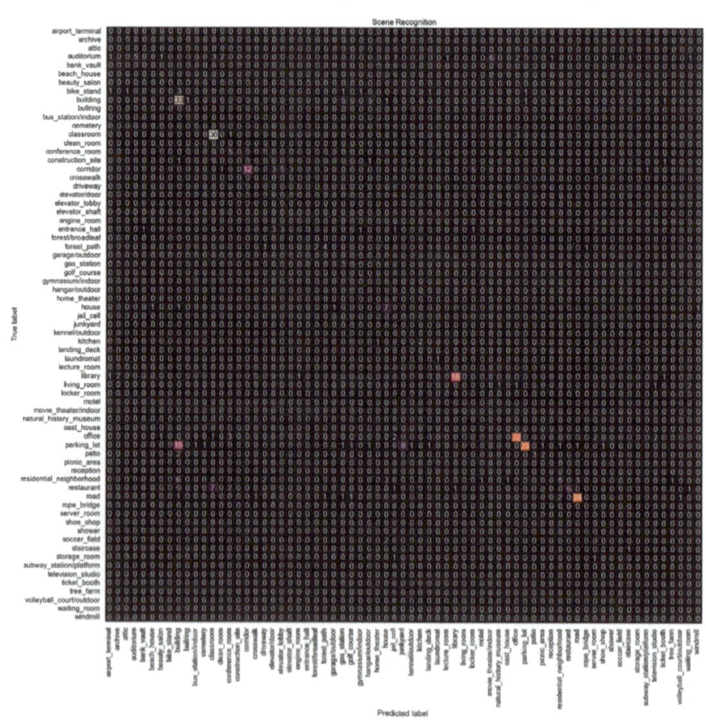

Figure A4. Confusion matrix of Configuration 4 on the scene recognition task.

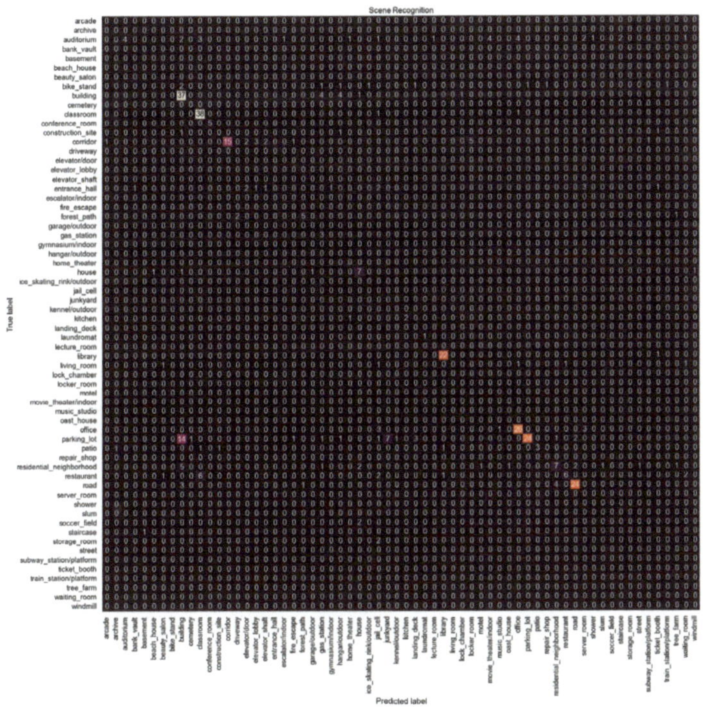

Figure A5. Confusion matrix of Configuration 5 on the scene recognition task.

Appendix B. Quantitative Results with Grayscale Constructed by Selecting other Bands

Table A1. Comparison of scene recognition accuracy and F1 score for grayscale images constructed by considering different spectral bands. K = 1 for configurations. Red text indicates the best value in the column.

Configuration	Acccuracy	F1 Score
Grayscale (699 nm)	0.5597	0.5896
Grayscale (656 nm)	0.5572	0.5874
Grayscale (614 nm)	0.5572	0.5972
Grayscale (425 nm)	0.5746	0.6069
Grayscale (PAN)	0.5697	0.6064
Grayscale (573 nm)	0.5597	0.6015
Grayscale (465 nm)	0.5696	0.5976
Grayscale (505 nm)	0.5696	0.5989
Grayscale (545 nm)	0.5672	0.6089

References

1. Xie, L.; Lee, F.; Liu, L.; Kotani, K.; Chen, Q. Scene recognition: A comprehensive survey. *Pattern Recognit.* **2020**, *102*, 107205. [CrossRef]
2. Zeng, D.; Liao, M.; Tavakolian, M.; Guo, Y.; Zhou, B.; Hu, D.; Pietikäinen, M.; Liu, L. Deep learning for scene classification: A survey. *arXiv* **2021**, arXiv:2101.10531.
3. Zhang, W.; Yu, X.; He, X. Learning Bidirectional Temporal Cues for Video-Based Person Re-Identification. *IEEE Trans. Circuits Syst. Video Technol.* **2018**, *28*, 2768–2776. [CrossRef]
4. Hou, J.; Zeng, H.; Zhu, J.; Hou, J.; Chen, J.; Ma, K.K. Deep Quadruplet Appearance Learning for Vehicle Re-Identification. *IEEE Trans. Veh. Technol.* **2019**, *68*, 8512–8522. [CrossRef]

5. Zhang, T.; Liu, S.; Xu, C.; Lu, H. Mining Semantic Context Information for Intelligent Video Surveillance of Traffic Scenes. *IEEE Trans. Ind. Inform.* **2013**, *9*, 149–160. [CrossRef]
6. Sreenu, G.; Durai, M.A.S. Intelligent video surveillance: A review through deep learning techniques for crowd analysis. *J. Big Data* **2019**, *6*, 48. [CrossRef]
7. Muhammad, K.; Ahmad, J.; Baik, S.W. Early fire detection using convolutional neural networks during surveillance for effective disaster management. *Neurocomputing* **2017**, *288*, 30–42. [CrossRef]
8. Nee, A.Y.C.; Ong, S.K.; Chryssolouris, G.; Mourtzis, D. Augmented reality applications in design and manufacturing. *Cirp Ann.-Manuf. Technol.* **2012**, *61*, 657–679. [CrossRef]
9. Vogel, J.; Schiele, B. Semantic Modeling of Natural Scenes for Content-Based Image Retrieval. *Int. J. Comput. Vis.* **2007**, *72*, 133–157. [CrossRef]
10. Zheng, L.; Yang, Y.; Tian, Q. SIFT Meets CNN: A Decade Survey of Instance Retrieval. *IEEE Trans. Pattern Anal. Mach. Intell.* **2016**, *40*, 1224–1244. [CrossRef] [PubMed]
11. Lapray, P.J.; Wang, X.; Thomas, J.B.; Gouton, P. Multispectral Filter Arrays: Recent Advances and Practical Implementation. *Sensors* **2014**, *14*, 21626–21659. [CrossRef]
12. Szummer, M.; Picard, R.W. Indoor-outdoor image classification. In Proceedings of the Proceedings 1998 IEEE International Workshop on Content-Based Access of Image and Video Database, Bombay, India, 3 January 1998; pp. 42–51.
13. Mao, J.; Jain, A.K. Texture classification and segmentation using multiresolution simultaneous autoregressive models. *Pattern Recognit.* **1992**, *25*, 173–188. [CrossRef]
14. Oliva, A.; Torralba, A. Modeling the shape of the scene: A holistic representation of the spatial envelope. *Int. J. Comput. Vis.* **2001**, *42*, 145–175. [CrossRef]
15. Fei-Fei, L.; Perona, P. A bayesian hierarchical model for learning natural scene categories. In Proceedings of the 2005 IEEE Computer Society Conference on Computer Vision and Pattern Recognition (CVPR'05), San Diego, CA, USA, 20–25 June 2005; Volume 2, pp. 524–531.
16. Lazebnik, S.; Schmid, C.; Ponce, J. Beyond Bags of Features: Spatial Pyramid Matching for Recognizing Natural Scene Categories. In Proceedings of the 2006 IEEE Computer Society Conference on Computer Vision and Pattern Recognition (CVPR'06), New York, NY, USA, 17–22 June 2006; Volume 2, pp. 2169–2178. [CrossRef]
17. Quattoni, A.; Torralba, A. Recognizing indoor scenes. In Proceedings of the 2009 IEEE Conference on Computer Vision and Pattern Recognition, Miami, FL, USA, 20–25 June 2009; pp. 413–420.
18. Boser, B.E.; Guyon, I.M.; Vapnik, V.N. A training algorithm for optimal margin classifiers. In Proceedings of the Fifth Annual Workshop on Computational Learning Theory, New York, NY, USA, 27–29 July 1992; pp. 144–152.
19. Krizhevsky, A.; Sutskever, I.; Hinton, G.E. Imagenet classification with deep convolutional neural networks. *Adv. Neural Inf. Process. Syst.* **2012**, *5*, 1106–1114. [CrossRef]
20. Zhou, B.; Lapedriza, A.; Xiao, J.; Torralba, A.; Oliva, A. Learning deep features for scene recognition using places database. *Adv. Neural Inf. Process. Syst.* **2014**, *27*. Available online: http://hdl.handle.net/1721.1/96941 (accessed on 4 February 2024).
21. Zhou, B.; Lapedriza, A.; Khosla, A.; Oliva, A.; Torralba, A. Places: A 10 million Image Database for Scene Recognition. *IEEE Trans. Pattern Anal. Mach. Intell.* **2017**, *40*, 1452–1464. [CrossRef]
22. Yang, S.; Ramanan, D. Multi-scale Recognition with DAG-CNNs. In Proceedings of the 2015 IEEE International Conference on Computer Vision (ICCV), Santiago, Chile, 7–13 December 2015; pp. 1215–1223. [CrossRef]
23. Zhou, B.; Khosla, A.; Lapedriza, A.; Oliva, A.; Torralba, A. Learning deep features for discriminative localization. In Proceedings of the IEEE Conference on Computer Vision and Pattern Recognition, Las Vegas, NV, USA, 27–30 June 2016; pp. 2921–2929.
24. Brown, M.A.; Süsstrunk, S. Multi-spectral SIFT for scene category recognition. In Proceedings of the CVPR 2011, Colorado Springs, CO, USA, 20–25 June 2011; pp. 177–184.
25. Lowe, D.G. Distinctive Image Features from Scale-Invariant Keypoints. *Int. J. Comput. Vis.* **2004**, *60*, 91–110. [CrossRef]
26. Xiao, Y.; Wu, J.; Yuan, J. mCENTRIST: A Multi-Channel Feature Generation Mechanism for Scene Categorization. *IEEE Trans. Image Process.* **2014**, *23*, 823–836. [CrossRef]
27. Wu, J.; Rehg, J.M. Centrist: A visual descriptor for scene categorization. *IEEE Trans. Pattern Anal. Mach. Intell.* **2010**, *33*, 1489–1501. [PubMed]
28. Sevo, I.; Avramović, A. Multispectral scene recognition based on dual convolutional neural networks. In Proceedings of the 10th International Symposium on Image and Signal Processing and Analysis, Ljubljana, Slovenia, 18–20 September 2017; pp. 126–130.
29. Elezabi, O.; Guesney-Bodet, S.; Thomas, J.B. Impact of Exposure and Illumination on Texture Classification Based on Raw Spectral Filter Array Images. *Sensors* **2023**, *23*, 5443. [CrossRef] [PubMed]
30. SILIOS CMS-C. 2023. Available online: https://www.silios.com/cms-series (accessed on 4 November 2023).
31. Li, Y.; Liao, N.; Bai, X.; Cheng, H.; Yang, W.; Deng, C. An on-line color defect detection method for printed matter based on snapshot multispectral camera. In Proceedings of the Advanced Optical Imaging Technologies, Beijing, China, 11–13 October 2018; SPIE: Philadelphia, PA, USA, 2018; Volume 10816, pp. 67–72.
32. IDS uEye Cockpit. 2023. Available online: https://en.ids-imaging.com/ids-software-suite.html (accessed on 4 November 2023).

33. He, K.; Zhang, X.; Ren, S.; Sun, J. Deep residual learning for image recognition. In Proceedings of the IEEE Conference on Computer Vision and Pattern Recognition, Las Vegas, NV, USA, 27–30 June 2016; pp. 770–778.
34. Howard, A.G.; Zhu, M.; Chen, B.; Kalenichenko, D.; Wang, W.; Weyand, T.; Andreetto, M.; Adam, H. Mobilenets: Efficient convolutional neural networks for mobile vision applications. *arXiv* **2017**, arXiv:1704.04861

Disclaimer/Publisher's Note: The statements, opinions and data contained in all publications are solely those of the individual author(s) and contributor(s) and not of MDPI and/or the editor(s). MDPI and/or the editor(s) disclaim responsibility for any injury to people or property resulting from any ideas, methods, instructions or products referred to in the content.

sensors

MDPI

Article

A Microlens Array Grating for Miniature Multi-Channel Spectrometers

Shuonan Shan [1,†], Jingwen Li [1,2,†], Peiyuan Liu [1], Qiaolin Li [1,*], Xiaohao Wang [1,2] and Xinghui Li [1,2,*]

1 Shenzhen International Graduate School, Tsinghua University, Shenzhen 518055, China; ssn21@mails.tsinghua.edu.cn (S.S.); jw-li20@mails.tsinghua.edu.cn (J.L.); liu-py19@tsinghua.org.cn (P.L.); wang.xiaohao@sz.tsinghua.edu.cn (X.W.)

2 Tsinghua-Berkeley Shenzhen Institute, Tsinghua University, Shenzhen 518055, China

* Correspondence: li.qiaolin@sz.tsinghua.edu.cn (Q.L.); li.xinghui@sz.tsinghua.edu.cn (X.L.)

† These authors contributed equally to this work.

Abstract: Most existing multi-channel spectrometers are constructed by physically stacking single-channel spectrometers, resulting in their large size, high weight, and limited number of channels. Therefore, their miniaturization is urgently needed. In this paper, a microlens array grating is designed for miniature multi-channel spectrometers. A transmissive element integrating microlens arrays and gratings, the MLAG, enables simultaneous focusing and dispersion. Using soft lithography, the MLAG was fabricated with a deviation of less than 2.2%. The dimensions are 10 mm × 10 mm × 4 mm with over 2000 available units. The MLAG spectrometer operates in the 400–700 nm wavelength range with a resolution of 6 nm. Additionally, the designed MLAG multi-channel spectrometer is experimentally verified to have independently valid cells that can be used in multichannel spectrometers. The wavelength position repeatability deviation of each cell is about 0.5 nm, and the repeatability of displacement measurements by the chromatic confocal sensor with the designed MLAG multi-channel spectrometer is less than 0.5 μm.

Keywords: microlens array; grating; fabrication; PDMS; soft lithography; miniature multi-channel spectrometer

Citation: Shan, S.; Li, J.; Liu, P.; Li, Q.; Wang, X.; Li, X. A Microlens Array Grating for Miniature Multi-Channel Spectrometers. *Sensors* **2023**, *23*, 8381. https://doi.org/10.3390/s23208381

Academic Editors: Nélia J. Alberto and Flavio Esposito

Received: 15 August 2023
Revised: 11 September 2023
Accepted: 19 September 2023
Published: 11 October 2023

Copyright: © 2023 by the authors. Licensee MDPI, Basel, Switzerland. This article is an open access article distributed under the terms and conditions of the Creative Commons Attribution (CC BY) license (https://creativecommons.org/licenses/by/4.0/).

1. Introduction

Spectrometers are used to obtain wavelength–light intensity relationships, which have important applications in many research fields, such as the spectral characteristics of different clinical conditions acquired by spectrometers including diagnosis of human eye diseases, monitoring of multiple gaseous air pollutants, chromatic confocal microscopes, and climate monitoring [1–5]. In recent years, spectrometer research has evolved towards broad spectral ranges, miniaturization, and ease of installation [6–8]. Meanwhile with the abundance of measurement scenarios, the demand for flat scanning and even three-dimensional information measurements has increased dramatically, which requires multi-channel spectrometers with multi-line processing capabilities, with requirements for measurement efficiency and data processing speed [9–12]. Thus, some novel methods for multi-spectrometers have been proposed in recent years, such as spectrometer arrays, which can be used to obtain and analyze multi-channel signals [13–16].

There are already a number of proven and reliable commercial multi-channel spectrometers available. The multi-channel spectrometer made by Ocean Optics has eight channels that can detect wavelengths from 180 to 1100 nm with a resolution of 0.1 nm and which weighs of 7 kg [17]. The multi-channel spectrometer made by HORIBA has 96 channels that can detect wavelengths from 360 to 780 nm with a resolution of less than 3.5 nm [18]. The dimensions and weights of these current products are too big and heavy due to the simple combination of the multi-channel spectrometers. Therefore, some novel products, in which the functions of focusing and dispersing are integrated into a small optical component, are proposed for the miniaturization of spectrometers [19,20]. The channel number is less than 100, which could not meet the requirement of industry manufacturing. Moreover, the multi-channel spectrometers here are constructed by physically stacking single-channel spectrometers, resulting in their large size, high weight, and limited number of channels. So, their miniaturization is urgently needed.

In existing research, there are several studies that propose some innovative multi-channel spectrometers. Silke Traut et al. used photoresist and holography to fabricate a composite structure with a microlens array on one side and a microlens array grating on the other, whose resolution is more than 10 nm [21]. In Tadayuki Hirano's research, the structure is designed as a blazed grating-air gap-microlens array and is fabricated by physical vapor deposition and electron beam direct writing. The element can resolve color images but with a low resolution [22]. Jun Shi et al. fabricated a microlens array grating structure by hot melting, self-assembly, and replication on the same surface. A resolution of up to 6.9 nm in the wavelength range of 450 nm~650 nm was achieved [23]. However, these fabrication methods require a high level of environmental cleanliness and materials. The lack of cleanliness during the production process and the curing shrinkage effect of the material can have a negative impact on the results [24,25].

In this paper, a microlens array grating (MLAG) structure [26–28], which has a microlens array on one side and a grating on the other side, was designed, fabricated, and verified [29,30]. Soft lithography was employed to fabricate the MLAG with PDMS [31–36]. The surface form of the fabricated component was inspected by AFM for the fabrication quality. The performance of the miniature multi-channel spectrometer with the fabricated microlens array grating was verified by some experiments. Furthermore, the application of this multi-channel spectrometer with the microlens array grating for the confocal sensor array was introduced, which can be developed for line-scanning measurements and/or confocal microscopy.

2. Theoretical Analysis

Conventionally, the lens and grating are employed in the spectrometer as two different elements for focusing and dispersing, respectively [37,38]. In this paper, the lens and grating are integrated into one component for the miniature spectrometer. As shown in Figure 1, a structure with many microlenses arrayed on one side and a grating on another side of the component surface, which is called microlens array grating, was designed for the multi-channel spectrometer.

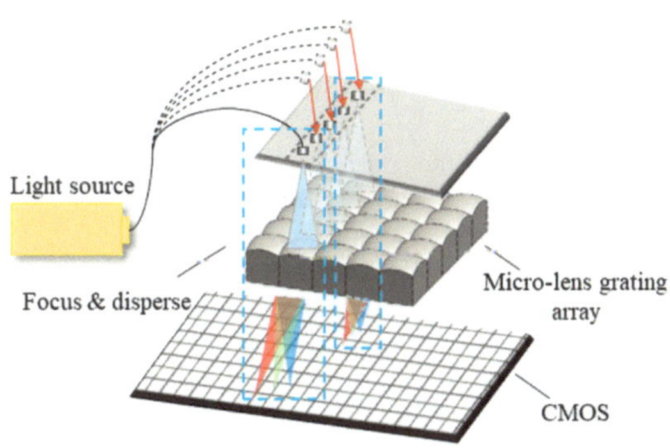

Figure 1. Schematic of a miniature multi-channel spectrometer with the microlens array grating.

2.1. Principle of the Microlens Array Grating

As shown in Figure 2a, a miniature multi-channel spectrometer can be realized since only a single unit of the microlens array grating can have the function of focusing and dispersing. A miniaturization of a conventional lens is employed as the microlens for focusing. Considering the complexity of the process and the requirements of the array design, the focusing of the core unit of the spectrometer is realized with a spherical convex lens, whose structure and main parameters are shown in Figure 2b, with f_{ef} as the effective front focal length, f_{eb} is the effective back focal distance, R_c is the radius of curvature, D is the lens diameter, and H is the unit thickness. A transmission grating, which can periodically modulate the amplitude and phase of the incident light, is employed for diffraction. The grating line density is 600 lines/mm.

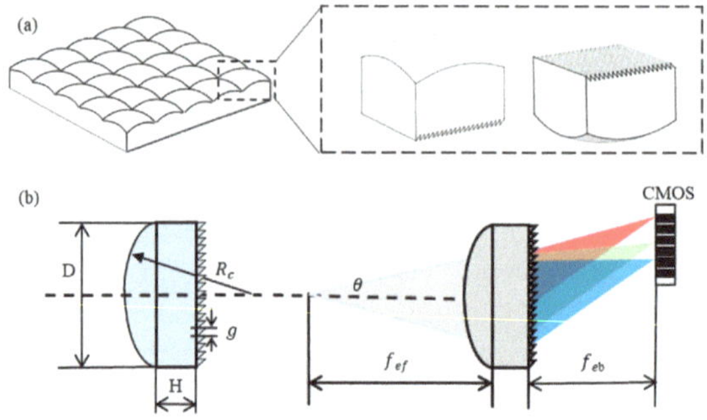

Figure 2. Design of a single unit of the microlens array grating: (**a**) the schematic of a single unit of the microlens array grating; (**b**) the principle of focusing and dispersing by the microlens grating.

2.2. Design of the Microlens Array Grating

Designing a microlens array grating (MLAG) spectrometer requires consideration of performance and dimensional goals and ensuring manufacturing feasibility. Following an analysis of research spectrometers, the target specifications were established as follows: physical dimensions below 1 cm × 1 cm × 1 cm, a measurement spectral range of 400–760 nm, and a resolution exceeding 10 nm. According to the fundamental imaging principle, the object distance roughly equals half the size of the spectrometer. Initially, a convex lens with an effective focal length (EFL) of 4 mm was selected to ensure an appropriate size. The grating period was set at 600 lines/mm, favoring a smaller grating period to achieve a larger diffraction angle and enhance optical performance. The preset MLAG parameters included a microlens diameter of 220 μm, microlens focal length of 4 mm, grating line density of 600 lines/mm, and microlens grating thickness (H) of 4 mm.

By utilizing an MLAG as its core component, the spectrometer model can be established with fiber optic arrays serving as the light source input and CMOS photodetectors as the signal reception. Through careful parameter design, simultaneous parallel analysis of various detection channels can be achieved, resulting in an array-type multi-channel micro-spectrometer.

The resolution of the spectrometer is significantly influenced by the pupil diameter. Theoretically, when the aperture diameter is excessively large, the image is magnified, resulting in reduced resolution. Conversely, if the aperture diameter is too small, there is a greater loss of available optical power, leading to reduced resolution as well. Therefore, after careful consideration, a pupil diameter of 10 μm was selected for the spectrometer. When light waves pass through the aperture, diffraction occurs. It is important to note that longer wavelengths result in larger divergence angles. Hence, the maximum divergence angle within the visible range (the maximum wavelength is 760 nm) is as shown in Equation (1) where θ_0 is the divergence angle, λ is wavelength D is entrance pupil diameter:

$$\theta_0 = 1.22\frac{\lambda}{D} \tag{1}$$

And the maximum incident half-angle width allowed by the preset microlens parameters is shown in Equation (2) where θ_0' is the maximum incident angle.

$$\theta = \theta_0' \tag{2}$$

As a consequence, the spot radius within this range will exceed the maximum limit permissible for the microlens surface element. To mitigate optical power loss and minimize the divergence angle of the light source, a three-piece objective lens set was implemented between the entry pupil and the MLAG. This integration ensures that the exit angle remains at or below 0.0125 rad.

2.3. The Simulation of the Designed Microlens Array Grating for the Miniature Multi-Channel Spectrometer

The simulation was carried out using ZEMAX for testing the performance of the spectrometer. For the spectrometer with the proposed microlens array grating, 11 groups of 22 wavelengths ranging from 400 nm to 700 nm were selected. Each group consisted of two wavelengths with similar values. The resolution of the spectrometer could be evaluated based on the relative positions of these two wavelengths in the image-plane spot.

A set of discrete sampling field-of-view points with different positions was established. The simulated fiber core diameter was 105 μm, with seven sampling points at horizontal positions (0, ±0.5 HFOV, ±0.707 HFOV, ±HFOV) and five sampling points at vertical positions (0, ±0.707 HFOV, ±HFOV). Subsequently, the spectrometer system simulation model was created using the lens data. The diffracted spectra from the MLAG, resulting from the passage of light waves through the lens group, were observed to fall on the image plane according to their respective wavelengths.

Table 1 presents the measured resolution of the MLAG spectrometer in each spectral band. The results demonstrate that the system achieved excellent resolution in the 400 nm to 700 nm band, with most positions achieving a resolution of 10 nm or less. In the lower frequency bands, the system achieved an optimal resolution of up to 6 nm.

Table 1. Resolution of the spectrometer at different wavelengths.

Wavelength/nm	Resolution/nm
400	13
450	12
470	11
550	9
600	9
625	7
650	6
675	8
700	9

This performance may not be as favorable compared to commercial spectrometers. However, it is important to note that the primary focus of the MLAG spectrometer is its size reduction and increased processing speed as a multi-channel, array-based miniature spectrometer. Therefore, the loss of resolution is within an acceptable range given these priorities.

3. Fabrication and Characterization

3.1. Fabrication Process

Soft lithography technology was chosen to fabricate microlens array gratings in a "sandwich" style. The main body of the microlens array grating consists of an elastic mold, while the grating and microlens array molds were used for printing on the two sides, respectively. It is worth noting that the imprinting of the microlens array involved the use of a negative mold, which was utilized to obtain the desired structure. The overall fabrication process is illustrated in Figure 3. This technique does not rely on expensive lithography equipment, and it enables the shaping of microstructures with nanometer resolution.

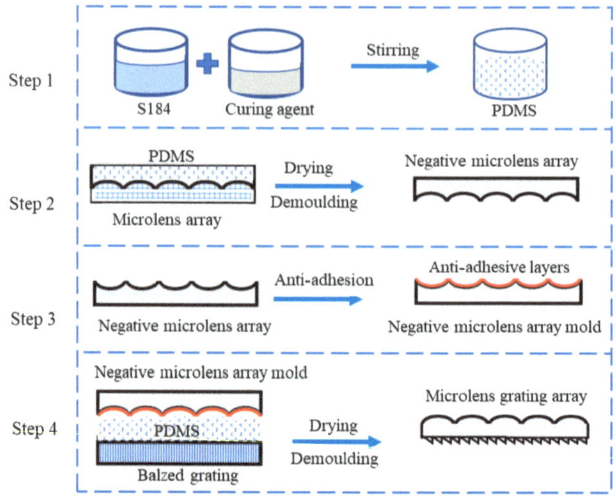

Figure 3. Fabrication process of microlens array grating (MLAG).

Prior to the fabrication process, the imprinting mold was selected according to the designed spectrometer unit. The microlens array was selected from the ML-S220-F4 lens array produced by Shanghai Microlight Technology Co., Shanghai, China. The lens is a square spherical mirror with a focal length of 4 mm, a subunit edge length of 220 μm, and a subunit size of φ16 mm × 2 mm with a size of 12.7 mm × 12.7 mm × 6 mm.

In the first step, PDMS was chosen as the elastic mold material due to its low interfacial separation energy (21.6 dyn/cm), chemical stability, thermal stability, homogeneity, isotropy, and high replication accuracy. The PDMS used was the Sylgard 184 two-component PDMS from Dow Corning, mixed in a ratio of 10:1 with the curing agent. The mixture was thoroughly stirred, and any air bubbles were eliminated using a vacuum to achieve transparency and readiness for use.

The primary replication process was designed to create a negative microlens array mold, as shown in Figure 4a. Special molds were used to pre-mount the microlens arrays, and a bubble-free PDMS mixture was poured slowly over the microlens arrays inside the replica molds. The assembly was then placed in a vacuum-drying oven and baked at 80 °C for 2 h to solidify the PDMS. The negative microlens array was obtained by demolding the solidified PDMS. Since both the negative microlens array and the MLAG are made of PDMS, they tend to adhere to each other. To prevent this, an anti-adhesive coating was applied in Step 3.

Molecular vapor deposition (MVD) was used for the anti-adhesive treatment of the negative microlens array. Two coating materials, Parylene C and Fluorine Nano, were selected. Both coatings had a thickness of 400 nm, which corresponds to approximately 12% of the microlens thickness of 3.32 μm. It can be assumed that the film layer has minimal impact on the microlens morphology, size, and focusing effect. Subsequently, separate tests were conducted on the MLAGs fabricated using the two different coatings.

After the negative microlens array was obtained, it and the grating could be used as molds for "sandwich-type" MLAG fabrication, as shown in Figure 4b. The process is similar to step 2, that is, the PDMS was poured into a special mold, the negative microlens array with anti-stick coating and blazed grating fixed in the mold of the upper and lower layers, and then the whole unit was placed in a vacuum oven, curing at 80 °C for 2 h. Then, the mold was removed, cooled, and demolded to obtain an MLAG with dimensions of 10 mm × 10 mm × 4 mm containing more than 2000 spectrometer units.

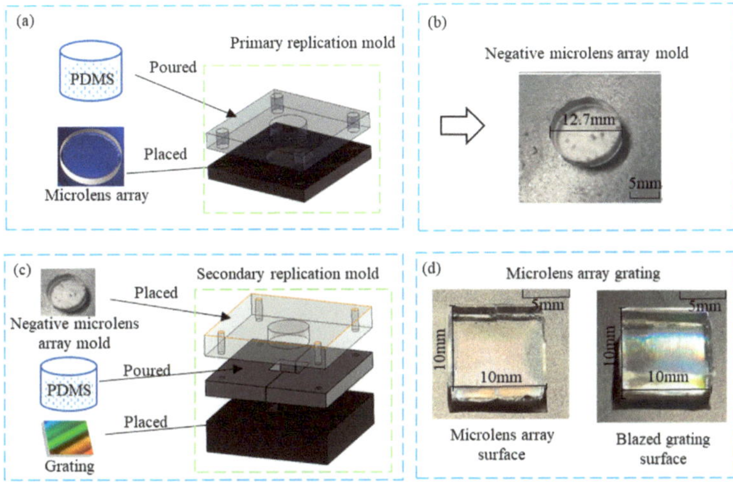

Figure 4. Soft lithography tools and results: (**a**) primary replication process; (**b**) inverted mold of the microlens array; (**c**) secondary replication process; (**d**) both sides of microlens array grating.

Dedicated molds were used in both steps 2 and 4 in order to complete the alignment and fabrication of the two micro-structured graphic planes, the two molds are shown in Figure 4a,b, respectively, and the specific parameters are shown in Tables 2 and 3.

Table 2. Parameters of primary replication mold.

Parameter	Upper Mold	Bottom Mold
Side length	40 mm	40 mm
Thickness	5 mm	4 mm
Groove shape	Circle	Circle
Groove diameter	12.7 mm	16.2 mm
Groove depth	5 mm	2 mm
Positioning hole diameter	4 mm	3 mm

Table 3. Parameters of secondary replication mold.

Parameter	Upper Mold	Middle Mold	Bottom Mold
Side length	40 mm	40 mm	40 mm
Thickness	4 mm	4 mm	10 mm
Groove shape	Circle	square	Square
Groove diameter	16.2 mm	10 mm	13.7 mm
Groove depth	2 mm	4 mm	6 mm
Positioning hole diameter	3 mm	2 mm	2 mm

3.2. Characterization of Microlens Array Grating

The prepared samples were investigated for surface morphology of the grating and microlens array using Bruker's Innova AFM at A. The probe lightly touched the grating surface, moving approximately 10 grating cycles away, and performed five scans on the same plane. The obtained results were flattened, and a cross-section was taken perpendicular to the grating lines to measure the cross-sectional characteristics of the grating groove shape, which provided data on grating height and period curves. These results are illustrated in Figure 5a–c, with corresponding morphological parameters presented in Table 4. Analysis of the test results indicates an average depth deviation of 2.2% and an average width deviation of 0.1% compared to the pristine grating template, aligning with expectations.

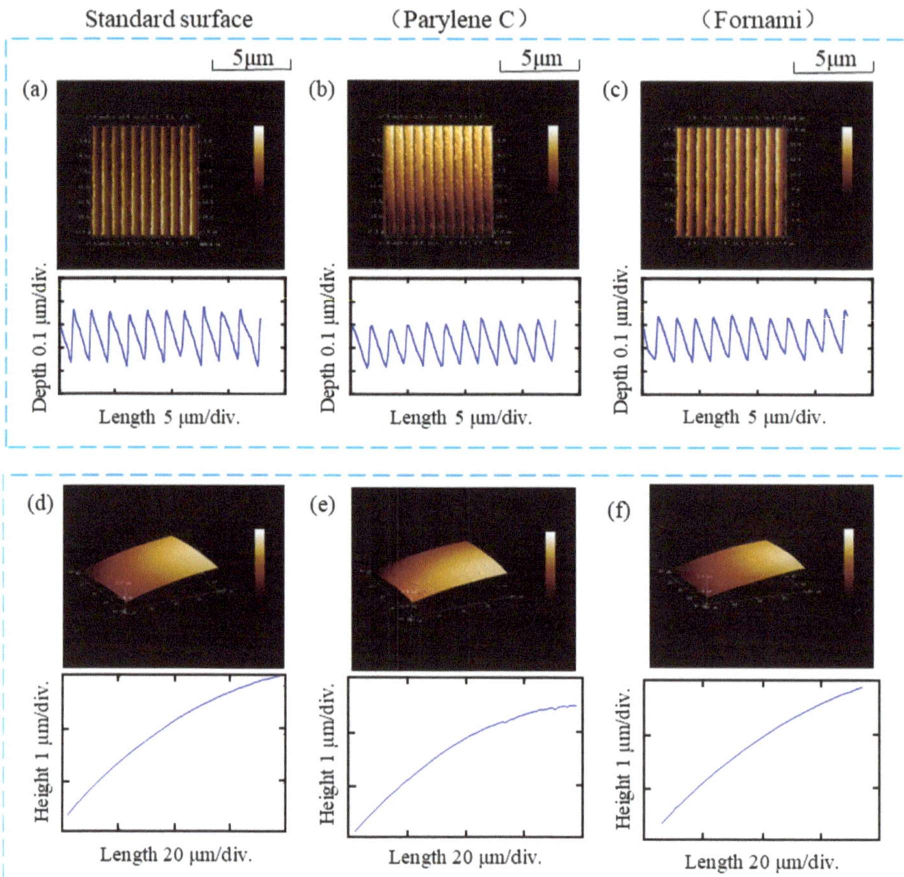

Figure 5. Characterization of the finished microlens array grating: (**a**) the standard grating surface with its depth and period; (**b**) the grating surface with its depth and period with Parylene is used as anti-adhesive coatings in the fabrication process; (**c**) the grating surface with its depth and period when Fluorine Nano is used as anti-adhesive coatings in the fabrication process; (**d**) the standard microlens surface; (**e**) the microlens when Parylene is used as anti-adhesive coatings in the fabrication process; (**f**) the microlens when Fluorine Nano is used as anti-adhesive coatings in the fabrication process.

Table 4. Grating surface morphology parameters.

Samples	Grating in MLGA	Grating
Width 8 cycles/μm	13.093	13.115
Average width/μm	1.637	1.639
Width deviation	0.10%	-
Average depth/nm	183.4	187.5
Depth deviation	2.20%	-

The lens surface was also measured using AFM to obtain its morphological information. The measurement results were flattened, and the surface's radius of curvature was fitted by selecting three random points on the surface. This process was repeated five times, and the average values were calculated to obtain the microlens height curve and determine its radius of curvature. Since the microlens surface is replicated using an inverted mold, which is associated with the anti-adhesive coating, the manufactured products with two different coatings, P (Parylene C) and F (Fluorine nano), were tested separately. The measurement results are depicted in Figure 5d–f, with the microlens morphology parameters presented in Table 5. The average radius of curvature for the microlens template was 1.768, while the microlens made with the P coating and F coating had average radii of curvature of 1.755 and 1.806, respectively. This represents a deviation of 0.7% for the former and 2.1% for the latter compared to the radius of curvature of the original microlens surface. It is apparent that the samples prepared with the P coating exhibit better quality and were therefore utilized in subsequent microlens–grating spectrometer experiments.

Table 5. Micro-lens surface morphology parameters.

Samples	Micro-Lens in MLGA		Micro-Lens
Coating material	P	F	-
Curvature/mm^{-1}	1/1.755	1/1.806	1/1.768
Deviation	0.70%	2.10%	-

4. Verification

4.1. Characterization of Spectrometer (Calibration)

An optical setup is shown in Figure 6a. The experimental system consists of the synthetic laser light source, three-piece objectives, CMOS detector (MV-CA060-10GC by HIKVISON), and computer. Laser beams of wavelengths of 632.8 nm (red), 473 nm (blue), and 405 nm (blue–violet) were combined and introduced to the fiber port. The light emitted from the fiber optic port was dispersive. After reducing the dispersion angle by a three-piece objective system, the light beam was focused and dispersed by the fabricated component. Light was imaged and recorded on the CMOS detector surface. Finally, the images were shown and analyzed by software on the computer. The image obtained by the CMOS detector is shown in Figure 6b. Since the grating is a diffractive element, the light spots appearing on the image are diffracted spots at various levels. The brightest spot is zero level spot and the other three spots are the first level diffraction spots. The centers of these spots are connected as the sampling axis. The light intensity on the sampling axis was recorded and is shown in Figure 6b. The intensity of zero level spot was higher than the intensity of the first level diffraction spot. According to the first level diffraction spots, the light beams with different wavelengths were distributed at different positions and recorded by pixels. The larger the wavelength is, the farther the center of the spot is from the zero-level spot. The mass center method was used to obtain the position of the spot center of first level diffraction spot. The real calculated method is shown in Equation (3) where P_{CM} is the focus pixel, P_n is the pixel number, $I(P_n)$ is light intensity at pixel number P_n.

$$P_{CM} = \frac{\sum P_n \cdot I(P_n)}{\sum I(P_n)} \tag{3}$$

Four surface elements were chosen at random and the images were recorded in order to test the performance of different surface elements. The results are shown in Table 6. The standard deviations of pixel positions were 0.81, 0.82, 0.95 for the different laser colors. The good consistency of different spectrometer units was verified.

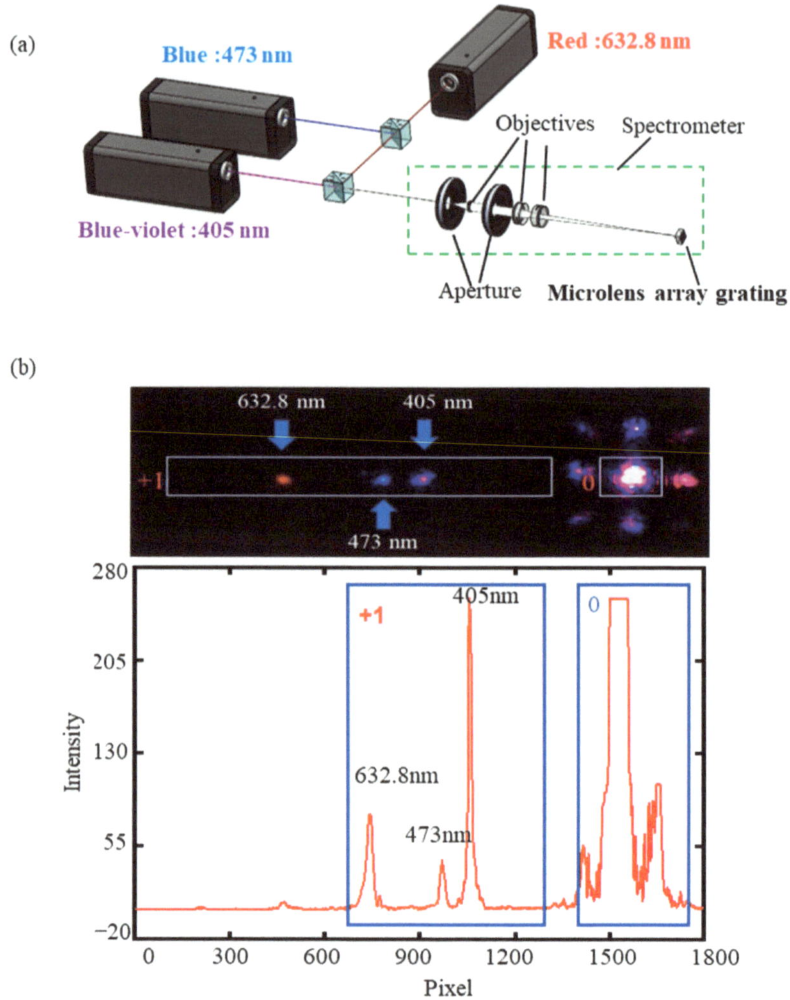

Figure 6. Spectral analysis ability testing system and results: (**a**) schematic diagram of the miniature spectrometer system; (**b**) results in a single microlens surface element.

Table 6. Results in a single microlens surface element.

No.	632.8 nm	473 nm	405 nm
1	773	541	458
2	772	541	457
3	771	542	456
4	772	540	456
Std (pixel)	0.81	0.82	0.95
Sta (nm)	0.48	0.64	0.81
Simulation (nm)		Around 0.5	

4.2. Application in Chromatic Confocal System

The feasibility of microlens array grating as the core component of the spectrometer has been verified in the previous section. In this section, the performance of the microlens array-grating spectrometer in a spectral confocal system is presented.

In the optical signals containing position information acquired by the chromatic confocal probe, the focused wavelength has the highest intensity, so it can be regarded as monochromatic light. The resolution of the chromatic confocal sensor depends on the wavelength resolution of the spectrometer to some extent.

As Figure 7a shows, the chromatic confocal sensor consists of the polychromatic light source, dispersive objectives, fiber coupler, and the spectrometer. Microlens array grating was used as the core element of the spectrometer in this system. In order to fit the small numerical aperture of microlens array grating, a white laser source, which was synthesized by three wavelengths, was used and compared with LED. The laser source has higher energy to make sure the optical signal can be noted by CMOS after the energy loss caused by the light path.

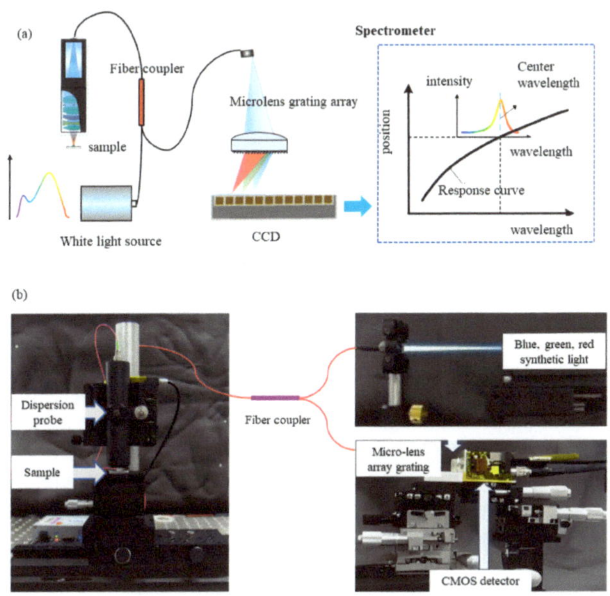

Figure 7. Application testing of microlens array grating spectrometer in spectral confocal measurement system: (**a**) schematic of the chromatic confocal system; (**b**) picture of the chromatic confocal system.

A z-stage was used to make the calibration experiments. Due to the principle of chromatic confocal sensors, different positions of samples correspond to different wavelengths of reflected light. As shown in Figure 8, the images can be recorded by controlling the z-stage. When the sample was in different positions, the color of the spot obtained was different. The curve of the pixel–wavelength relationship can be obtained by calibration according to the previous section. In the chromatic confocal system, there was a correspondence between the sample position and the spot position in the whole system since there is a functional relationship between the position and the focused wavelength. In Figure 8a, the light spot positions of three measurements are P1, P2, and P3, respectively. The displacement stage was controlled to reach the corresponding position after recording the position of the z-stage and the pixel position at the center of the spot for each measurement. Figure 8 shows the results of the experiments, which were carried out 20 times. The standard deviations of pixel positions were 0.31, 0.15, and 0.13, respectively. In conclusion, the spectrometer showed good stability in the chromatic confocal system.

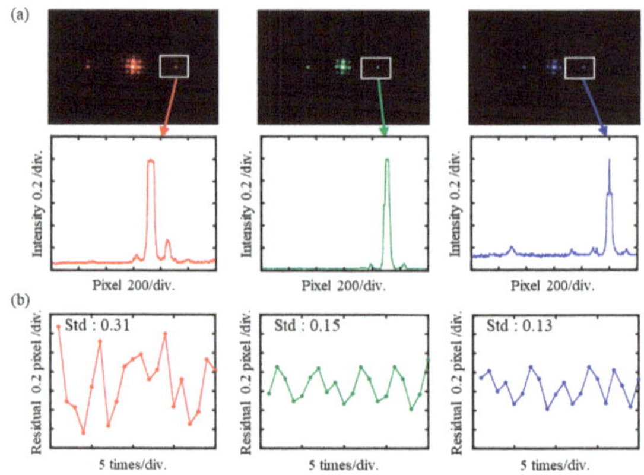

Figure 8. Experimental results: (**a**) optical signal detected by CMOS and its intensity (**b**) residual error.

5. Conclusions

A microlens array gating (MLAG), which has the functions of focusing and dispersion, is proposed for the miniature multi-channel spectrometer. The single microlens with a bin size of 0.22 mm and focal length of 4 mm was arrayed on one side of the component with a dimension of 10 mm × 10 mm × 4 mm, namely, there are more than 2000 microlenses on the surface. The other side of the component was a blazed grating with a density of 600 lines/mm. The fabrication of the microlens array grating was achieved by soft lithography, which consisted of four steps: PDMS preparation, primary replication process, anti-adhesive treatment process, and secondary replication process. The quality of the fabricated MLAG was confirmed through characterization using AFM microscopy. To verify the performance of the developed miniature multi-channel spectrometer utilizing the proposed MLAG, the centroid position repeatability of the diffractive light spot was detected by the centroid method. The results demonstrate that the centroid position repeatability of the diffractive light spot was kept within 1 pixel, and the performance of different microlens grating panels was consistent. The experiments involving spectral confocal measurement were carried out to verify the application of a miniature multi-channel spectrometer with an MLAG. The results reveal that the repeatability of the spot centroid position was within 0.5 pixels at the three typical positions within the corresponding 400 μm measuring range, which is expected to bring a measurement resolution of 0.5 μm for the spectral confocal system.

Author Contributions: Conceptualization, S.S. and J.L.; Methodology, S.S. and J.L.; Software, P.L.; Formal analysis, Q.L.; Investigation, J.L.; Resources, X.W.; Data curation, P.L.; Writing—original draft, S.S.; Writing—review & editing, Q.L.; Visualization, P.L.; Supervision, Q.L. and X.W.; Project administration, X.L.; Funding acquisition, X.L. All authors have read and agreed to the published version of the manuscript.

Funding: This work was supported in part by the Shenzhen Stable Supporting Program under grant WDZC20200820200655001, in part by the National Natural Science Foundation of China under Grant 62275142, in part by the Basic and Applied Basic Research Foundation of Guangdong Province under Grant 2021B1515120007, and in part by the Tsinghua University under Grant QD2020001N.

Conflicts of Interest: The authors declare that they have no known competing financial interests or personal relationships that could have appeared to influence the work reported in this paper.

References

1. Gao, W. (Ed.) *Metrology*; Precision Manufacturing; Springer: Singapore, 2019; ISBN 978-981-10-4937-8.
2. Sato, R.; Li, X.; Fischer, A.; Chen, L.-C.; Chen, C.; Shimomura, R.; Gao, W. Signal Processing and Artificial Intelligence for Dual-Detection Confocal Probes. *Int. J. Precis. Eng. Manuf.* **2023**. [CrossRef]

3. Schifano, L.; Berghmans, F.; Dewitte, S.; Smeesters, L. Optical Design of a Novel Wide-Field-of-View Space-Based Spectrometer for Climate Monitoring. *Sensors* **2022**, *22*, 5841. [CrossRef] [PubMed]
4. Fathy, A.; Gnambodoe-Capochichi, M.; Sabry, Y.M.; Anwar, M.; Ghoname, A.O.; Saeed, A.; Leprince-Wang, Y.; Khalil, D.; Bourouina, T. Potential of a Miniature Spectral Analyzer for District-Scale Monitoring of Multiple Gaseous Air Pollutants. *Sensors* **2023**, *23*, 6343. [CrossRef]
5. Dastgeer, G.; Shahzad, Z.M.; Chae, H.; Kim, Y.H.; Ko, B.M.; Eom, J. Bipolar Junction Transistor Exhibiting Excellent Output Characteristics with a Prompt Response against the Selective Protein. *Adv. Funct. Mater.* **2022**, *32*, 2204781. [CrossRef]
6. Zhao, Q.; Tong, L.; Gao, B. Spectral Analysis of Stationary Signals Based on Two Simplified Arrangements of Chirp Transform Spectrometer. *Electronics* **2021**, *10*, 65. [CrossRef]
7. Feng, L.; He, X.; Li, Y.; Wei, L.; Nie, Y.; Jing, J.; Zhou, J. Compact Shortwave Infrared Imaging Spectrometer Based on a Catadioptric Prism. *Sensors* **2022**, *22*, 4611. [CrossRef]
8. Zhang, X.; Li, B.; Jiang, X.; Gu, G.; Li, H.; Wang, X.; Lin, G. Design of a Prism-Grating Wide Spectral Range Transmittance Imaging Spectrometer. *Sensors* **2023**, *23*, 5050. [CrossRef]
9. Zhou, Z.; Liu, Q.; Fu, Y.; Xu, X.; Wang, C.; Deng, M. Multi-Channel Fiber Optical Spectrometer for High-Throughput Characterization of Photoluminescence Properties. *Rev. Sci. Instrum.* **2020**, *91*, 123113. [CrossRef] [PubMed]
10. Li, Z.; Li, H.; Zhu, X.; Peng, Z.; Zhang, G.; Yang, J.; Wang, F.; Zhang, Y.F.; Sun, L.; Wang, R.; et al. Directly Printed Embedded Metal Mesh for Flexible Transparent Electrode via Liquid Substrate Electric-Field-Driven Jet. *Adv. Sci.* **2022**, *9*, 2105331. [CrossRef]
11. Yu, F.; Yu, S.; Li, C.; Li, Z.; Song, F.; Xu, Z.; Zhu, Y.; Dai, C.; Cao, X.; Zhang, Z.; et al. Molecular Engineering of Biomimetic Donor-Acceptor Conjugated Microporous Polymers with Full-Spectrum Response and an Unusual Electronic Shuttle for Enhanced Uranium(VI) Photoreduction. *Chem. Eng. J.* **2023**, *466*, 143285. [CrossRef]
12. Zhu, X.; Liu, M.; Qi, X.; Li, H.; Zhang, Y.F.; Li, Z.; Peng, Z.; Yang, J.; Qian, L.; Xu, Q.; et al. Templateless, Plating-Free Fabrication of Flexible Transparent Electrodes with Embedded Silver Mesh by Electric-Field-Driven Microscale 3D Printing and Hybrid Hot Embossing. *Adv. Mater.* **2021**, *33*, 2007772. [CrossRef]
13. Yang, Z.; Albrow-Owen, T.; Cai, W.; Hasan, T. Miniaturization of Optical Spectrometers. *Science* **2021**, *371*, eabe0722. [CrossRef] [PubMed]
14. Walter, P.; Kamalov, A.; Gatton, A.; Driver, T.; Bhogadi, D.; Castagna, J.C.; Cheng, X.; Shi, H.; Obaid, R.; Cryan, J.; et al. Multi-Resolution Electron Spectrometer Array for Future Free-Electron Laser Experiments. *J. Synchrotron Radiat.* **2021**, *28*, 1364–1376. [CrossRef] [PubMed]
15. Gaigalas, A.K.; Wang, L.; He, H.J.; DeRose, P. Procedures for Wavelength Calibration and Spectral Response Correction of CCD Array Spectrometers. *J. Res. Natl. Inst. Stand. Technol.* **2009**, *114*, 215. [CrossRef] [PubMed]
16. Danz, N.; Höfer, B.; Förster, E.; Flügel-Paul, T.; Harzendorf, T.; Dannberg, P.; Leitel, R.; Kleinle, S.; Brunner, R. Miniature Integrated Micro-Spectrometer Array for Snap Shot Multispectral Sensing. *Opt. Express* **2019**, *27*, 5719–5728. [CrossRef]
17. MX2500 Series, Ocean Optics. Available online: www.oceaninsight.cn (accessed on 12 July 2023).
18. HORIBA OEM, HORIBA. Available online: www.horiba.com (accessed on 12 July 2023).
19. Wang, Y.; Qu, Y.; Zhao, H.; Fan, X. Construction, Spectral Modeling, Parameter Inversion-Based Calibration, and Application of an Echelle Spectrometer. *Sensors* **2023**, *23*, 6630. [CrossRef] [PubMed]
20. Wang, Q.; Shen, H.; Liu, W.; Zhang, J.; Meng, L. Design of Compact Mid-Infrared Cooled Echelle Spectrometer Based on Toroidal Uniform-Line-Spaced (TULS) Grating. *Sensors* **2022**, *22*, 7291. [CrossRef]
21. Traut, S. Holographically Recorded Gratings on Microlenses for a Miniaturized Spectrometer Array. *Opt. Eng.* **2000**, *39*, 290–298. [CrossRef]
22. Hirano, T.; Shimatani, N.; Kintaka, K.; Nishio, K.; Awatsuji, Y.; Ura, S. Combined Blazed Grating and Microlens Array for Color Image Sensing. *Jpn. J. Appl. Phys.* **2014**, *53*, 032501. [CrossRef]
23. Shi, J.; Huang, Y.S.; Peng, L.N.; Ni, Z.J.; Zhang, D.W. Grating/Microlens Arrays Fabricated by Hot-Melting, Self-Assembly and Replication. *Opt. Mater.* **2020**, *104*, 109733. [CrossRef]
24. Dastgeer, G.; Afzal, A.M.; Jaffery, S.H.A.; Imran, M.; Assiri, M.A.; Nisar, S. Gate Modulation of the Spin Current in Graphene/WSe2 van Der Waals Heterostructure at Room Temperature. *J. Alloys Compd.* **2022**, *919*, 165815. [CrossRef]
25. Elahi, E.; Suleman, M.; Nisar, S.; Sharma, P.R.; Iqbal, M.W.; Patil, S.A.; Kim, H.; Abbas, S.; Chavan, V.D.; Dastgeer, G.; et al. Robust Approach towards Wearable Power Efficient Transistors with Low Subthreshold Swing. *Mater. Today Phys.* **2023**, *30*, 100943. [CrossRef]
26. Zhang, L.; Yi, A.Y.; Yan, J. Flexible Fabrication of Fresnel Micro-Lens Array by off-Spindle-Axis Diamond Turning and Precision Glass Molding. *Precis. Eng.* **2022**, *74*, 186–194. [CrossRef]
27. Zhou, Q.; Li, X.; Geng, M.; Hu, H.; Ni, K.; Zhong, L.; Yan, P.; Wang, X. Economic Fabrication of a Novel Hybrid Planar Grating/Fresnel Lens for Miniature Spectrometers. *Opt. Express* **2018**, *26*, 6079–6089. [CrossRef]
28. Li, X.; Zhang, J.; Zhou, Q.; Ni, K.; Pang, J.; Tian, R. Design of a Variable-Line-Spacing Grating Pattern for Spectrometers Based on a Grating Fresnel Device. *Opt. Lett.* **2016**, *41*, 1470–1473. [CrossRef]
29. Ma, L.; Lin, J.; Ma, Y.; Jin, P.; Tan, J. Subwavelength Focusing of Micro Grating-Fresnel Lens. *Opt. Commun.* **2013**, *298–299*, 242–245. [CrossRef]
30. Lee, C.L.; Dawson, M.D.; Gu, E. Diamond Double-Sided Micro-Lenses and Reflection Gratings. *Opt. Mater.* **2010**, *32*, 1123–1129. [CrossRef]

31. Qin, D.; Xia, Y.; Whitesides, G.M. Soft Lithography for Micro- and Nanoscale Patterning. *Nat. Protoc.* **2010**, *5*, 491. [CrossRef] [PubMed]
32. Sun, R.; Yang, H.; Rock, D.M.; Danaei, R.; Panat, R.; Kessler, M.R.; Li, L. Manufacturing PDMS Micro Lens Array Using Spin Coating under a Multiphase System. *J. Micromechan. Microeng.* **2017**, *27*, 055012. [CrossRef]
33. Zhou, Q.; Li, X.; Ni, K.; Tian, R.; Pang, J. Holographic Fabrication of Large-Constant Concave Gratings for Wide-Range Flat-Field Spectrometers with the Addition of a Concave Lens. *Opt. Express* **2016**, *24*, 732–738. [CrossRef] [PubMed]
34. Li, X.; Ni, K.; Zhou, Q.; Yan, P.; Pang, J.; Wang, X. Improved Master-Replica Separation Process for Fabrication of a Blazed Concave Grating by Using a Combination-Type Convex Grating. *Appl. Opt.* **2017**, *56*, 298–302. [CrossRef]
35. Zhao, C.; Cheung, C.F.; Xu, P. High-Efficiency Sub-Microscale Uncertainty Measurement Method Using Pattern Recognition. *ISA Trans.* **2020**, *101*, 503–514. [CrossRef] [PubMed]
36. Jiang, W.; Wang, H.; Xie, W.; Qu, Z. Lithography Alignment Techniques Based on Moiré Fringe. *Photonics* **2023**, *10*, 351. [CrossRef]
37. Gao, W.; Shimizu, Y. *Optical Metrology for Precision Engineering*; De Gruyter: Vienna, Austria, 2021; ISBN 9783110542363.
38. Scheeline, A. How to Design a Spectrometer. *Appl. Spectrosc.* **2017**, *71*, 2237–2252. [CrossRef] [PubMed]

Disclaimer/Publisher's Note: The statements, opinions and data contained in all publications are solely those of the individual author(s) and contributor(s) and not of MDPI and/or the editor(s). MDPI and/or the editor(s) disclaim responsibility for any injury to people or property resulting from any ideas, methods, instructions or products referred to in the content.

sensors

MDPI

Article

Gaussian Decomposition vs. Semiclassical Quantum Simulation: Obtaining the High-Order Derivatives of a Spectrum in the Case of Photosynthetic Pigment Optical Properties Studying

Andrei P. Razjivin [1,*], Vladimir S. Kozlovsky [1], Aleksandr A. Ashikhmin [2] and Roman Y. Pishchalnikov [3,*]

1 Belozersky Research Institute of Physico-Chemical Biology, Moscow State University, 119992 Moscow, Russia; wladask@gmail.com
2 Institute of Basic Biological Problems, Russian Academy of Sciences, 142290 Pushchino, Russia; ashikhminaa@gmail.com
3 Prokhorov General Physics Institute of the Russian Academy of Sciences, 119991 Moscow, Russia
* Correspondence: razjivin@belozersky.msu.ru (A.P.R.); rpishchal@kapella.gpi.ru (R.Y.P.); Tel.: +7-4995038777 (R.Y.P.)

Abstract: In this paper, a procedure for obtaining undistorted high derivatives (up to the eighth order) of the optical absorption spectra of biomolecule pigments has been developed. To assess the effectiveness of the procedure, the theoretical spectra of bacteriochlorophyll ***a***, chlorophyll ***a***, spheroidene, and spheroidenone were simulated by fitting the experimental spectra using the differential evolution algorithm. The experimental spectra were also approximated using sets of Gaussians to calculate the model absorption spectra. Theoretical and model spectra can be differentiated without smoothing (high-frequency noise filtering) to obtain high derivatives. Superimposition of the noise track on the model spectra allows us to obtain test spectra similar to the experimental ones. Comparison of the high derivatives of the model spectra with those of the test spectra allows us to find the optimal parameters of the filter, the application of which leads to minimal differences between the high derivatives of the model and test spectra. For all four studied pigments, it was shown that smoothing the experimental spectra with optimal filters makes it possible to obtain the eighth derivatives of the experimental spectra, which were close to the eighth derivatives of their theoretical spectra.

Keywords: chlorophyll; bacteriochlorophyll; carotenoid; gaussian decomposition; absorption spectrum; multimode brownian oscillator model; differential evolution

Citation: Razjivin, A.P.; Kozlovsky, V.S.; Ashikhmin, A.A.; Pishchalnikov, R.Y. Gaussian Decomposition vs. Semiclassical Quantum Simulation: Obtaining the High-Order Derivatives of a Spectrum in the Case of Photosynthetic Pigment Optical Properties Studying. *Sensors* **2023**, *23*, 8248. https://doi.org/10.3390/s23198248

Academic Editors: Qing Yu, Ran Tu, Ting Liu and Lina Li

Received: 24 August 2023
Revised: 25 September 2023
Accepted: 28 September 2023
Published: 5 October 2023

Copyright: © 2023 by the authors. Licensee MDPI, Basel, Switzerland. This article is an open access article distributed under the terms and conditions of the Creative Commons Attribution (CC BY) license (https://creativecommons.org/licenses/by/4.0/).

1. Introduction

The use of derivatives to increase the spectral resolution was first proposed by Lord Rutherford and was also initially applied in the work of his colleagues [1]. To date, more than a thousand papers have been published in which the term "derivative spectroscopy" is mentioned in the title. The theory and practice of derivative spectroscopy are covered in detail in many books [2–4], reviews [5–11], and studies [7,12–20].

Mostly the first or the second derivatives are used, although in the software of modern commercial spectrophotometers it is possible to obtain derivatives of spectra up to four orders. The accuracy of recording spectra is limited by the noise characteristics of the electronics of the spectrophotometers. In modern spectrophotometers, the amplitude of the noise track is usually close to 0.0001 o.u. If, for instance, the long-wavelength absorption peak of chlorophyll *a* (Chl) is about one optical unit, then the signal is 10,000 times greater than the noise. However, a single differentiation of the spectrum decreases this ratio to ~100. With a twofold differentiation, the signal will be equal or so to the noise. Therefore, more or less satisfactory spectrum derivatives are obtained only for the first derivative (without smoothing)

In all cases, when it comes to derivatives above the first, different methods of filtering (smoothing) the data are used in order to suppress high-frequency noise. Among the smoothing methods, the most commonly mentioned are adjacent averaging, the Savitsky–Golay algorithm, FFT-filter, and others. Filtering (smoothing) the spectra makes it possible to obtain a higher derivative, but at the same time this derivative turns out to be distorted—the signal bands are broadened, their amplitude drops, and wavy parasitic bands appear. It is easy to show the distortions of the derivatives due to filtering the spectra by modeling, but it is not possible to take into account and correct the distortions. This is especially true for high derivatives, for example, of the eighth order [21–24]. We have previously used the eighth derivative to reveal the upper exciton band of the bacteriochlorophyll ***a*** (BChl) dimer of the photosynthetic reaction center of purple bacteria [25]. In general, it is difficult to formalize the procedure of filtering (smoothing) and differentiating to obtain high derivatives [26].

If we ignore the formal side of obtaining derivative spectra, we can ask how the spectra of polyatomic dye molecules (pigments, chromophores) and their derivatives should look like, based on physical considerations. Molecular spectra in the near UV, visible, and near IR regions consist of bands corresponding to electronic transitions. In addition to electronic transitions, so-called phonon wings may be present in the spectrum. These bands are due to the interaction of molecular vibrations with the electronic states. In most cases, the contribution of rotational components is too small when pigments are studied in solvents, and their traces are not detected on the spectra.

Ab initio simulations of absorption spectra of polyatomic dye molecules require substantial computational resources, and when it comes to the relatively fast simulation of the obtained spectroscopy data, this type of simulation is not promising. The use of semi-classical quantum theories describing the interaction of the electric field with matter can be considered promising in this case [27,28]. The optical properties of photosynthetic pigments such as chlorophylls, bacteriochlorophylls, and carotenoids, as well as the pigment–protein complexes containing them, are the focus of intensive theoretical research [27,29,30]. The basic concept of semi-classical calculations is the spectral density function, which carries information about the vibronic structure of the molecule and about the intensity of interaction of different vibrational modes with electronic excitation [28]. However, modeling with semi-classical theories has one drawback, which is the need to adjust microparameters that cannot be measured directly, and their values can only be estimated be fitting experimental spectra. A successful solution to this problem was the application of a multiparametric optimization algorithm—differential evolution [31,32]. The development of quantum models of the $|S_0\rangle \rightarrow |S_2\rangle$ electronic transition of carotenoids and the $|S_0\rangle \rightarrow |Q_y\rangle$ transition of chlorophylls and bacteriochlorophylls, as well as the fitting of experimental spectra, were implemented by using differential evolution [33–35].

It is clear that even semi-classical calculations are time consuming, and it can be difficult to use them directly with spectrometric instruments. One of the possible simplifications is to apply an approach when the optical spectra of dye molecules are approximated by a set of bell-shaped Gaussian curves [12,14,23,36–38]. It must be stressed that from the point of view of quantum theory, such presentation can be considered as a rough approximation. Thus, the 0-0 electronic transition band of Chl (BChl) in solution at room temperature in the frequency domain is described by a rather narrow Gaussian function of large amplitude. The phonon wing components give a broad band of smaller amplitude lying towards higher frequencies (higher energy). Doppler broadening and other interactions lead to some deviation of the shape of the electronic transition band from Gaussian. Moreover, this shape deviation can easily be taken into account by adding one more or, in extreme cases, two Gaussian bands. Yet, more Gaussian bands are required to fit the electronic transition of carotenoids than in the case of electronic transition of chlorophylls. Thus, we are faced with the main problem of the decomposition into Gaussian bands: the approximation of the spectrum by Gaussian curves is ambiguous and the success of the decomposition depends on the correct initial choice of the position, width, and amplitude of each Gaussian curve [19].

Thus, the main goal of our work is to demonstrate that Gaussian decomposition of experimental absorption spectra of photosynthetic pigments can be used a suitable approximation of the shape of fitted spectra (for further calculation of high order undistorted derivatives) instead of simulating the linear optical response within the framework of semi-classical quantum theory. Moreover, taking into account the computational cost of data processing, Gaussian decomposition is much faster than modeling the absorption profile using semi-classical quantum theories; it can then be argued that this method of data approximation can be used directly in spectroscopic devices (remote sensing) for on-the-fly data processing (Figure 1).

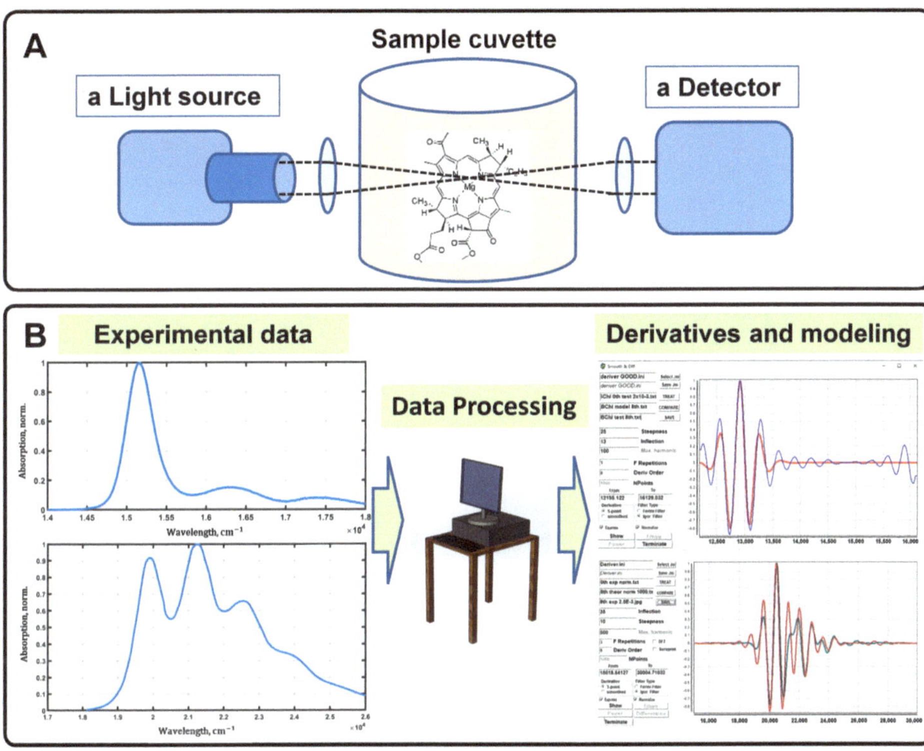

Figure 1. General scheme of a spectrophotometric setup, data processing, and modeling. The light source, sample cuvette, and detector, which are the essential elements of any spectrophotometer, allow for measuring the optical properties of pigment molecules in solvents as a function of wavelength (**A**). The measured spectra are subjected to the proper processing, in particular, signal smoothing is performed, which is necessary for its further analysis and modeling (**B**).

2. Materials and Methods

2.1. Absorption Spectra of Photosynthetic Pigments

To perform the optical properties simulations, absorption spectra of Chl, BChl, spheroidene, and spheroidenone were taken from the published studies [35,39]. All spectra were measured at room temperature; Chl and BChl were diluted in dimethyl ether and taken in a range from 300 to 800 nm with a step of 1 nm, while spheroidene and spheroidenone were diluted in a 7/2 (v/v) acetone/methanol mixture and in a range from 450 to 550 nm (Figure 2).

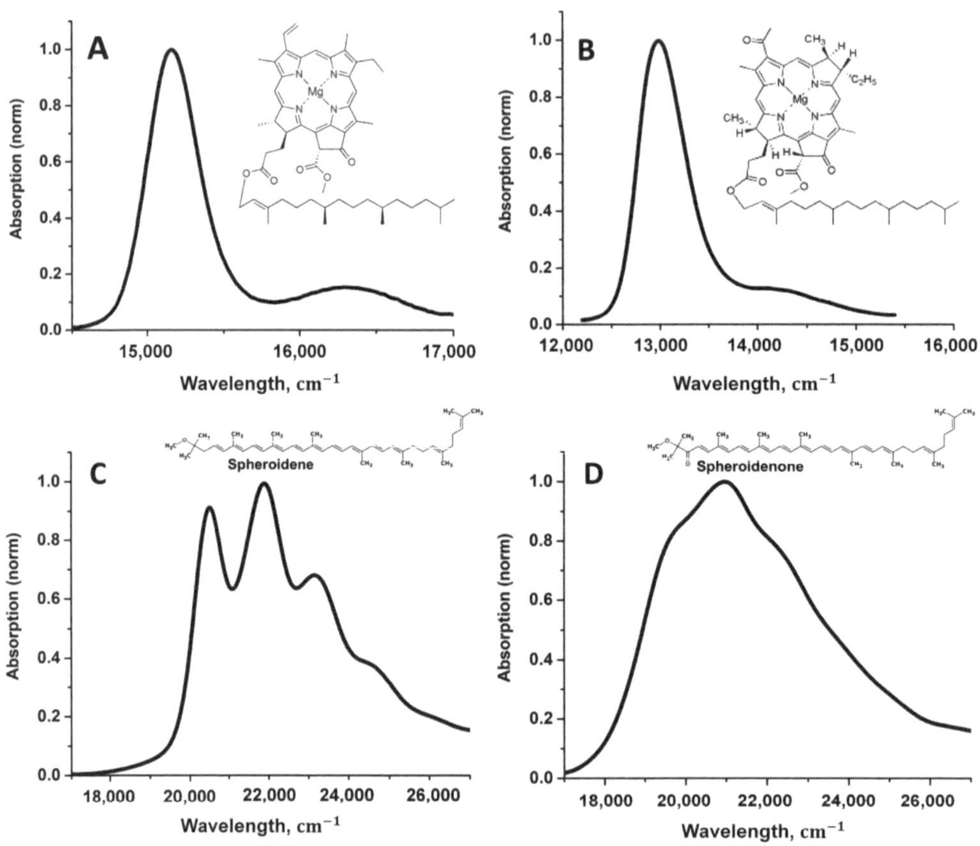

Figure 2. Experimentally measured absorption spectra of Chl (**A**), BChl (**B**), spheroidene (**C**), and speroidenone (**D**) in solvents at room temperature.

2.2. Multimode Brownian Oscillators Model

The bands of different intensities and widths observed in the absorption spectrum of an organic pigment in the visible wavelength range are the result of the transition from one electronic state to another. Assuming that the molecule can be in the ground $|g\rangle$ and excited $|e\rangle$ electronic states, it is then possible to simulate the absorption spectrum profile using a semi-classical theory called the multimode Brownian oscillator model (Figure 3A).

Within the framework of the applied theory, the number of vibronic modes associated with an electronic state is determined by the spectral density function (Figure 3B–D). Moreover, it is assumed that these vibrations of the molecular skeleton interact with the bath modes, which allows us to take into account the influence of the local surroundings. Each vibronic mode is represented by three parameters, namely the frequency ω_j, the relaxation rate γ_j, and the Huang–Rhys factor S_j. Depending on the ratio between ω_j and γ_j, different types of nuclear motions can be represented. S_j is the effective interaction energy of jth mode with the electronic state. The expression for the spectral density function is

$$C''(\omega) = \sum_j \frac{2S_j\omega_j^3\omega\gamma_j}{\left(\omega_j^2 - \omega^2\right)^2 + \omega^2\gamma_j^2}, \tag{1}$$

$C''(\omega)$ is a real function that is used to calculate the complex correlation function of the electronic energy gap between $|g\rangle$ and $|e\rangle$ states [28]. The temperature dependent correlation function is then written as

$$g(t) = \frac{1}{2\pi}\int_{-\infty}^{\infty} d\omega \frac{1-\cos\omega t}{\omega^2}\coth(\beta\hbar\omega/2)C''(\omega) - \frac{i}{2\pi}\int_{-\infty}^{\infty} d\omega \frac{\sin(\omega t)-\omega t}{\omega^2}C''(\omega), \quad (2)$$

This function is also called the line-shape function and is used in the expression for the absorption spectrum lineshape:

$$\sigma_{abs}(\omega) = \frac{1}{\pi}Re\int_0^{\infty} dt\, e^{i(\omega-\Omega_{eg})t}e^{-g(t)}e^{-\frac{1}{2}(\Delta t)^2}, \quad (3)$$

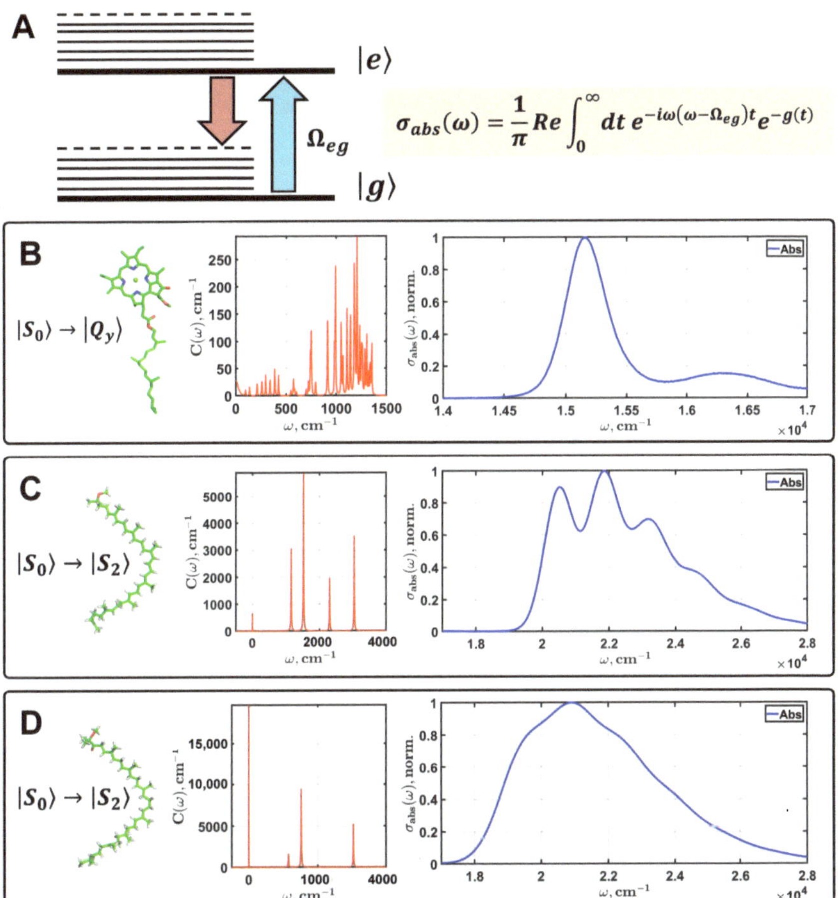

Figure 3. Theoretical modeling of the linear optical response of photosynthetic pigments within the framework of multimode Brownian oscillator theory. A diagram of the electronic transition from the ground $|g\rangle$ to the excited $|e\rangle$ state is shown in plot (**A**). Thick lines are the electronic states; thin lines are the manifold of vibronic states of a pigment interacting with the electronic states. Ω_{ge} is the electronic energy gap. $\sigma_{abs}(\omega)$ is an integral expression for the absorption spectrum depending on the lineshape function $g(t)$. The simulated spectra of the $|S_0\rangle \rightarrow |Q_y\rangle$ transition of Chl (**B**) and the $|S_0\rangle \rightarrow |S_2\rangle$ transition of shperoidene (**C**) and spheroidenone (**D**) are presented. Red lines are the spectral densities of pigments; the linear absorption spectra are blue.

Here, Ω_{eg} is the 0-0 frequency of the electronic transition $|g\rangle \rightarrow |e\rangle$. $\Delta = FWHM/2\sqrt{2 \cdot ln2}$ is the inhomogeneous broadening parameter. The inhomogeneous broadening of absorption is an effect due to the influence of the immediate surroundings on the electronic transition of the pigment (the effect of the solvent). $FWHM$ is the full width at the half maximum of Gaussian distribution.

Thus, to simulate a realistic absorption spectrum according to (3), we need to optimize the following parameters of the quantum model: Ω_{eg}, $FWHM$, and $\{\omega_j, S_j, \gamma_j\}$, where $j \in [1, \ldots, n]$, n is the number of vibronic modes specific for a pigment.

2.3. The Eighth Derivative Calculation Procedure

Original software was developed to process the experimental spectra. It allows us to upload and download data files, visualize the spectra, as well as smooth the original data and calculate derivatives of any order. The procedure of gaining high-order derivatives of the pigment absorption spectrum using Gaussian decomposition involves several steps. Considering the long-wavelength absorption band of chlorophyll a as an example, let us explain the series of operations that was used in our calculations. Initially, the Chl spectrum was measured in diethyl ether at room temperature in a range from 300 to 800 nm with a step of 1 nm. Only the part of the spectrum from 600 to 800 nm (the so-called Qy band) was taken for further processing (Figure 2A). This spectral region is used for Gaussian decomposition. The main electronic peak of the Qy band is shifted to the center of the figure to avoid edge effects during differentiation.

The input parameters of the procedure of approximation of absorption spectra by Gaussians are the number of Gaussians, their intensities, the width at half maximum, and the frequency shift. In general, the fitting can be optimized by using any multiparametric optimization [19], but specifically for Chl and carotenoid spectra, acceptable results can be obtained without optimization. It should be emphasized that the approximated spectrum is a smooth curve, for which it is easy to obtain the derivative of any order. However, the experimental absorption spectrum is an array of values measured to an accuracy ± 0.0001 (if the maximum of intensities is normalized to 1.0) or worse. When recorded on a spectrophotometer, this measurement error gives a "noise track", the amplitude of which can vary due to the different sensitivity of the instrument at different wavelengths (Figure 4A,B).

To simulate the effect of real measured data, a noise track taken on a spectrophotometer with an empty cuvette was used. This straightened noise array is added to the model spectrum and the test spectrum is obtained (Figure 4A). The test spectrum and the experimental spectrum contain approximately the same noise component. Therefore, the test spectrum can be used to find the parameters of the best filter (Figure 4C,D). By applying filtering (smoothing) and differentiation of the test spectrum, you can obtain, for example, its eighth derivative. However, the exact form of this eighth derivative is already known (this is the eighth derivative of the model spectrum). Comparing the eighth derivatives of the test spectrum and the model spectrum, one can choose the filter parameters that give the best match for these derivatives. This filter will be called the optimal filter. The use of this optimal filter to obtain the eighth derivative of the experimental Chl spectrum should ensure good agreement between the eighth derivatives of the experimental and theoretical Chl spectra (Figure 4E,F).

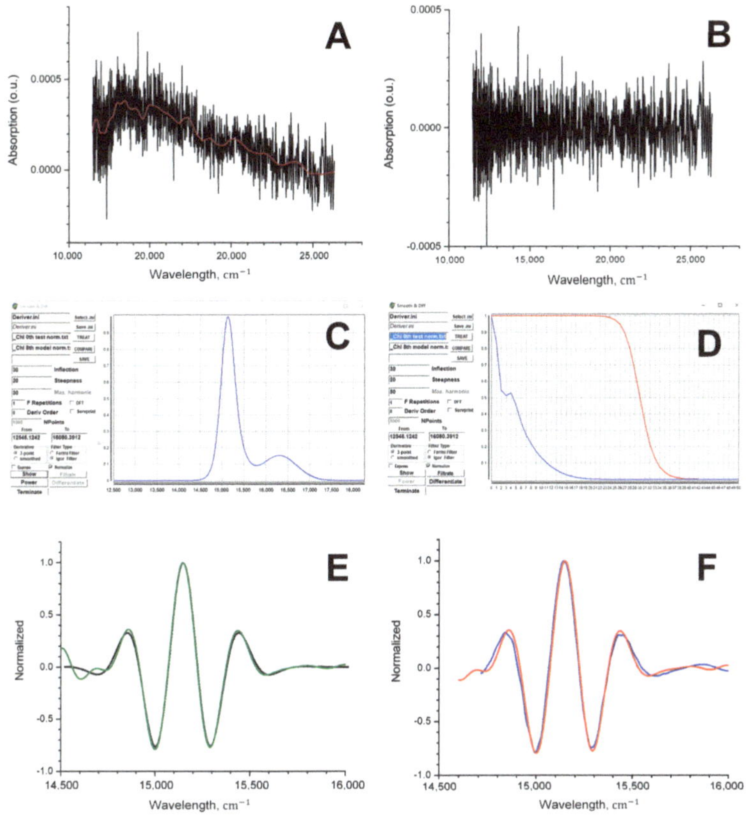

Figure 4. The noise line obtained on the spectrophotometer when registering without a cuvette in the sample compartment; the red curve is a smoothed noise line (**A**). The straightened noise obtained from the noise line by subtracting the smoothed noise line (**B**). Search for optimal FFT filter parameters using Chl as an example. The workspace of Deriver602green program: (**C**)—processed spectrum, (**D**)—the frequency representation of the spectrum (blue) and the filter curve (red). (**E**) Normalized eighth derivatives of the test (green) and model (black) Chl spectra. (**F**) Normalized eighth derivatives of the experimental (red) and theoretical (blue) Chl spectra.

3. Results

It is well known that the shape of absorption spectra of photosynthetic pigments is determined by the intensity of the nuclear motions of molecular skeleton [28,37]. Applying the multimode Brownian oscillator model, these types of motions can be taken into account by evaluating the spectral density function [33,34]. In this study we will distinguish between spectra calculated theoretically using spectral density and spectra obtained using Gaussian decomposition.

Parameters of the multimode Brownian oscillator model for simulation of the theoretical absorption spectra of Chl (Figure 3B) and BChl were taken from our previous studies [35,40]. Carotenoid spectra were simulated especially for this study (Figure 3C,D). Parameters of the spectral density, the electronic energy gap, and the full width at half maximum are listed in Table A1 in Appendix A. The time and frequency scales of simulated spectra contain $2^{12} = 4096$ points. This is due to the use of the fast Fourier transform to process the spectra, which requires 2^n point arrays. All the spectra were saved with an accuracy of 16 digits. Numerical differentiation of such simulated data makes it possible to obtain derivatives up to the eighth order without preliminary smoothing, namely without distortion. It has to be stressed that the eighth derivative of the spectrum of any dye is the natural limit for a standard personal computer when using double precision. When calculating derivatives of a higher order, there are problems related to the representation of numbers in the RAM of the computer.

Spectra obtained by Gaussian decomposition are much easier to simulate. Using the procedures described in the previous section and applying our software, we can approximate the experimental spectra with a high degree of fidelity (Figure 5A,C,E,J). The key point is to find the optimal filter parameters. The use of test spectra for smoothing allowed us to adjust the filter finely and we obtained the smallest discrepancies between the spectrum of an ideal higher derivative (for example, of the eighth order) and the corresponding model derivative.

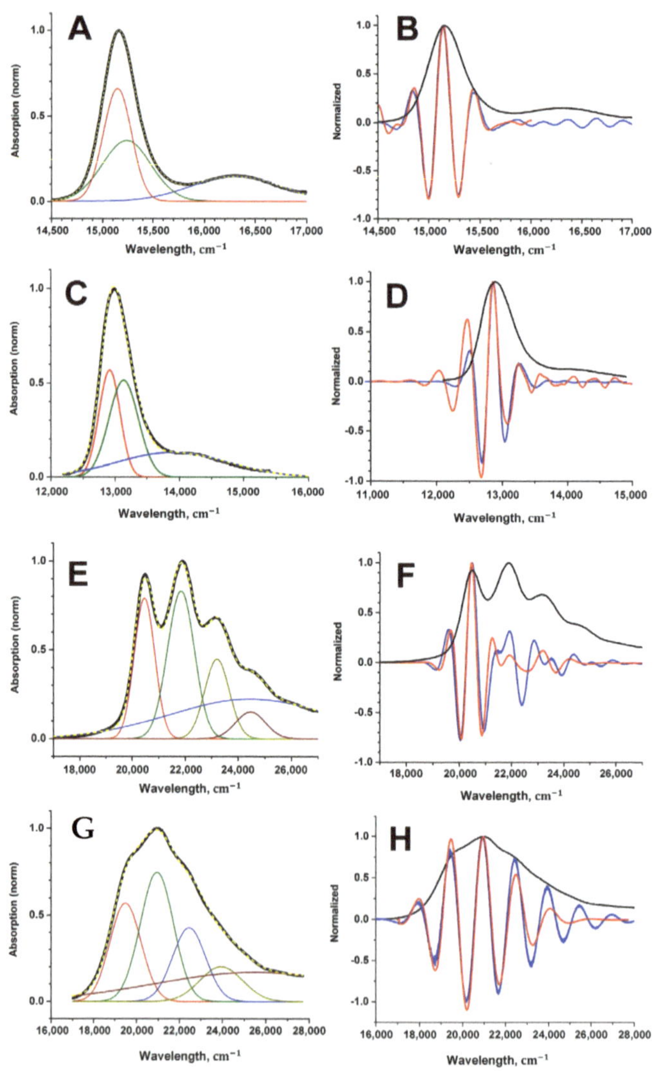

Figure 5. Experimental absorption spectra of Chl (**A**), BChl (**C**), (black curves) and its decomposition into three Gaussian bands (red, green, and blue), as well as experimental spectra of spheroidene (**E**) and spheroidenone (**G**) and its decomposition into five Gaussian bands. The resulting spectra, which are the sum of Gaussians, are shown by the yellow dashed lines. Normalized eighth derivatives of the pigment experimental spectra (red curves) and those of theoretical spectra (blue curves) are shown on the graphs (**B**,**D**,**F**,**H**).

To check that the proposed optimal filter parameter selection technique works, we applied it to obtain the eighth derivatives of the experimental spectra of four photosynthetic pigments for which the theoretical spectra were calculated (Chl, BChl, spheroidene, and spheroidenone). The results are shown in Figure 5. It can be seen that the proposed technique makes it possible to obtain high derivatives of the experimental spectrum, which are quite close to the theoretical ones.

4. Discussion

Application of the Gaussian decomposition has a long tradition, particularly in astronomy [19,41,42]. The optical data obtained by astronomers are essentially different from the spectroscopy of biological pigments and proteins. Astronomers measure and analyze the intensities of atomic transitions, whereas the object of study for biophysicists is the electronic transitions of multi-atomic organic molecules, complicated by the influence of the immediate environment (solvents or proteins). Electronic transitions in atoms and small inorganic molecules in space mostly have a Gaussian shape, therefore decomposition into Gaussians, in addition to a trivial smoothing of spectral curves and noise removal, allows astronomers to extract from the absorption and emission spectra parameters of physical processes (for instance, the Doppler shift of frequencies, the full width at half maximum) occurring in the interstellar space. The computational implementation of modern advances in artificial intelligence enables efficient automation of astronomical data processing [13,19,43].

Unlike the spectra of astronomical objects, the spectra of biological pigments are much more difficult to interpret. In general, the shape of their absorption profile is not Gaussian and is determined by the integral expression (3). This expression can turn to the Gaussian function only at a certain relation between the parameters of the theory, which are in fact never fulfilled for organic pigments at room temperature [28,30,37]. Experimental spectra of photosynthetic pigments are shown in Figure 2. It should be noted that the Gaussian decomposition in the study of organic and photosynthetic pigments has its own history [44–47]. However, in spite of the achieved success, one cannot but note the fact that the parameters of Gaussians obtained after decomposition have no physical meaning and do not reflect real physical processes occurring in pigments when absorbing light quanta. On the other hand, the high derivatives of the Chl and BChl spectra in these studies are characterized by a large number of narrow peaks, which were attributed to the absorption bands of certain transitions, which in fact can be interpret as insufficient filtering (smoothing) of the original spectra. Interestingly, in addition to decomposition into Gaussians or Lorentzians, the Padé transform was used to estimate the shape of the derivative spectrum, which, however, also excludes the possibility of physical interpretation of the obtained data [48].

The absorption spectra of Chl and BChl in the range of 12,000–17,000 cm^{-1} (830–590 nm) and carotenoids in the range of 18,000–26,000 cm^{-1} (560–380 nm) were not arbitrary chosen. It is in these ranges that the absorption bands of strictly defined electronic transitions are found. Thus, for this spectral region it is allowed to use the multimode Brownian oscillator model, which will reproduce the true absorption profile with high accuracy. With the help of the differential evolution algorithm, we have fitted the given spectra finely [35], however, from a practical point of view, if we are talking about automation of the processes of analysis and registration of optical data, carrying out such modelling will not allow us to analyze the spectra on the spot. Moreover, if we consider the problem more generally, it is obvious that the set of vibronic modes of any organic pigment cannot be known exactly. This circumstance fundamentally complicates the application of the multimode Brownian oscillator model and the use of this theory becomes irrational if the purpose of modeling is to obtain high-order derivatives. It is worth noting that the application of semi-classical theories to on-the-fly modeling of optical properties has been discussed before [15,16,18].

Consequently, to calculate higher order derivatives, we replace the approximation of the experimental spectrum based on a physical model (discussed above) with the approximation of the experimental spectrum by the sum of Gaussians. It is clear that such an envelope is an analytic curve for which it is easy to obtain accurate high derivatives. We call the envelope the model curve, and its derivatives, the model derivatives. By adding the noise track to the model spectrum, we obtain a test spectrum that approximates the experimental one in its shape and noise characteristics.

Smoothing (filtering) the test spectrum with the selected filter, we obtain its eighth derivative and compare it with the model eighth derivative. The process continues until it is possible to obtain the best match of these two derivatives and thereby find the parameters of the optimal filter. Since it is assumed that the noise characteristics of the test spectrum do not differ from the noise characteristics of the experimental one, the use of the optimal filter for processing the experimental spectrum should allow one to obtain its "true" eighth derivative. For the four compounds that are considered in this work, it is possible to check whether their eighth derivatives obtained using the proposed method are really "true". As can be seen from the comparison of the derivatives obtained by smoothing/differentiating the experimental spectra and the theoretical derivatives (Figure 5), the agreement is quite good.

5. Conclusions

It was shown that the noisy experimental spectra of photosynthetic pigments can be processed by Gaussian decomposition in order to obtain the undistorted high-order derivatives. The spectra of the studied pigments (chlorophyll, bacteriochlorophyll, and carotenoids) were chosen in a way that their frequency range allowed the use of a semi-classical theory called the multimode Brownian oscillator model to simulate the shape of the absorption spectrum. The calculated true theoretical spectra were used as reference spectra to generate higher derivatives, in particularly the eighth one. At the same time, the absorption spectra were modeled using Gaussian decomposition: three Gaussians were sufficient to obtain the envelope curve of the spectrum for Chl and BChl, and five Gaussians were needed in the case of carotenoids. Overlaying a noise track on the model spectra allowed us to synthesize the test experimental data roughly, as it would be read from a spectrophotometer. By comparing the model derivatives of the eighth order and those of the test spectra, the optimal settings of the smoothing filter were found, where the spectra of the high derivatives match perfectly.

Thus, it was demonstrated that Gaussian decomposition in modeling the optical response of biological pigments allows us to obtain undistorted high-order derivatives with good precision. Considering that the computational costs of Gaussian decomposition is much less than optimization of spectra modeling by semi-classical quantum theory, this method can be used in on-the-fly spectral data analysis with minimal computational costs. It is also worth noting that in the future it is reasonable to consider the possibility of optimizing the Gaussian decomposition with the help of differential evolution. The algorithm of multiparametric optimization has already shown high efficiency in quantum calculations of the optical response of photosynthetic pigments [33–35].

Author Contributions: Supervision, conceptualization, writing—review and editing, A.P.R.; validation, formal analysis, visualization, investigation, V.S.K.; validation, formal analysis, investigation, A.A.A.; conceptualization, methodology, software, visualization, writing—original draft preparation, writing—review and editing, R.Y.P. All authors have read and agreed to the published version of the manuscript.

Funding: This work was supported by the Russian Science Foundation (RSF # 22-21-00905, https://rscf.ru/en/project/22-21-00905/, accessed on 10 February 2023).

Institutional Review Board Statement: Not applicable.

Informed Consent Statement: Not applicable.

Data Availability Statement: Not applicable.

Conflicts of Interest: The authors declare no conflict of interest. The funders had no role in the design of the study; in the collection, analyses, or interpretation of data; in the writing of the manuscript, or in the decision to publish the results.

Appendix A

Table A1. Parameters of the multimode Brownian oscillator model for simulation of spheroidene (Figure 3C) and spheroidenone (Figure 3D) absorption spectra. $\nu_1 = 1524$ cm^{-1}, $\nu_2 = 1158$ cm^{-1}, $\nu_3 = 1006$ cm^{-1}, $\nu_4 = 965$ cm^{-1} are the characteristic vibronic modes of carotenoids; $2\nu_1$ and $2\nu_2$ are the overtones of the corresponding frequencies; $\nu_1 + \nu_2$ is sum of the frequencies; S_j are Huang-Rhys factors, where $j = \{\nu_1, \nu_2, \nu_3, \nu_4,\ 2\nu_1, 2\nu_2, \nu_1 + \nu_2\}$; Ω_{eg} is the electronic energy gap; $FWHM_\Omega$ is the full width at half maximum.

	Ω_{eg}	$FWHM_\Omega$	ω_{low}	S_{low}	γ_{low}	S_{ν_1}	S_{ν_2}	S_{ν_3}	S_{ν_4}	$2\nu_2$	$2\nu_1$	$\nu_1 + \nu_2$
Spheroidene	22,631.0	1007.6	4.9	3.0	42.7	0.63	0.57	7.3×10^{-3}	1.5×10^{-3}	9.3×10^{-2}	6.9×10^{-8}	9.7×10^{-2}
Spheroidenone	21,785.8	1668.3	1.0	3.2×10^{-3}	450.4	1.02	0.30	3.2×10^{-11}	2.5×10^{-11}	1.1×10^{-11}	1.2×10^{-11}	0.15

References

1. Dymond, E. On the measurement of the critical potentials of gases. In Proceedings of the Mathematical Proceedings of the Cambridge Philosophical Society, Cambridge, UK, 20 September 1924; pp. 405–408.
2. Talsky, G. *Derivative Spectrophotometry: Low and Higher Order*; VCH: Weinheim, Germany; Basel, Switzerland; Cambridge, UK; New York, NY, USA; Tokyo, Japan, 1994.
3. Dubrovkin, J. *Derivative Spectroscopy*; Cambridge Scholars Publishing: Newcastle upon Tyne, UK, 2020.
4. Saakov, V.S.; Krivchenko, A.I.; Rozengart, E.V.; Danilova, I.G. *Derivative Spectrophotometry and PAM-Fluorescence in Comparative Biochemistry*; Springer: Berlin/Heidelberg, Germany, 2015.
5. Karpinska, J. Basic principles and analytical application of derivative spectrophotometry. In *Macro to Nano Spectroscopy*; Jamal, U., Ed.; InTech Open: London, UK, 2012; pp. 253–256. [CrossRef]
6. Karpińska, J. Derivative spectrophotometry—Recent applications and directions of developments. *Talanta* **2004**, *64*, 801–822. [CrossRef] [PubMed]
7. Czarnecki, M.A. Resolution enhancement in second-derivative spectra. *Appl. Spectrosc.* **2015**, *69*, 67–74. [CrossRef] [PubMed]
8. Ojeda, C.B.; Rojas, F.S. Recent applications in derivative ultraviolet/visible absorption spectrophotometry: 2009–2011: A review. *Microchem. J.* **2013**, *106*, 1–16. [CrossRef]
9. Abdel-Aziz, Y.; El-sayed, N.; El-Salem, A. Recent developments of derivative spectrophotometry and their analytical application. *Anal. Sci.* **2005**, *21*, 595–614. [CrossRef]
10. Craven, P.G.; Fairhurst, S.A.; Sutcliffe, L.H. A simple approach to derivative spectroscopy. *Spectrochim. Acta Part A-Mol. Biomol. Spectrosc.* **1988**, *44*, 539–545. [CrossRef]
11. Belkić, D.; Belkić, K. Review of recent applications of the conventional and derivative fast Padé transform for magnetic resonance spectroscopy. *J. Math. Chem.* **2019**, *57*, 385–464. [CrossRef]
12. Kovalenko, A.V.; Vovk, S.M.; Plakhtii, Y.G. Sum Decomposition Method for Gaussian Functions Comprising an Experimental Photoluminescence Spectrum. *J. Appl. Spectrosc.* **2021**, *88*, 357–362. [CrossRef]
13. Riener, M.; Kainulainen, J.; Beuther, H.; Henshaw, J.; Orkisz, J.; Wang, Y. Autonomous Gaussian decomposition of the Galactic Ring Survey-I. Global statistics and properties of the 13CO emission data. *Astron. Astrophys.* **2020**, *633*, A14. [CrossRef]
14. Fellows, A.P.; Casford, M.T.L.; Davies, P.B. Spectral analysis and deconvolution of the amide I band of proteins presenting with high-frequency noise and baseline shifts. *Appl. Spectrosc.* **2020**, *74*, 597–615. [CrossRef]
15. Begusic, T.; Vanicek, J. On-the-fly ab initio semiclassical evaluation of vibronic spectra at finite temperature. *J. Chem. Phys.* **2020**, *153*, 024105. [CrossRef]
16. Begusic, T.; Vanicek, J. On-the-fly ab initio semiclassical evaluation of third-order response functions for two-dimensional electronic spectroscopy. *J. Chem. Phys.* **2020**, *153*, 184110. [CrossRef]
17. Shao, X.; Cui, X.; Wang, M.; Cai, W. High order derivative to investigate the complexity of the near infrared spectra of aqueous solutions. *Spectrochim. Acta Part A Mol. Biomol. Spectrosc.* **2019**, *213*, 83–89. [CrossRef] [PubMed]
18. Patoz, A.; Begusic, T.; Vanicek, J. On-the-Fly Ab Initio Semiclassical Evaluation of Absorption Spectra of Polyatomic Molecules beyond the Condon Approximation. *J. Phys. Chem. Lett.* **2018**, *9*, 2367–2372. [CrossRef] [PubMed]
19. Lindner, R.R.; Vera-Ciro, C.; Murray, C.E.; Stanimirovic, S.; Babler, B.; Heiles, C.; Hennebelle, P.; Goss, W.M.; Dickey, J. Autonomous Gaussian Decomposition. *Astron. J.* **2015**, *149*, 138. [CrossRef]
20. Klein, M.P.; Dratz, E.A. Derivative spectroscopy with recording spectrometers. *Rev. Sci. Instrum.* **1968**, *39*, 397–399. [CrossRef]
21. Ichimura, K. Procedures for Converting Electronic Absorption Spectra into Higher-Order Derivatives to Examine Photoinduced Spectral Changes. *Bull. Chem. Soc. Jpn.* **2016**, *89*, 549–564. [CrossRef]
22. Ichimura, K. Photoisomerisation of azobenzene crystals in aqueous dispersions examined by higher order derivative spectra. *Phys. Chem. Chem. Phys.* **2015**, *17*, 2722–2733. [CrossRef]

23. Kovalenko, A.V.; Plakhtiy, E.G.; Vovk, S.M. Application of derivative spectroscopy method to photoluminescence in ZnS: Mn nanocrystals. *Ukr. J. Phys. Opt.* **2018**, *19*, 133–138. [CrossRef]
24. Zhukovskii, Y.G.; Saakov, V.S. Evaluation of the heterogeneity and specificity of promising antitumoral preparations by means of high-order derivative spectroscopy. *Dokl. Biol. Sci.* **2002**, *386*, 440–444. [CrossRef]
25. Mikhailyuk, I.K.; Knox, P.P.; Paschenko, V.Z.; Razjivin, A.P.; Lokstein, H. Analysis of absorption spectra of purple bacterial reaction centers in the near infrared region by higher order derivative spectroscopy. *Biophys. Chem.* **2006**, *122*, 16–26. [CrossRef]
26. Ichimura, K. The Reproducibility and Reliability of UV–vis Higher-Order Derivative Spectroscopy for Quantitative Analysis of Spectral Changes. *Bull. Chem. Soc. Jpn.* **2017**, *90*, 411–418. [CrossRef]
27. Jang, S.J.; Mennucci, B. Delocalized excitons in natural light-harvesting complexes. *Rev. Mod. Phys.* **2018**, *90*, 035003. [CrossRef]
28. Mukamel, S. *Principles of Nonlinear Optical Spectroscopy*; Oxford University Press: New York, NY, USA; Oxford, UK, 1995; Volume 6, p. 543.
29. Renger, T. Semiclassical Modified Redfield and Generalized Forster Theories of Exciton Relaxation/Transfer in Light-Harvesting Complexes: The Quest for the Principle of Detailed Balance. *J. Phys. Chem. B* **2021**, *125*, 6406–6416. [CrossRef]
30. Mirkovic, T.; Ostroumov, E.E.; Anna, J.M.; van Grondelle, R.; Govindjee; Scholes, G.D. Light Absorption and Energy Transfer in the Antenna Complexes of Photosynthetic Organisms. *Chem. Rev.* **2017**, *117*, 249–293. [CrossRef] [PubMed]
31. Pant, M.; Zaheer, H.; Garcia-Hernandez, L.; Abraham, A. Differential Evolution: A review of more than two decades of research. *Eng. Appl. Artif. Intell.* **2020**, *90*, 103479. [CrossRef]
32. Storn, R.; Price, K. Differential evolution-A simple and efficient heuristic for global optimization over continuous spaces. *J. Glob. Optim.* **1997**, *11*, 341–359. [CrossRef]
33. Pishchalnikov, R.Y.; Yaroshevich, I.A.; Zlenko, D.V.; Tsoraev, G.V.; Osipov, E.M.; Lazarenko, V.A.; Parshina, E.Y.; Chesalin, D.D.; Sluchanko, N.N.; Maksimov, E.G. The role of the local environment on the structural heterogeneity of carotenoid β-ionone rings. *Photosynth. Res.* **2022**, *156*, 3–17. [CrossRef]
34. Chesalin, D.D.; Kulikov, E.A.; Yaroshevich, I.A.; Maksimov, E.G.; Selishcheva, A.A.; Pishchalnikov, R.Y. Differential evolution reveals the effect of polar and nonpolar solvents on carotenoids: A case study of astaxanthin optical response modeling. *Swarm Evol. Comput.* **2022**, *75*, 101210. [CrossRef]
35. Pishchalnikov, R. Application of the differential evolution for simulation of the linear optical response of photosynthetic pigments. *J. Comput. Phys.* **2018**, *372*, 603–615. [CrossRef]
36. Antonov, L.; Nedeltcheva, D. Resolution of overlapping UV–Vis absorption bands and quantitative analysis. *Chem. Soc. Rev.* **2000**, *29*, 217–227. [CrossRef]
37. Zucchelli, G.; Jennings, R.C.; Garlaschi, F.M.; Cinque, G.; Bassi, R.; Cremonesi, O. The calculated in vitro and in vivo chlorophyll a absorption bandshape. *Biophys. J.* **2002**, *82*, 378–390. [CrossRef] [PubMed]
38. Malkova, V.I.; Makotchenko, E.V. Analysis of composite bands in the electronic absorption spectra of compounds in solutions. *Russ. J. Gen. Chem.* **2008**, *78*, 1161–1166. [CrossRef]
39. Ashikhmin, A.; Makhneva, Z.; Bolshakov, M.; Moskalenko, A. Incorporation of spheroidene and spheroidenone into light-harvesting complexes from purple sulfur bacteria. *J. Photochem. Photobiol. B-Biol.* **2017**, *170*, 99–107. [CrossRef]
40. Pishchalnikov, R.; Shubin, V.; Razjivin, A. Single Molecule Fluorescence Spectroscopy of PSI Trimers from Arthrospira platensis: A Computational Approach. *Molecules* **2019**, *24*, 822. [CrossRef] [PubMed]
41. Krčo, M.; Goldsmith, P.F.; Brown, R.L.; Li, D. An Improved Technique for Measurement of Cold H I in Molecular Cloud Cores. *Astrophys. J.* **2008**, *689*, 276. [CrossRef]
42. Stutzki, J.; Guesten, R. High Spatial Resolution Isotopic CO and CS Observations of M17 SW: The Clumpy Structure of the Molecular Cloud Core. *Astrophys. J.* **1990**, *356*, 513. [CrossRef]
43. Hampton, E.J.; Medling, A.M.; Groves, B.; Kewley, L.; Dopita, M.; Davies, R.; Ho, I.T.; Kaasinen, M.; Leslie, S.; Sharp, R.; et al. Using an artificial neural network to classify multicomponent emission lines with integral field spectroscopy from SAMI and S7. *Mon. Not. R. Astron. Soc.* **2017**, *470*, 3395–3416. [CrossRef]
44. Fell, A.F. Biomedical applications of derivative spectroscopy. *TrAC Trends Anal. Chem.* **1983**, *2*, 63–66. [CrossRef]
45. Whitten, W.B.; Nairn, J.A.; Pearlstein, R.M. Derivative absorption spectroscopy from 5—300 K of bacteriochlorophyll a-protein from Prosthecochloris aestuarii. *Biochim. Biophys. Acta (BBA)-Bioenerg.* **1978**, *503*, 251–262. [CrossRef]
46. Talsky, G.; Mayring, L.; Kreuzer, H. High-resolution, higher-order UV/VIS derivative spectrophotometry. *Angew. Chem. Int. Ed. Engl.* **1978**, *17*, 785–799. [CrossRef]
47. Butler, W.L.; Hopkins, D.W. Higher derivative analysis of complex absorption spectra. *Photochem. Photobiol.* **1970**, *12*, 439–450. [CrossRef]
48. Belkić, D.; Belkić, K. Exact quantification by the nonparametric fast Padé transform using only shape estimation of high-order derivatives of envelopes. *J. Math. Chem.* **2018**, *56*, 268–314. [CrossRef]

Disclaimer/Publisher's Note: The statements, opinions and data contained in all publications are solely those of the individual author(s) and contributor(s) and not of MDPI and/or the editor(s). MDPI and/or the editor(s) disclaim responsibility for any injury to people or property resulting from any ideas, methods, instructions or products referred to in the content.

sensors

Article

A Spectral Encoding Simulator for Broadband Active Illumination and Reconstruction-Based Spectral Measurement

Peng Jiang [1,2], Xiaoxu Wang [1], Zihui Zhang [1], Guochao Gu [1], Jifeng Li [1], Heng Wu [1,2], Limin He [1,2] and Guanyu Lin [1,*]

1 Changchun Institute of Optics, Fine Mechanics and Physics, Chinese Academy of Sciences, Changchun 130033, China
2 University of Chinese Academy of Sciences, Beijing 100049, China
* Correspondence: linguanyu1976@163.com

Abstract: Spectral reflectance or transmittance measurements provide intrinsic information on the material of an object and are widely used in remote sensing, agriculture, diagnostic medicine, etc. Most reconstruction-based spectral reflectance or transmittance measurement methods based on broadband active illumination use narrow-band LEDs or lamps combined with specific filters as spectral encoding light sources. These light sources cannot achieve the designed spectral encoding with a high resolution and accuracy due to their low degree of freedom for adjustment, leading to inaccurate spectral measurements. To address this issue, we designed a spectral encoding simulator for active illumination. The simulator is composed of a prismatic spectral imaging system and a digital micromirror device. The spectral wavelengths and intensity are adjusted by switching the micromirrors. We used it to simulate spectral encodings according to the spectral distribution on micromirrors and solved the DMD patterns corresponding to the spectral encodings with a convex optimization algorithm. To verify the applicability of the simulator for spectral measurements based on active illumination, we used it to numerically simulate existing spectral encodings. We also numerically simulated a high-resolution Gaussian random measurement encoding for compressed sensing and measured the spectral reflectance of one vegetation type and two minerals through numerical simulations. We reconstructed the spectral transmittance of a calibrated filter through an experiment. The results show that the simulator can measure the spectral reflectance or transmittance with a high resolution and accuracy.

Keywords: spectral measurement; spectral encoding; active illumination; compressed sensing

Citation: Jiang, P.; Wang, X.; Zhang, Z.; Gu, G.; Li, J.; Wu, H.; He, L.; Lin, G. A Spectral Encoding Simulator for Broadband Active Illumination and Reconstruction-Based Spectral Measurement. *Sensors* **2023**, *23*, 4608. https://doi.org/10.3390/s23104608

Academic Editor: Anna Chiara De Luca

Received: 7 March 2023
Revised: 25 April 2023
Accepted: 5 May 2023
Published: 10 May 2023

Copyright: © 2023 by the authors. Licensee MDPI, Basel, Switzerland. This article is an open access article distributed under the terms and conditions of the Creative Commons Attribution (CC BY) license (https://creativecommons.org/licenses/by/4.0/).

1. Introduction

Every object has its own unique spectral characteristics; as a result, the spectrum is regarded as the "fingerprint" of the object. The spectral reflectance or transmittance of an object contains considerable inherent physical information that can be applied in object classification [1,2], remote sensing [3–6], medical diagnosis [7–9], and image reproduction [10,11]. Spectral measurement methods based on active illumination address the constraints between the spectral resolution and signal-to-noise ratio of traditional spectral measurement methods and have gained attention due to their rapidity and accuracy [12]. Active illumination is an important component of computational optical imaging. This improves the imaging resolution and environmental adaptability by encoding space, time, polarization, or spectral information on the light source side. When the spectral signal is collected under broadband active illumination, the spectral reflectance or transmittance of the target object can be measured. In recent years, researchers have performed many studies on active-illumination-based spectral measurement techniques. Park et al. used a set of LEDs in a multiplexed sequence to illuminate a scene and applied a simple empirical linear model to measure the spectral reflectance [13]; however, although the measurement algorithm was accurate and fast, the combined spectrum of the LEDs was not continuously tunable due to the fixed spectrum of each LED. In actual use, this light source led to fitting difficulties with spectral encodings with a high resolution and accuracy. In addition to using LEDs as active illumination sources, Chi et al. proposed a new active illumination approach. They selected 16 channels using a set of 228 possible filters and used an optimized broadband filtered illumination to obtain multispectral reflectance information [14]. However, they used tungsten and xenon lamps with specific filters as light sources for active illumination. While this approach is simple, the spectral encoding of the nearly optimal illumination set, which was based on the designed Hadamard code, could not be fitted because the equivalent broadband filters were not physically available. The above traditional illumination methods are inflexible. Han et al. measured the reflectance of a scene using a digital light processing (DLP) projector as a light source with spectrally distinct illuminations and a high-speed camera [15]. DLP projectors equipped with RGB color wheels have only three fixed illumination spectra. Although high-speed spectrum switching can be achieved for dynamic scenes, accurate spectral measurements are difficult to attain. For a more accurate reconstruction of the spectral reflectance, Zhang et al. developed a deeply learned broadband encoding stochastic hyperspectral camera with an active measurement mode that used a broadband light source and random filter to illuminate the target object. With this approach, accurate and dynamic measurements of the spectra in the entire field of view were obtained by detecting the target spectral reflectance [16]; however, the random filter used in the directional design was difficult to manufacture.

To address the various deficiencies in the existing active illumination light sources, we designed a spectral encoding simulator (SES) based on spectral tuning technology. In reconstruction-based spectral measurements, there are various methods for spectral encoding. A commonly used method is performed at the receiver side, usually using devices such as filters for encoding. One of the other methods is performed at the light source side, encoding the spectrum of the initial light source into a specific spectrum (the premise is that active illumination is applied to spectral measurements), and our SES is an active encoding light source.

In recent years, an increasing number of scholars have applied digital micromirror devices (DMDs) in spectral tuning. These devices control the wavelength and intensity of the spectrum through two dimensions, namely, the DMD rows and columns. For example, in 2005, MacKinnon et al. proposed a spectral programmable engine (SPLE) that used a DMD to output a specific spectral distribution (SPD), which has been used in the fields of biochemistry and biomedicine [17]. This system was an early prototype of a programmable spectral light engine based on a DMD, which has inspired various spectral light engines that have been applied in different fields. In 2006, Chuang et al. proposed a DMD pattern-scanning calibration method and a digital programmable light spectrum synthesis system, which was applied to the synthesis of various infrared C-band (1530–1565 nm) spectral profiles [18]. In 2017, Luo et al. proposed a programmable light source in the visible range based on the combination of a prism and an echelle grating [19]. Rice et al. developed a visible hyperspectral image projector (HIP). This projector has a double-DMD symmetrical structure, with the DMDs used as the spectral engine and spatial engine. The HIP outperforms conventional digital light processing (DLP) projectors because it has hyperspectral image projection capabilities [20]. Hirai et al. proposed a projector with a programmable light source (Optronic Laboratories OL490) and a DMD. The projection principle was the same as that of DLP projectors, and the proposed projector achieved spectral image projection similar to the HIP [21]. In addition to the abovementioned techniques, spectral light engines have been developed in various fields, such as stellar simulation [22], lasers [23], and spectral calibration [24–26]. In addition, there are some commercially available spectral light engines, such as Optronic Laboratories OL490 and the recently developed Chromatiq Spectral Engine (CSE) from Energetic (Hamamatsu). These spectral light engines use gratings to achieve good and smooth spectral matching; however, they cannot achieve a broad spectral range or high brightness. While the abovementioned spectral light engines are all based on DMDs, there are some problems in using them for spectral measurements. Active illumination sources for spectral measurements require a broad spectral range, high spectral encoding accuracy, and high optical throughput. Thus, the light source is better suited for using a prismatic spectral imaging system.

The main work of this study is to present a method of applying the SES to active-illumination-based spectral measurement and to verify the feasibility of the method using numerical simulations. We designed the SES based on spectral tuning technology. The SES uses the flexible spatial light modulation capability of the DMD to simulate spectral encoding in the broadband range of 300–1100 nm, which provides a basis for broadband active illumination and the reconstruction of spectral measurements. The SES uses a prismatic spectral imaging system to image the spectral image onto the DMD. The spectral dimension of the image corresponds to the dimension of the DMD micromirror columns, and the spatial dimension corresponds to the dimension of the DMD micromirror rows. Therefore, the on–off state of the DMD micromirror columns determines the wavelength of the output spectrum, and the number of micromirrors in the on state in each column determines the intensity of this wavelength. After the hardware structure was determined, we verified the high optical throughput of the SES through numerical calculations and analysis. To accurately simulate spectral encoding, a spectral encoding simulation method based on a convex optimization algorithm is used to calculate the optimal state of each DMD micromirror, and the controller controls the on–off state of 1920×1080 micromirrors according to the calculated optimal state. Furthermore, we theoretically analyzed the spectral encoding simulation capability and performance of the SES. These are the preconditions for its application in the field of spectral measurement. To verify the applicability of the SES in spectral measurements, we used the SES to simulate some existing spectral encoding schemes. Moreover, we also used the SES to perform a Gaussian random spectral encoding, achieving compressed-sensing-based spectral measurements. The theoretical analysis and numerical simulation results show that the SES can be used to achieve accurate complex spectral encoding simulations, and the compressed-sensing-based spectral measurements using this light source show a priority in the spectral reconstruction accuracy, because the SES compensates for the low accuracy in the spectral encoding simulations near the high end of the wavelength tuning range. The proposed SES can be used not only for high accuracy active-illumination-based spectral measurements but also as a simulation and verification platform for various spectral encoding schemes because of its ability to achieve high-resolution and accurate spectral encoding simulations.

The main structure of the paper is as follows: Section 1 is the introduction. In Section 2, the principle (Section 2.1) and structure (Section 2.2) of the SES are introduced, and the optical throughput is numerically simulated (Section 2.3). In Section 3, the spectral encoding simulation principle (Section 3.1) is introduced, and, after the position–wavelength function is numerical simulated, the model of the spectral response function is established (Section 3.2). Then, the spectral resolution of the SES is numerically simulated; after that, the spectral encoding simulation method is finally introduced (Section 3.3). In Section 4, the spectral encoding simulation capability and performance is numerically simulated. In Section 5, the applicability of the SES in spectral measurement is verified by the numerical simulation of the spectral encoding designed by various groups (Section 5.1) and the numerical simulation of spectral measurement based on compressed sensing (Section 5.2). In Section 6, the performance of the SES is verified by spectral encoding simulation capability experiments (Section 6.1) and compressed sensing experiments (Section 6.2). Section 7 is the conclusion.

2. System Principle and Structures

2.1. System Principle

The hardware principle of the simulated system is shown in Figure 1. The DMD controller controls the state of the 1920×1080 micromirrors according to the target SPD to encode the spectrum and then outputs the spectrum through the integrating sphere. The output of the integrating sphere is measured by the fiber optic spectrometer to evaluate the spectral encoding simulation accuracy.

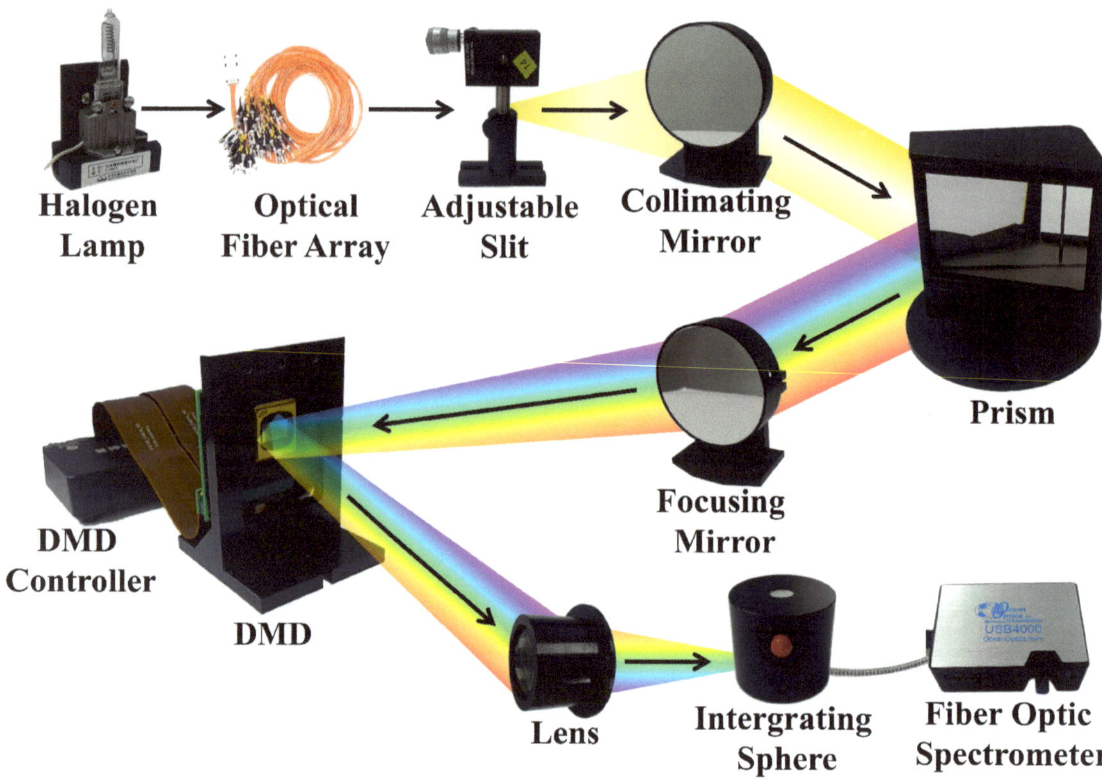

Figure 1. The hardware principle of the system.

2.2. System Structure

The main body of the SES uses the Czerny–Turner structure. The ray of the initial light sources (1000 W halogen lamp) is converged to the slit through a closely arranged fiber array and passes through an off-axis parabolic mirror (collimating mirror), a prism (dispersion element), and a spherical mirror (focusing mirror), imaging on the DMD. The spatial light modulation characteristics of the DMD are used to modulate the spectral image. The light is finally output through the converging lens and integrating sphere. The slit is 0.1×6 mm, the numerical aperture of the object is 0.12, the vertical magnification is 1.83, and the image height is 10.98 mm. In order to achieve a wide operating spectral range of 300–1100 nm and high optical throughput, the prismatic spectral imaging system and the fiber array are needed. The fiber array consists of 48 fibers with a diameter of 105 microns and a numerical aperture of 0.12. They are closely arranged in a 1×48 array to illuminate the slit. It is just enough to fill the entire 0.1×6 mm slit. The fibers on the other side are dispersed to couple more energy.

The design of the optical path of the system is shown in Figure 2a, and the design software is zemax 14.2. Figure 2b shows the design of the mechanical structure, and the design software is ug 10.0.

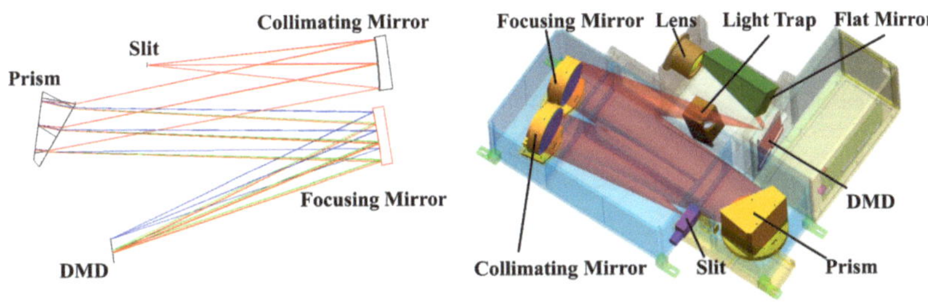

(a) Optical path (b) Mechanical structure

Figure 2. The optical–mechanical structure of the system: (**a**) the design of the optical path of the system; (**b**) the design of the mechanical structure.

2.3. Optical Throughput of the System

As an active illumination light source, the SES needs to have a high optical throughput to achieve spectral measurements with high signal-to-noise ratios. In order to improve the optical throughput of the SES and broaden the operating band, we used a prism made of JGS2 quartz. In addition, all reflective surfaces were simulated as coated with aluminum, and the front surface of the prism was coated with MgF2.

The use of the fiber array can also improve the optical throughput. The 48 optical fibers are equipped with 3 mm diameter grin lenses and placed at a distance of 50 mm from the 1000 W halogen lamp. In an ideal situation, the energy entering the fiber array is the product of the irradiance of the halogen lamp at 50 mm, the total area of all the grin lenses, and the coupling efficiency. For the same light source and coupling mode, the illuminance and coupling efficiency can be considered equal, so the optical energy gain is the ratio between the area of the fiber array and the area of the slit directly irradiated by the light source. This ratio is equal to approximately 300.

In the software simulation (the simulation software is lighttools 6.0), we set the light source spectrum as a 1000 W halogen lamp spectrum. The energy entering the slit was 1.94 W, and all the working surfaces were linked with the corresponding film characteristics. The simulation results are shown in Figure 3. Figure 3a shows the irradiance distribution on the DMD surface and Figure 3b shows the spectral power distribution on the DMD surface. The light energy utilization efficiency through the slit was 68.72%. However, the average diffraction efficiency with blazed or concave gratings was less than 50%, and they cannot be used with a broad spectral range. If other optical components are included, the light energy utilization rate decreases. For example, the initial light source of the CSE is directly used to illuminate the slit, and the CSE uses a concave grating, which operates in the range of 380–780 nm. The results of spectral measurements measured by the CSE have a low signal-to-noise ratio.

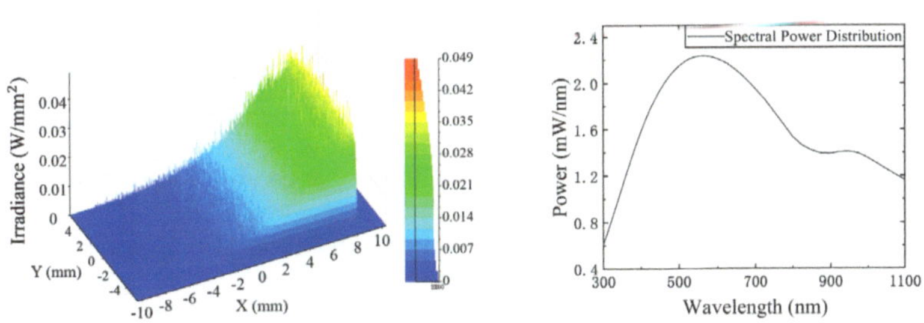

(a) Irradiance on DMD (b) Spectral power distribution on DMD

Figure 3. The optical throughput of the system: (**a**) the irradiance on DMD; (**b**) the spectral power distribution on DMD.

3. Spectral Encoding Simulation Method and Model

3.1. Spectral Encoding Simulation Principle

The spectral encoding simulation principle, which is also the main process of the spectral measurement simulation, is shown in Figure 4.

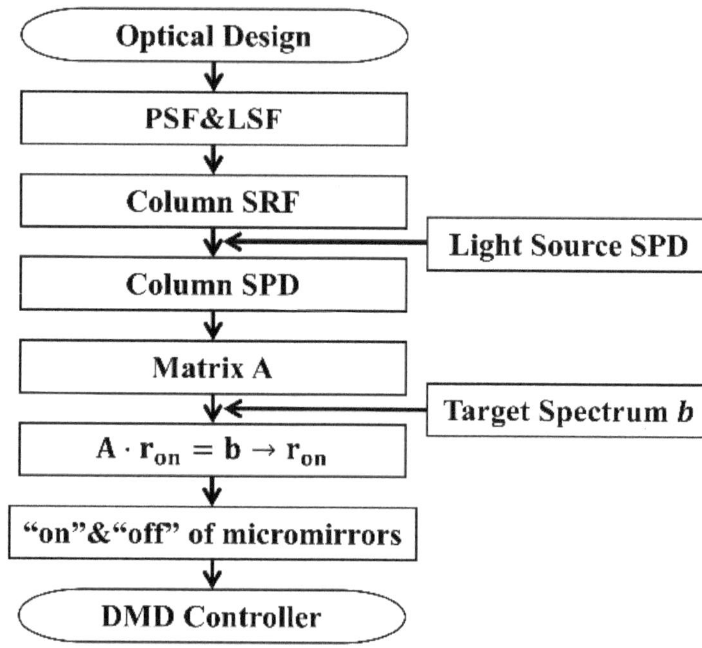

Figure 4. Flow chart of the spectral encoding simulation method. The PSF and LSF can be obtained by optical design, and the DMD column SPD can be calculated. The spectral matrix A is combined with all the SPDs. After the target spectrum is set, the state of DMD can be calculated. The state of DMD is fed into the controller.

As shown in the flow chart in Figure 4, based on the optical design, we can obtain the point spread function (PSF) and line spread function (LSF). The PSF describes the response of the imaging system to a point light source. Figure 5a shows the PSF in the on-axis field of view at 700 nm (the software is zemax 14.2). The LSF, which can be obtained by integrating the PSF in the slit length direction, describes the response of the imaging system to a line light source. The LSF can be used to obtain the spectral response function (SRF) of each DMD micromirror column. The product of the SRF and the initial light source SPD is the column SPD. All DMD column SPDs are divided into column vectors according to the central wavelength of the corresponding DMD micromirror column. These vectors are combined into a matrix known as spectral matrix A. Therefore, the row vector in matrix A represents the intensity of the corresponding wavelength in all DMD columns. The column vector in matrix A represents the SPD in the whole spectral band corresponding to the DMD column. Figure 5b shows a contour map of a part of matrix A. The horizontal axis represents DMD micromirror columns 1 to 128. The vertical axis represents wavelengths ranging from 300 nm to 308.54 nm. The blue parts in Figure 5b are all 0, which indicates that the micromirrors in this column do not receive light with the corresponding wavelength.

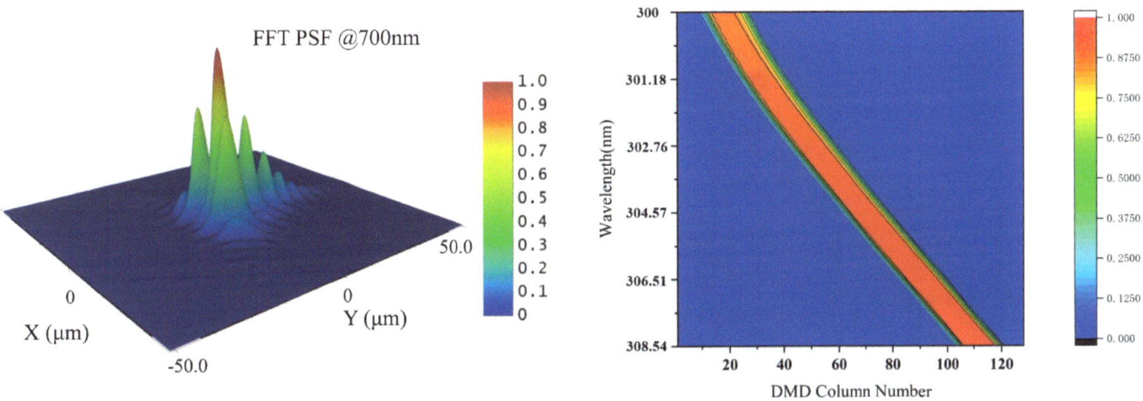

(a) Point spread function @700nm

(b) A part of the matrix **A**

Figure 5. The spectral encoding simulation models: (**a**) the point spread function exported from optical design software; (**b**) contour map of a part of the matrix A.

Spectral matrix A and a convex optimization algorithm can be used to determine the state of each DMD micromirror, leading to accurate spectral encoding simulations. The detailed encoding simulation steps are described in Sections 3.2 and 3.3.

3.2. Model of Spectral Response Function

As shown in the flow chart in Figure 4, the SPD on each micromirror must be known to modulate the spectral image on the image plane by the DMD. The SPD on each micromirror is determined by the product of the initial light source SPD (discussed in Section 4.1) and the spectral response function (SRF) on that micromirror. The SRF is an inherent property of the prismatic spectral imaging system, which is mainly determined by the spectroscopic principle and the efficiency of the optical system. Spectrometers have inherent spectral smile effects. Through monochromatic ray tracing, we find that the spectral smile effect of the SES system is not significant, and the edge and center positions of the monochromatic image with the most severe smile effects are only three micromirrors apart. However, it will be a great challenge to find the SRFs of all micromirrors on the DMD, and the simulation accuracy is not significantly improved by calculating the SRFs of all micromirrors. Therefore, we assume that the micromirror SRFs in each DMD micromirror column are the same and that these SRFs are uniformly distributed in the spatial dimensional direction. Therefore, it is necessary to calculate only the sum of micromirror SRFs in each single DMD micromirror column. The SRF of the prismatic spectral imaging system is shown in Equation (1).

$$\mathrm{SRF}(\lambda) = \mathrm{rect}\left(\frac{\mathrm{x}(\lambda)}{\mathrm{b}}\right) * \left[\mathrm{LSF}(\mathrm{x}(\lambda)) * \mathrm{rect}\left(\frac{\mathrm{x}(\lambda)}{\mathrm{a}}\right)\right]\tau(\lambda), \tag{1}$$

where * represents the convolution operation, a is the slit width, b is the micromirror width, and Rect is a rectangular function. When the independent variable is between −0.5 and 0.5, it equals 1; otherwise, it equals 0. The former rectangular function represents the response function of the micromirror, and the latter represents the slit function. For each micromirror, its center is defined as zero, and x is the position in this micromirror coordinate system. The spectral power received by the micromirror can be regarded as the response of the micromirror to the translation and superposition of each monochromatic image of the incident slit, and this process is equivalent to the convolution of the micromirror response function and the response function of the incident slit. $\tau(\lambda)$ is the system optical efficiency, and $\mathrm{LSF}(\mathrm{x}(\lambda))$ is the LSF at the position corresponding to the DMD micromirror column, which can be obtained by integrating the PSF in the slit length direction after the optical design has been determined. $x(\lambda)$ is the position coordinate along the dispersion direction on the image plane. There is a mapping relationship with the wavelength in the spectral dimension. Using the method of monochromatic ray tracing, the monochromatic image position coordinates can be simulated (the simulation software is lighttools 6.0). As shown in Figure 6a, the distribution of the monochromatic images from left to right is 300 nm, 400 nm, 500 nm, 600 nm, 700 nm, 800 nm, 900 nm, 1000 nm, and 1100 nm on the image plane. Figure 6b shows the correlation of x to wavelength.

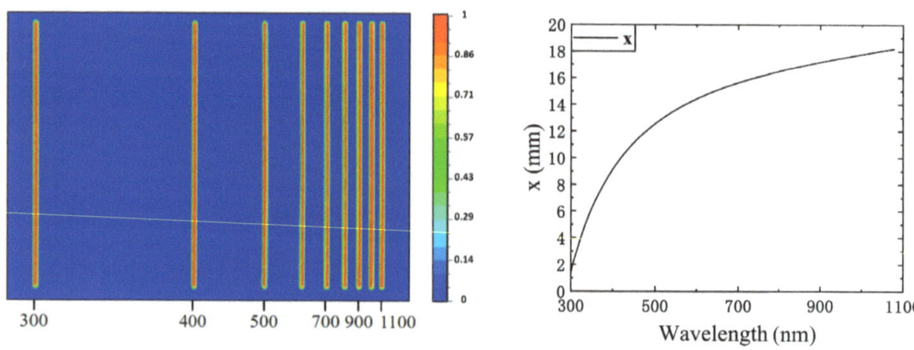

(a) Monochromatic images (b) Position-wavelength function

Figure 6. The positions of monochromatic images at different wavelengths on the image plane: (**a**) the distribution of monochromatic images at 9 wavelengths (the distance between 2 adjacent wavelengths is 100 nm) on the image plane; (**b**) the correlation of position to wavelength.

When the position–wavelength mapping relationship is applied in the SRF equation, the SRF curve can be obtained as shown in Figure 7. A total of 1575 DMD micromirror columns are covered in the range of 300–1100 nm, so a total of 1575 SRF curves are needed. The normalized SRF curves of the four DMD micromirror columns with center wavelengths of 325.45 nm, 457.15 nm, 694.07 nm, and 1078.27 nm are shown in Figure 7.

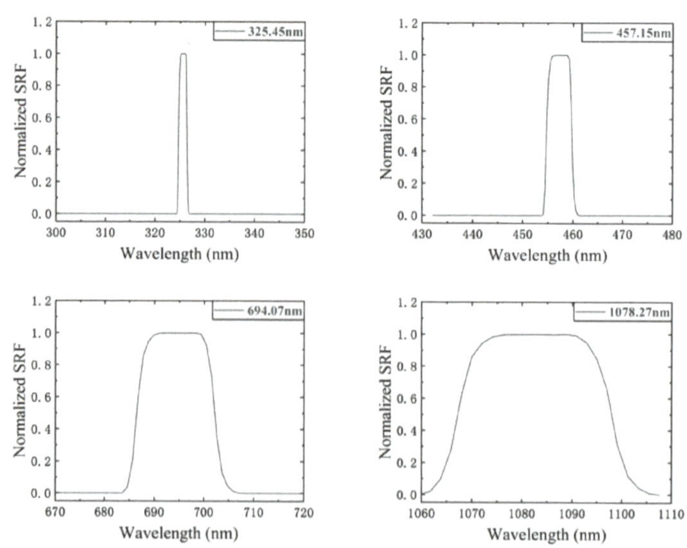

Figure 7. The normalized SRF curves of the four DMD columns with center wavelengths of 325.45 nm, 457.15 nm, 694.07 nm, and 1078.27 nm.

The full width at half maximum (FWHM) of the SRF also represents the spectral resolution, as seen in Figures 7 and 8. The FWHM varies at different locations and increases with increases in the wavelength. This places a constraint on the spectral encoding simulation performance of the SES. The specific constraint relationship is discussed in Section 4.2.

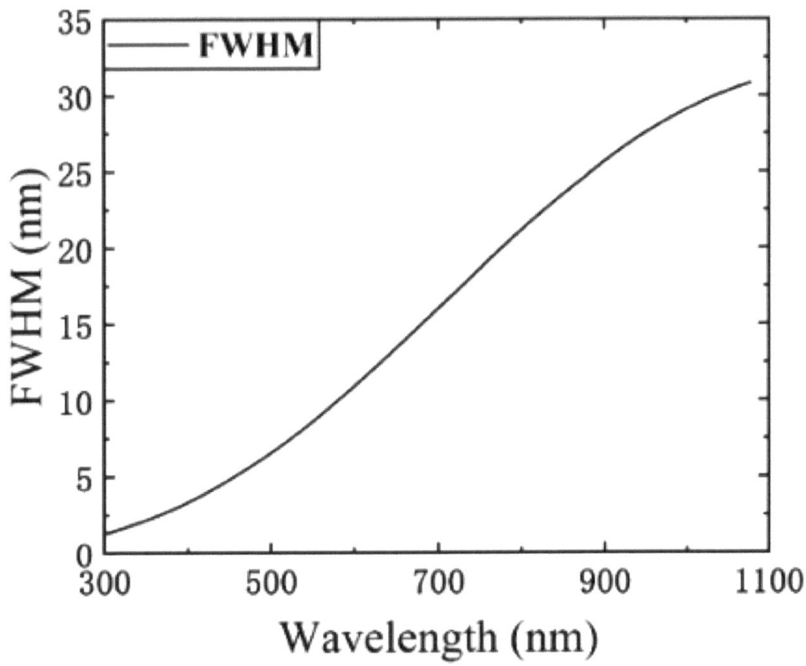

Figure 8. The spectral resolution of SES (numerical simulation).

3.3. Spectral Encoding Simulation Method

The SRFs of the 1575 DMD micromirror columns are discussed in Section 3.2. These SRFs are multiplied by the initial source SPD to obtain the SPDs of the 1575 DMD micromirror columns. Each SRF is discretized into a 1575-dimensional column vector according to the central wavelength of the corresponding DMD micromirror column and then combined into a 1575 × 1575 matrix known as SPD matrix A. The spectral encoding simulation process is shown in Figure 9. Figure 9a shows matrix A, which consists of the SPDs of the 1575 DMD micromirror columns. A halogen lamp with a 5000 K color temperature is used as the initial light source. Figure 9b shows the simulation process of a random target SPD. The simulation process can be modeled as follows:

$$\mathbf{A} \cdot \mathbf{r}_{on} = \mathbf{b}, \tag{2}$$

where $\mathbf{r}_{\text{on}}$ and $\mathbf{b}$ are 1575-dimensional column vectors. $\mathbf{r}_{\text{on}}$ is the unknown to be determined and represents the proportion of the micromirrors in the "on" state in each DMD micromirror column, which has a value between 0 and 1. A value of 0 means that the micromirrors in the column are all off, while a value of 1 means that the micromirrors in the column are all on. $\mathbf{b}$ is obtained by dividing the target SPD by the spectral reflectance of the micromirrors, the transmittance of the lens and the reflectance of the integrating sphere diffuse reflector.

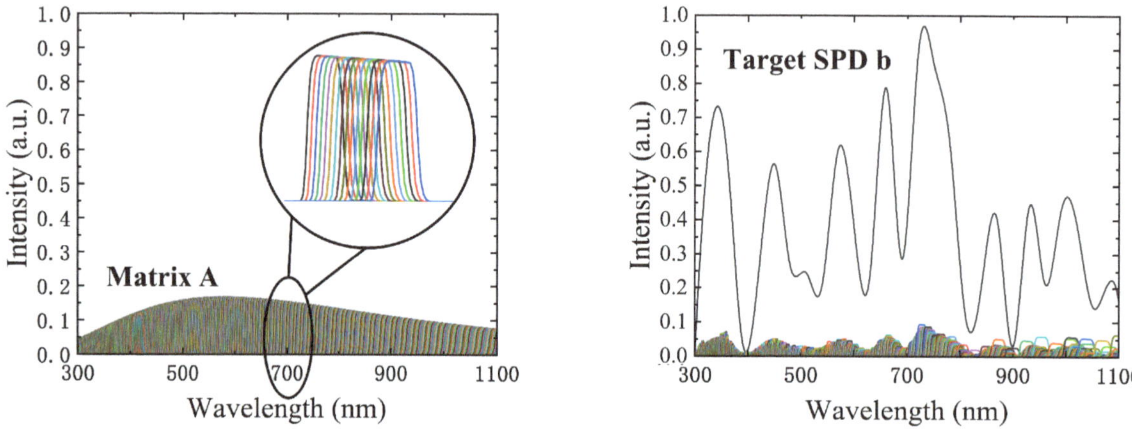

(a) Matrix **A**

(b) Simulation process

Figure 9. The spectral encoding simulation process: (**a**) the 1575 × 1575 SPD matrix A (expanded by column); (**b**) the simulation process of a random target SPD (the sum of all the modulated columns is close to the target SPD b).

To solve for $\mathbf{r}_{on}$, the 2-norm is used as the evaluation function to calculate $\mathbf{r}_{on}$, and the result of $\mathbf{r}_{on}$ is used to guide the switching of each DMD micromirror column. The equation then becomes a convex optimization problem with 1575 constraints, as shown below:

$$\widehat{\mathbf{r}_{on}} = \underset{\mathbf{r}_{on}}{\operatorname{argmin}}(\|\mathbf{A} \cdot \mathbf{r}_{on} - \mathbf{b}\|_2), \text{ s.t. } \quad 0 \leq \mathbf{r}_{on} \leq 1. \tag{3}$$

We use a convex optimization algorithm to solve this equation [27,28], which can calculate the proportion of the micromirrors that should be in the on state in each column with minimum deviation. As a toolkit, cvx is simple to use. By entering the code according to Equation (3) and importing A and b, the result can be easily obtained. $\mathbf{r}_{on}$ is multiplied by the number of DMD micromirror rows and rounded to the nearest integer to determine the number of micromirrors that should be in the on state in each column. To minimize the error caused by the spectral smile mentioned in Section 3.2, the code is designed so that each column of DMD micromirrors turns on from the middle to both ends.

4. Spectral Encoding Simulation Capability and Performance

Matrix A and the target SPD b in the algorithm described in the previous section affect the spectral encoding simulation capability and spectral encoding simulation performance of the SES, respectively. These effects are discussed in the following sections. (All of the following programs were written in python).

4.1. Spectral Encoding Simulation Capability

Matrix A determines the spectral encoding simulation capability, and the product of the SPD of the initial light source and the SRFs of the SES determines matrix A. For different types of initial light sources, the spectral encoding simulation capability is different. A halogen lamp (with a color temperature of 5000 K) and a xenon lamp were used for simulations. The SPDs of these two initial light sources are shown in Figure 10a,d, and a random target SPD was simulated with two initial light sources. The numerical simulation results are shown in Figure 10b,e. The simulation capability of the SES with xenon lamps with characteristic spectra was inferior to that of the SES with halogen lamps in the range of the characteristic spectral bands. Therefore, to improve the simulation capability of the SES, the initial light source should be halogen lamps without characteristic spectra. Figure 10c,f show the DMD state when the halogen lamp and xenon lamp are used for simulations, respectively (black pixels represent the off state and white pixels represent the on state).

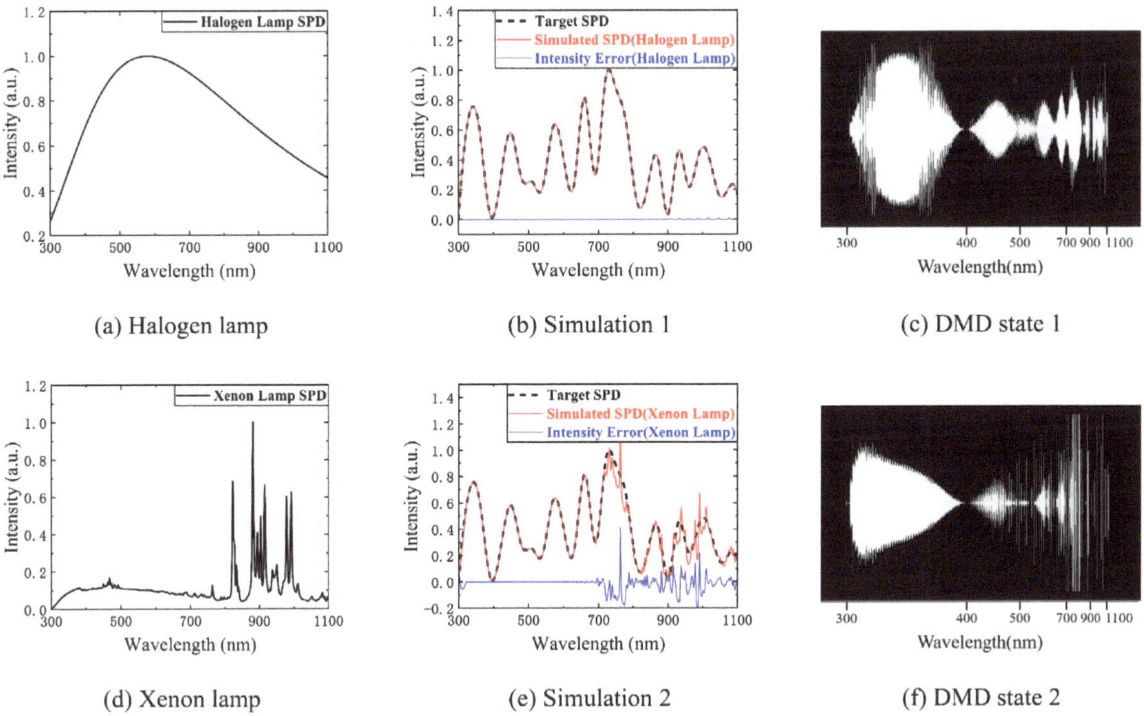

Figure 10. The spectral encoding simulation capability: (**a**) the halogen lamp SPD; (**b**) the numerical simulation result using the halogen lamp; (**c**) the DMD state when the halogen lamp is used for simulation (black pixels represent the off state and white pixels represent the on state); (**d**) the xenon lamp SPD; (**e**) the numerical simulation result using the xenon lamp; (**f**) the DMD state when the xenon lamp is used for simulation.

4.2. Spectral Encoding Simulation Performance

The target SPD b determines the spectral encoding simulation performance. Due to the nonlinear spectroscopic characteristics of the prism, the spectral resolution (FWHM of the SRF) decreases with increases in the wavelength. Figure 8 shows the trend of the change in the FWHM of the SRF with the wavelength. The maximum FWHM is 30 nm in the range of 300–1100 nm. Therefore, the resolution of the target SPD b should be greater than 30 nm to obtain an accurate spectral encoding simulation. Clearly, the resolution of b can increase with a decrease in the range of wavelengths.

Figure 11a–c show extreme cases of 30 nm, 20 nm, and 10 nm resolutions, respectively. Figure 11d–f show the DMD states of the three numerical simulation cases. The root mean square errors (RMSEs) of the three cases are 1.22%, 8.05%, and 13.83%, respectively. Since our numerical simulations are in extreme states, the determined RMSE corresponds to the maximum deviation of the three cases.

The above numerical simulation results show that the error increases with increases in the wavelength, and the main reason is that the SES uses a prismatic spectral imaging system. The broad bandwidth near the high end of the wavelength tuning range could limit the simulation capability and performance.

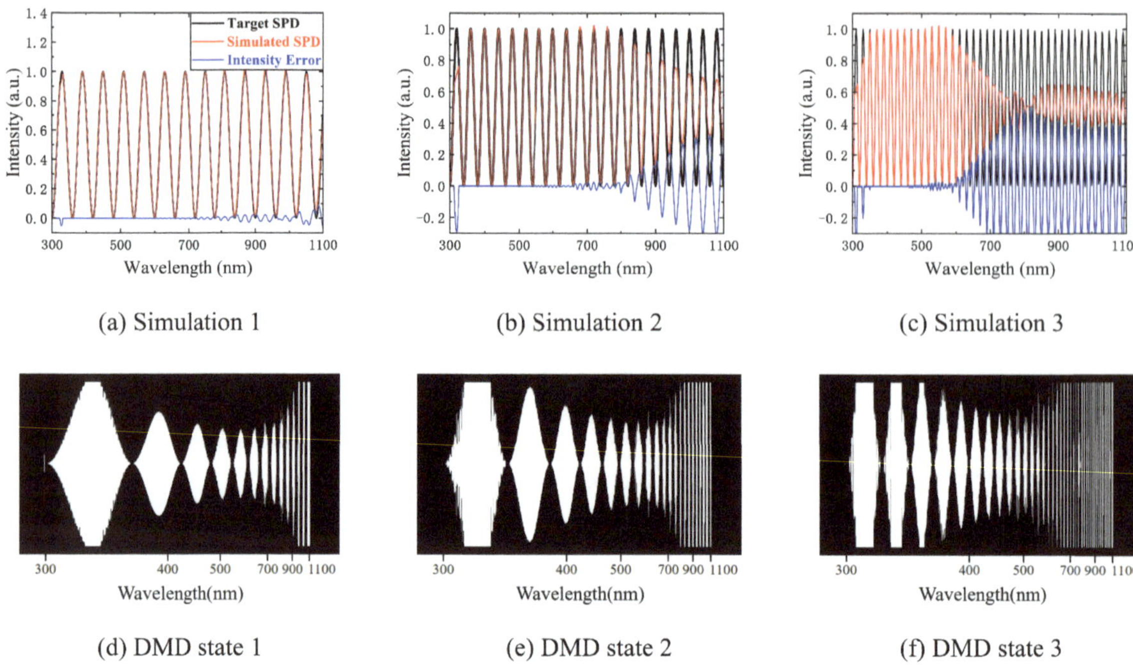

Figure 11. The spectral encoding simulation performance: (**a**) the numerical simulation result of the extreme case of 30 nm resolution; (**b**) the numerical simulation result of the extreme case of 20 nm resolution; (**c**) the numerical simulation result of the extreme case of 10 nm resolution; (**d**) the DMD state of the 30 nm resolution case; (**e**) the DMD state of the 20 nm resolution case; (**f**) the DMD state of the 10 nm resolution case.

5. Applicability of the SES in Spectral Measurement

5.1. Simulation of Spectral Encoding

The SES can be used to simulate spectral encoding schemes in existing spectral measurement methods based on active illumination. We numerically simulated spectral encoding schemes designed by various groups.

Chi et al. designed the spectral encoding of a nearly optimal illumination set based on the Hadamard code. Since physically perfect narrow bandpass filters do not exist, they used the SPD information of a real narrow bandpass interference filter (Edmund Optics) to simulate nearly optimal illumination [14]. Our SES simulated spectral encoding based on the Hadamard code numerically, which cannot be achieved with light sources and filters, and the numerical simulation results are shown in Figure 12a.

Han et al. used a DLP projector as an illumination source and combined it with a high-speed camera to measure spectral reflectance [15]. This approach has the advantage of being able to obtain spectral reflectance measurements in dynamic scenes by using the high-speed switching illumination SPD of the DLP projector, whose core component is a DMD. Our SES can achieve the same SPD switching speed as the DLP projector. When combined with a high-speed camera, it can also obtain the spectral reflectance measurements of dynamic scenes. Figure 12b shows the performance of the SES in numerically simulating the green spectral illumination channel in the RGB of the DLP projector.

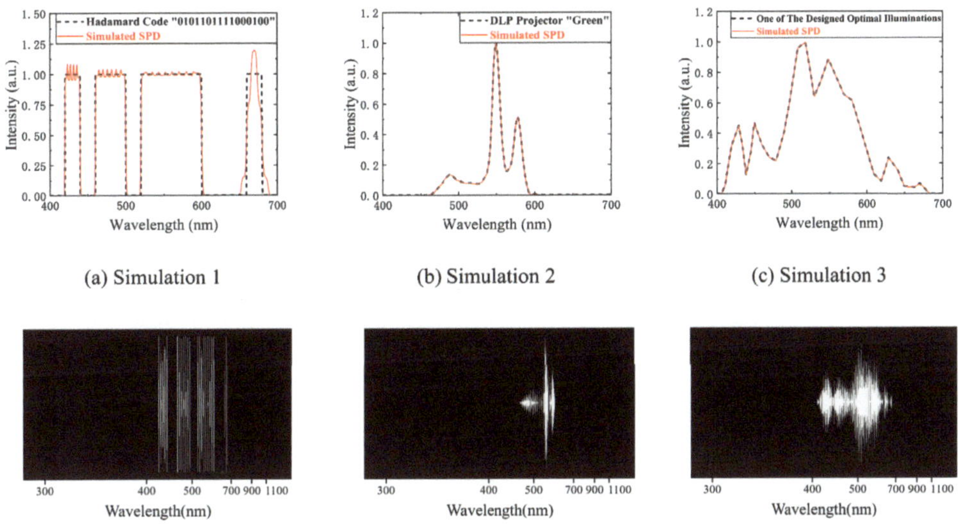

(a) Simulation 1 (b) Simulation 2 (c) Simulation 3

(d) DMD state 1 (e) DMD state 2 (f) DMD state 3

Figure 12. The numerical simulation of spectral encoding designed by various groups: (**a**) the numerical simulation result of spectral encoding based on Hadamard code; (**b**) the numerical simulation result of the green spectral illumination channel in the RGB of the DLP projector; (**c**) the numerical simulation result of one of the optimal illumination SPDs; (**d**) the DMD state when simulating spectral encoding based on Hadamard code; (**e**) the DMD state when simulating the green spectral illumination channel; (**f**) the DMD state when simulating one of the optimal illumination SPDs.

Fu et al. presented a simple and efficient convolutional neural network (CNN)-based spectral reflectance recovery method with optimal illumination. They designed an illumination optimization layer to optimally multiplex illumination spectra in a given dataset or to design the optimal one under physical restrictions [29]. We numerically simulated one of the designed optimal illumination SPDs for spectral encoding. The numerical simulation results are shown in Figure 12c.

In summary, our SES achieves good numerical simulation for the illumination spectral encoding used by existing active illumination methods.

Compared to the spectral measurements achieved by Zhang et al. using a light source and random filter combined with their deeply learned broadband encoding stochastic hyperspectral camera [16], we used the SES to replace the light source and random filter and simulated the measurement of spectral reflectance or transmittance by the compressed sensing method numerically, as discussed in the Section 5.2.

5.2. Spectral Measurement Based on Compressed Sensing

Compressed sensing (CS) is a sampling theory that obtains measurements under a specific measurement basis $\mathbf{\Phi}$ and represents the signal to be sampled sparsely in the time domain by means of a sparse basis $\mathbf{\Psi}$. The sparse signal can be finally recovered by various algorithms. CS can reduce the number of samples while maintaining signal integrity, which can be applied to the measurement of spectral reflectance or transmittance. The principle of CS is shown in the following equation:

$$y = \mathbf{\Phi}\mathbf{\Psi}\mathbf{S}, \tag{4}$$

where y is the $N_1 \times 1$ observation vector formed by M measurements ($N_1 = M$), $\mathbf{\Phi}$ is the $N_1 \times N_2$ observation matrix, $\mathbf{\Psi}$ is the sparse matrix of $N_2 \times N_2$, and $\mathbf{S}$ is the sparse signal of $N_2 \times 1$. According to compressive sensing theory, the key to measuring whether **S** can be compressively losslessly sampled is the restricted isometry property (RIP). The independent and identically distributed Gaussian random measurement matrix can be the universal compressive sensing measurement basis. The sensing matrix formed by the Gaussian random measurement matrix and most sparse bases can satisfy the RIP; therefore, we choose the Gaussian random matrix as the measurement basis. In addition, we use the overcomplete dictionary as the sparse basis and construct the sparse basis with a Gaussian function as the parent function, which can be well sparsed for the spectral reflectance or transmittance data.

As shown in Equation (5) in the Gaussian function, the central wavelength is μ, and the FWHM is $\sqrt{8\ln 2}$. This is equal to the spectral sampling interval λ_s. After stretching the Gaussian function 2^m times (m is an integer), we can obtain a new Gaussian function with a stretched FWHM. N times (n is an integer) the stretched FWHM is the center wavelength offset to yield the new Gaussian function that covers the entire spectral range. The set of functions forms a sparse basis $\boldsymbol{\Psi}$, as shown in Equation (6).

$$g(\lambda;\mu,\sigma) = \frac{1}{\sqrt{2\pi}\cdot\sigma}\exp\left[-\frac{(\lambda-\mu)^2}{2\sigma^2}\right], \tag{5}$$

$$\left\{\begin{array}{l} \boldsymbol{\Psi}_{m,n}(\lambda) = \sqrt{2^m}\cdot g\left(\lambda; 2^m n\lambda_s, \frac{2^m\lambda_s}{\sqrt{8\ln 2}}\right); \\ 1 \le 2^m \le \frac{\lambda_{max}-\lambda_{min}}{\lambda_s}\ ;\ \frac{\lambda_{min}}{\lambda_s 2^m} \le n \le \frac{\lambda_{max}}{\lambda_s 2^m}\ ;\ m,n \in \mathbf{Z} \end{array}\right\}. \tag{6}$$

Finally, the convex optimization algorithm is used to measure the spectral reflectance or transmittance. The algorithm is shown in Equation (7).

$$\hat{\mathbf{S}} = \underset{S}{\operatorname{argmin}}(\|\boldsymbol{\Phi}\boldsymbol{\Psi}\mathbf{S} - \mathbf{y}\|_1 + 0.05\|\mathbf{S}\|_1). \tag{7}$$

After the measurement basis $\boldsymbol{\Phi}$, the sparse basis $\boldsymbol{\Psi}$, and the algorithm are determined, the spectral measurement process is proposed, as shown in Figure 13. The N_1 row vectors in $\boldsymbol{\Phi}$ are regarded as the spectral encoding. The SES simulates this spectral encoding as the active illumination SPD, and the camera is used to acquire M reflectance or transmittance signals y. The sparse spectral reflectance or transmittance signal **S** is measured by the convex optimization algorithm (cvx) [27,28], and $\boldsymbol{\Psi}\mathbf{S}$ is the measured spectral reflectance or transmittance signal.

For the illumination, the integrating sphere can be replaced with a fiber coupler to improve energy utilization. In addition, light guides, which have larger diameters than optical fibers and higher energy utilization than optical fibers and integrating spheres, can also be used as illumination elements for the SES. The commercially available quartz light guides and liquid light guides can directly couple the light from the SES. The use of a fiber coupler or light guides for illumination can meet spectral measurement requirements.

The most important and difficult part of compressed sensing is the physical implementation of the measurement basis $\boldsymbol{\Phi}$. Our SES can easily implement a Gaussian random encoding because of its flexibility. We choose a spectral sampling interval of 10 nm (N_2= 80) to generate a Gaussian random matrix. This is regarded as the spectral encoding. The numerical simulation results are shown in Figure 14. Four of the numerical simulations are shown, with the bottom right figure showing the worst numerical simulation results.

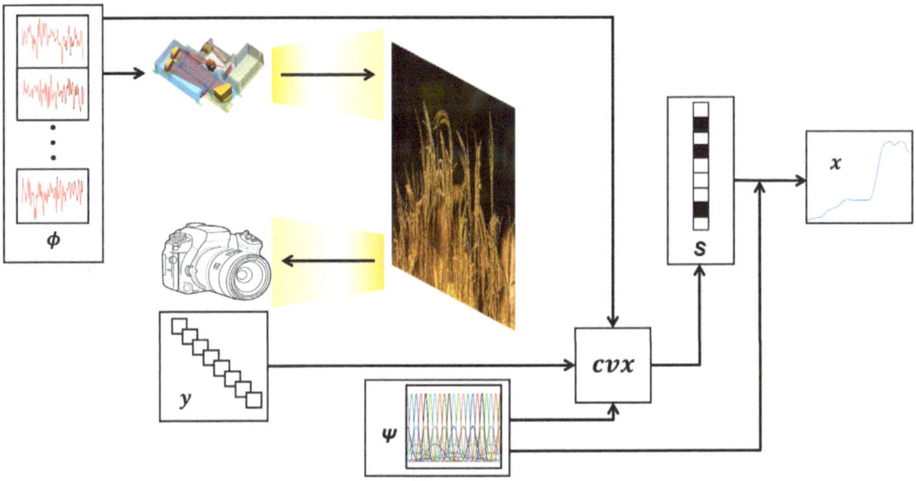

Figure 13. The process of spectral reflectance measurement. Φ is the measurement basis, which is simulated by SES, y is the signal obtained by the detector, and Ψ is the sparse basis. The above three are used as input quantities, and the sparse signal S is obtained by the *cvx*. x is the measured spectral reflectance or transmittance signal obtained by multiplying Ψ and S.

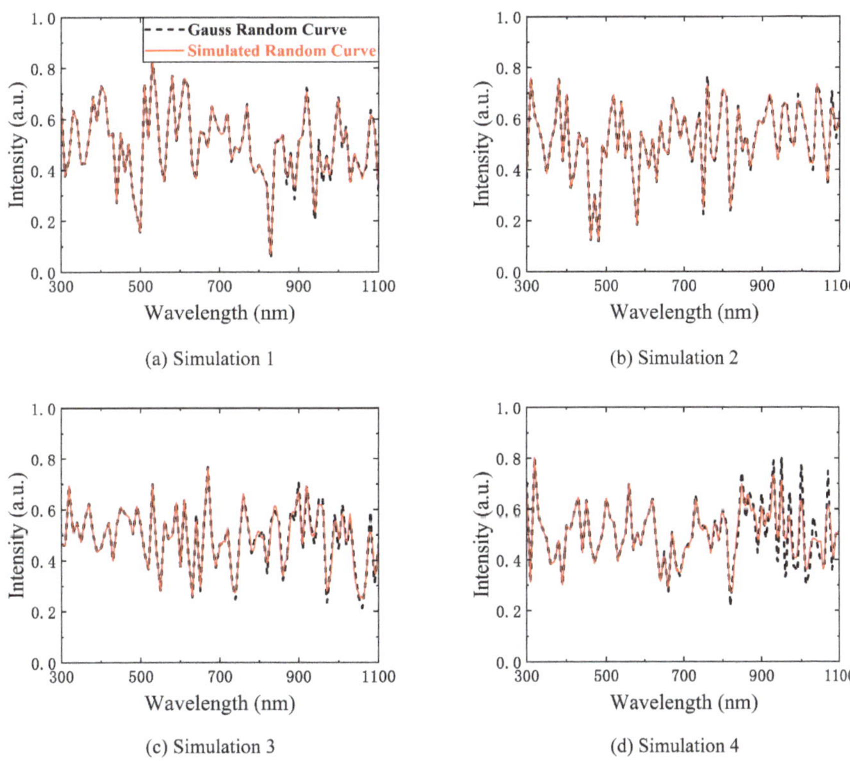

Figure 14. Four of the numerical simulations of Gauss random encoding. The black curves are Gauss random curves and the red curves are simulated random curves. The simulation performance is shown in the four numerical simulations.

According to the RIP, all column vectors in the Gaussian random matrix should be approximately orthogonal, that is, with weak correlation. It is calculated that the mean of the absolute value of the cross-correlation coefficient of the generated Gaussian random matrix is 0.087 and the maximum value is 0.428. The mean value of the absolute value of the cross-correlation coefficient of the simulated matrix is 0.0901 and the maximum value is 0.434, which can meet the requirement of weak correlation. Figure 15 shows the numerical simulation of Grass_dry.4+.6green from the US Geological Survey (USGS) spectral reflectance database [30] with the real Gaussian random matrix and the SES simulated matrix as the measurement encoding. The values for the number of measurements M are 60, 50, 40, and 30, and the product of the measurement encoding and the spectral reflectance is regarded as the measurement value y.

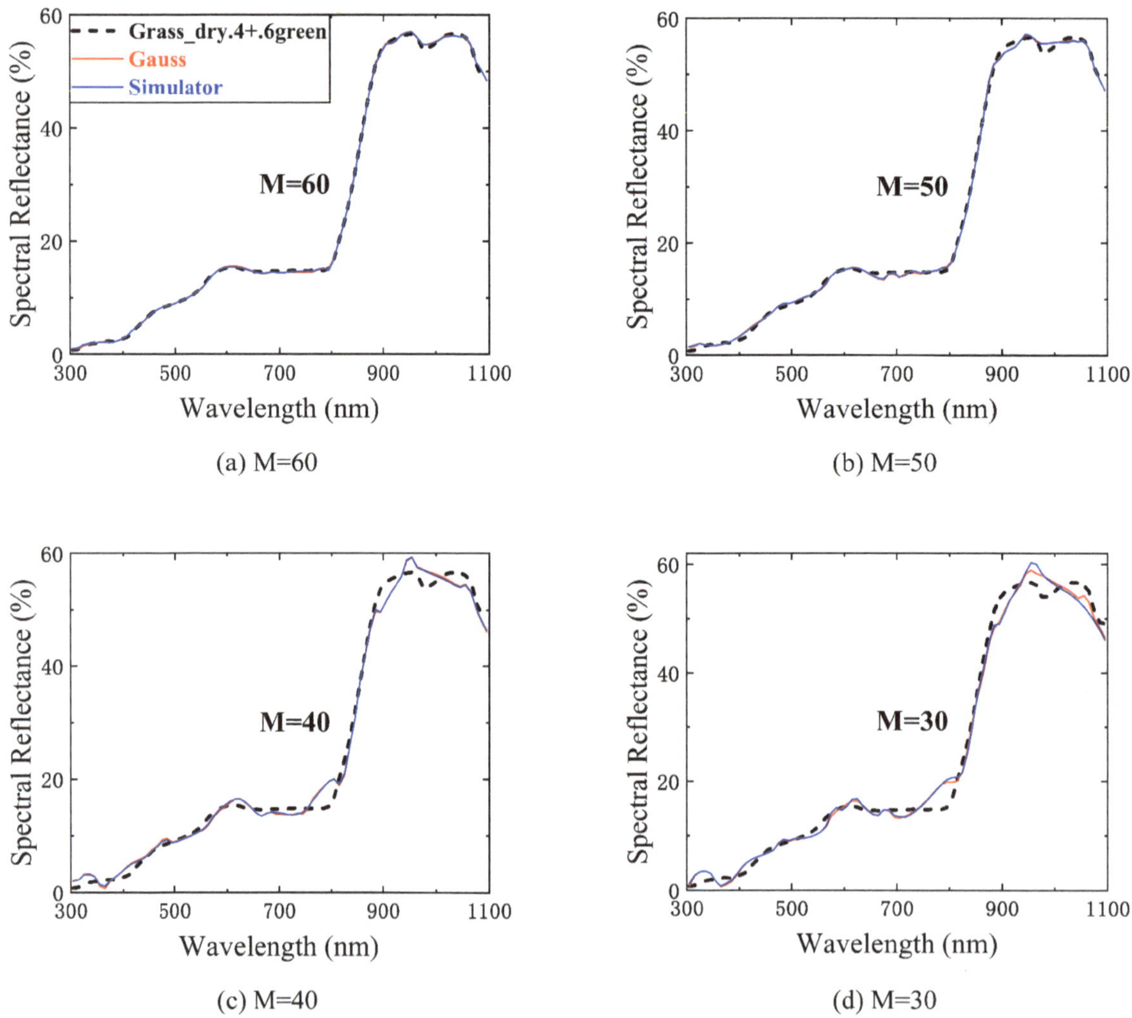

(a) M=60 (b) M=50 (c) M=40 (d) M=30

Figure 15. The numerical simulation of Grass_dry.4+.6green spectral reflectance with the real Gaussian random matrix and the SES simulated matrix as the measurement encoding (M is 60, 50, 40, and 30).

The numerical simulation results show that the reconstruction ability of the two measurement bases gradually diverges (the deviation between the blue line and the red line in the figure) as M decreases. The following table (Table 1) lists the RMSEs of the spectral measurements obtained with the two measurement matrices using different numbers of measurements. The simulated matrix and the real Gaussian random matrix have basically the same performance on the spectral measurement, and there is no apparent superiority or inferiority relationship between them. This indicates that the simulation error on the Gaussian random matrix does not affect the measurement accuracy of spectral reflectance or transmittance.

Table 1. The RMSEs of spectral measurement with two measurement matrices.

M	RMSE of Gauss (%)	RMSE of SES (%)
80	0.181	0.185
70	0.232	0.211
60	0.29	0.275
50	0.599	0.619
40	1.636	1.584
30	1.773	2

In the USGS database, we randomly selected the spectral reflectance of two minerals (Axinite_HS342.3B & Goethite_WS222) [30] and measured them at a sampling rate of 50% (M = 40). The results are shown in Figure 16a,b. Figure 16c shows the change in the RMSE for three spectral reflectance measurements with measurement time M. The measurement deviation has a significant downward trend as the measurement time increases and tends to stabilize after reaching 58% (M = 47). The RMSE is less than 1%.

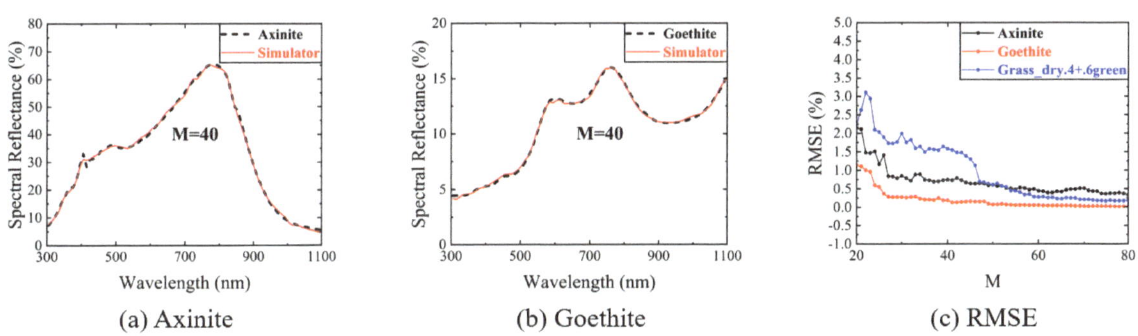

Figure 16. The spectral reflectance measurement of two minerals at a sampling rate of 50% (M = 40): (**a**) the spectral reflectance measurement of Axinite_HS342.3B; (**b**) the spectral reflectance measurement of Goethite_WS222; (**c**) the change in RMSE for three spectral reflectance measurements with measurement time M.

In summary, with the flexible spectral encoding simulation capability, the SES can simulate a variety of complex spectral encoding, such as Gaussian random measurement encoding, spectral encoding based on deep learning, and Hadamard-based spectral encoding, so as to achieve accurate spectral reconstruction based on active illumination.

6. Validation Experiments

To verify the performance of the SES, we designed two validation experiments to verify its spectral encoding simulation capability and applicability of compressed sensing, respectively.

6.1. Experiments to Verify the Capability of Spectral Encoding Simulation

To verify the spectral encoding simulation capability and performance of SES, we mounted the experimental platform as shown in Figure 17.

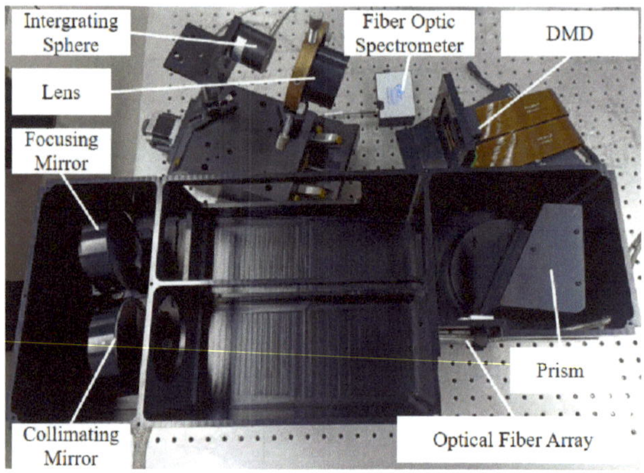

Figure 17. The drawing of the experimental platform.

The initial light source is a 1000 W spectral irradiance standard lamp 210,701 (National Institute of Metrology, Beijing, China). The optical fiber array is RH-48-127-8-L (Ruihe, Beijing, China). The prism, the collimating mirror, the focusing mirror, and the lens were specialized manufactured by the Changchun Institute of Optics, Fine Mechanics and Physics, China. The DMD is the Discovery F4110 (Jinhua Fldiscovery, Jinhua, China), and the core DMD chip is the DLP9500 (Texas Instruments, Dallas, TX, USA). The DMD control program is F4110DMDControlerV2.0, which is provided by Jinhua Fldiscovery. The integrating sphere is the JFQ-25 (Hangxin, Guangzhou, China). The fiber optic spectrometer is the USB4000 (Ocean Optics, Orlando, FL, USA). All other control and data acquisition software were written in Python 3.6.

The USB4000 (Ocean Optics) has a high response at 450–850 nm, so our experiments were operated at 450–850 nm.

After the experimental platform was mounted, we tested the main performance of the SES, including the SPDs received by each micromirror column and the spectral resolution of the system. We tested the SPDs of the micromirror columns by opening the single column. Figure 18a shows the result of several SPDs. The tested SPDs were also used to calibrate the SES. Based on the SPDs, we solved the spectral resolution of the SES as shown in Figure 18b.

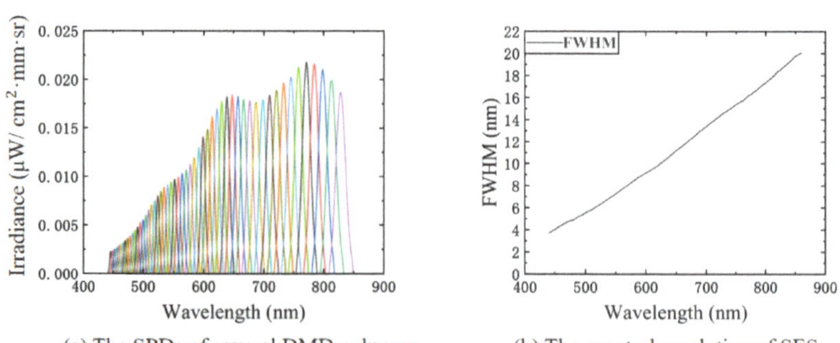

Figure 18. The tests of basic SES performance: (**a**) the DMD column SPDs of SES; (**b**) the spectral resolution of SES.

We experimentally simulated random spectra with a 10 nm resolution at 450–850 nm after the experimental platform was mounted and calibrated, and the results are shown in Figure 19a,b. The average relative errors of the two experiments were 3.07 and 3.81%. The RMSE values were 0.0065 μW/cm^2·mm·sr and 0.0043 μW/cm^2·mm·sr. The main reason for the errors was that the spectral and radiometric calibrations of the experimental platform were not precise enough.

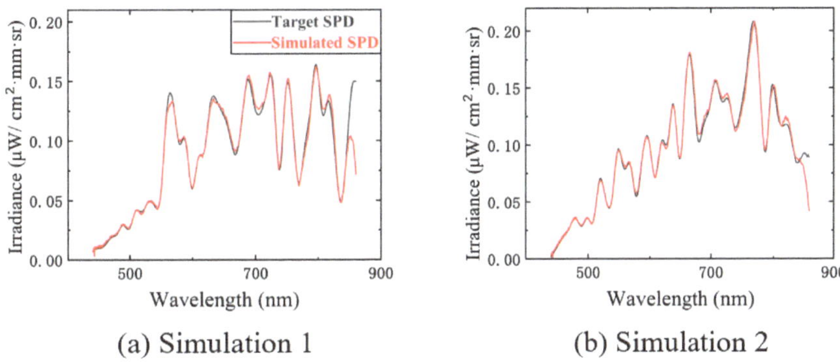

Figure 19. The experimental simulations of two random spectral encodings: (**a**) the experimental simulation result (red simulated SPD in the figure) of spectral encoding 1 (black target SPD in the figure); (**b**) the experimental simulation result of spectral encoding 2.

6.2. Experiments to Verify the Applicability of Compressed Sensing

To verify the applicability of compressed sensing, we used an existing calibrated filter as an experimental sample and pasted the filter in front of the integrating sphere, as shown in Figure 20.

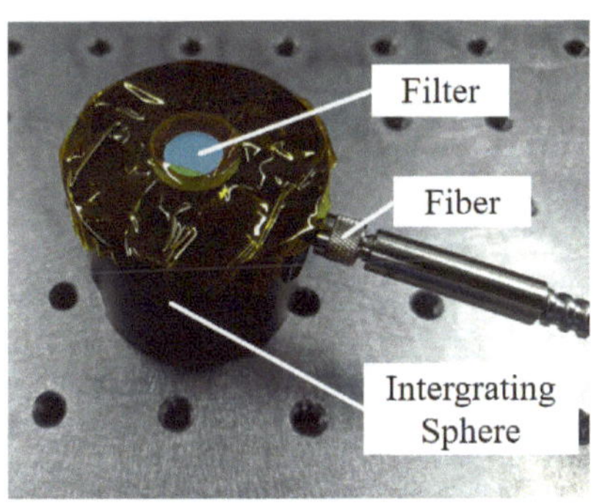

Figure 20. The drawing of the experimental sample.

We used M′ random spectra (one random spectrum corresponds to one measurement, so the M′ is equal to the measurement time M) to illuminate the filter with a sampling resolution of 10 nm, which is the same as the sampling resolution of the filter calibration result, and then used a single-point detector to collect the energy signal. We finally used the compressed sensing algorithm to reconstruct the filter spectral transmittance. The results are shown in Figure 21a–d, which show the reconstruction effect for M values of 40, 35, 30, and 25, respectively. Since compressed sensing requires known random spectra, we still used the USB4000 to measure the random spectra, so we reconstructed the spectral transmittance at 450–850 nm.

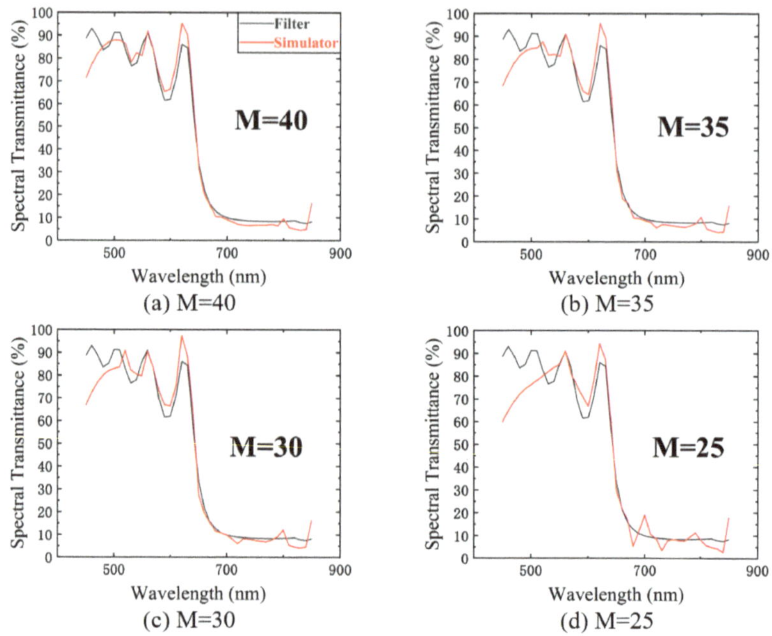

Figure 21. The transmittance measurements with different measurement time M (M = 40, 35, 30, and 25). The black curves in the figures are the calibrated transmittance of the filter. The red curves are the measurement results obtained from transmittance reconstruction experiments.

The causes of the reconstruction errors in the figure are the detector measurement errors, especially the time drift error and the detector response error. The former leads to a gradual increase of the error when the number of measurements increases, and the latter leads to a poorer reconstruction in the low response band (450–550 nm). Figure 22 shows the change in the RMSE for the spectral transmittance measurements with measurement time M.

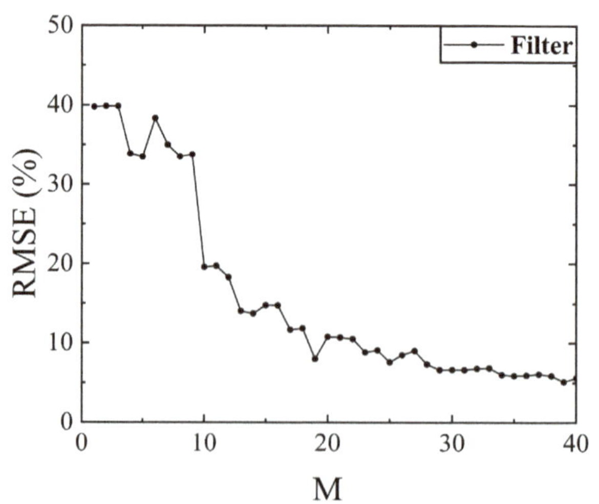

Figure 22. The RMSE for the spectral transmittance measurements with measurement time M.

7. Conclusions

Since the light sources used in existing methods of spectral measurement cannot effectively implement the designed active illumination spectral encoding, we designed a spectral encoding simulator for the broadband of 300–1100 nm and calculated the spectral response function on the image plane of the prismatic spectral imaging system and the spectral distribution on the DMD. According to the spectral distribution on the DMD, we used a convex optimization algorithm to design a spectral encoding simulation method. In addition, we explored the impact of different initial light sources on the spectral encoding simulation capability and the impact of different target spectral resolutions on the spectral encoding simulation performance. The simulation data show that the use of tungsten halogen lamps without characteristic spectra can improve the spectral encoding simulation capability. The SES can simulate spectral encoding at 30 nm, 20 nm, and 10 nm resolutions in the extreme case with RMSEs of less than 1.22%, 8.05%, and 13.83%, respectively. We also simulated the spectral encoding for most of the existing active illumination methods. The results show that the SES can achieve the spectral encoding designed by other research groups. Finally, based on compressed sensing theory, we simulated a 10 nm resolution Gaussian random measurement encoding and measured the spectral reflectance or transmittance. The simulation results show that the goodness of fit of the spectral measurement is higher than 0.998 when the sampling rate is larger than 50%. The recovery rate (success rate) of the system is high because the cvx we use is stable, and the accuracy is acceptable.

Depending on the spectral modulation characteristics of the DMD, the main advantage of our SES is that it can flexibly simulate various spectral encodings and combine various algorithms to measure spectral reflectance or transmittance for active illumination. The main disadvantage of our SES is that the optical throughput of the initial light source is reduced due to the use of the slit. To compensate for this defect, we use a prismatic spectral imaging system. We are exploring an optical design without slits to improve the optical throughput, and this problem is expected to be resolved in future work. Our SES also has other potential applications, such as in star simulators [22], calibration light sources [24–26], and other scenes requiring tunable spectra. The SES can even be used as a simulation and verification platform for various forms of spectral encoding schemes and spectral reconstruction verification.

Author Contributions: Conceptualization, Z.Z. and G.L.; data curation, P.J.; formal analysis, L.H.; investigation, P.J., X.W. and G.L.; methodology, P.J. and X.W.; project administration, G.L.; resources, X.W. and G.L.; software, J.L. and H.W.; supervision, G.L.; validation, Z.Z. and G.G.; writing—original draft, P.J.; writing—review and editing, P.J. and X.W. All authors have read and agreed to the published version of the manuscript.

Funding: This research was funded by the National Natural Science Foundation of China (62205330, 62005268) and Key Science and Technology R & D program of Jilin Province, China (No. 20220203195SF).

Institutional Review Board Statement: Not applicable.

Informed Consent Statement: Not applicable.

Data Availability Statement: Data underlying the results presented in this paper are available in Refs. [27,28,30].

Acknowledgments: Thanks for the help and technical support from other authors, and thanks for the financial support from the foundation.

Conflicts of Interest: The authors declare no conflict of interest.

References

1. Cao, X.; Zhou, F.; Xu, L.; Meng, D.; Xu, Z.; Paisley, J. Hyperspectral Image Classification with Markov Random Fields and a Convolutional Neural Network. *IEEE Trans. Image Process.* **2018**, *27*, 2354–2367. [CrossRef] [PubMed]
2. Qi, B.; Zhao, C.; Youn, E.; Nansen, C. Use of weighting algorithms to improve traditional support vector machine based classifications of reflectance data. *Opt. Express* **2011**, *19*, 26816–26826. [CrossRef] [PubMed]
3. Thuillier, G.; Zhu, P.; Snow, M.; Zhang, P.; Ye, X. Characteristics of solar-irradiance spectra from measurements, modeling, and theoretical approach. *Light Sci. Appl.* **2022**, *11*, 79. [CrossRef] [PubMed]
4. Bioucas-Dias, J.M.; Plaza, A.; Camps-Valls, G.; Scheunders, P.; Nasrabadi, N.; Chanussot, J. Hyperspectral Remote Sensing Data Analysis and Future Challenges. *IEEE Geosci. Remote Sens. Mag.* **2013**, *1*, 6–36. [CrossRef]
5. Gillis, D.B.; Bowles, J.H.; Montes, M.J.; Moses, W.J. Propagation of sensor noise in oceanic hyperspectral remote sensing. *Opt. Express* **2018**, *26*, A818–A831. [CrossRef] [PubMed]
6. Loghmari, M.A.; Naceur, M.S.; Boussema, M.R. A Spectral and Spatial Source Separation of Multispectral Images. *IEEE Trans. Geosci. Remote Sens.* **2006**, *44*, 3659–3673. [CrossRef]
7. Randeberg, L.L.; Kollias, N.; Baarstad, I.; Zeng, H.; Choi, B.; Løke, T.; Kaspersen, P.; Malek, R.S.; Wong, B.J.; Svaasand, L.O.; et al. Hyperspectral imaging of bruised skin. In Proceedings of the Photonic Therapeutics and Diagnostics II, San Jose, CA, USA, 21–26 January 2006.
8. Nakariyakul, S.; Casasent, D.P. Fast feature selection algorithm for poultry skin tumor detection in hyperspectral data. *J. Food Eng.* **2009**, *94*, 358–365. [CrossRef]
9. Stamatas, G.N.; Balas, C.; Kollias, N. Hyperspectral image acquisition and analysis of skin. In Proceedings of the Spectral Imaging: Instrumentation, Applications, And Analysis II, San Jose, CA, USA, 25–31 January 2003; pp. 77–82.
10. Lam, A.; Sato, I. Spectral Modeling and Relighting of Reflective-Fluorescent Scenes. In Proceedings of the 2013 IEEE Conference on Computer Vision and Pattern Recognition, Portland, OR, USA, 23–28 June 2013; pp. 1452–1459.
11. Oh, S.W.; Brown, M.S.; Pollefeys, M.; Kim, S.J. Do It Yourself Hyperspectral Imaging with Everyday Digital Cameras. In Proceedings of the 2016 IEEE Conference on Computer Vision and Pattern Recognition (CVPR), Las Vegas, NV, USA, 27–30 June 2016; pp. 2461–2469.
12. Lam, A.; Subpa-Asa, A.; Sato, I.; Okabe, T.; Sato, Y. Spectral Imaging Using Basis Lights. In Proceedings of the 24th British Machine Vision Conference 2013, Bristol, UK, 9–13 September 2013; pp. 41.1–41.11.
13. Park, J.I.; Lee, M.H.; Grossberg, M.D.; Nayar, S.K. Multispectral Imaging Using Multiplexed Illumination. In Proceedings of the 2007 IEEE 11th International Conference on Computer Vision, Rio de Janeiro, Brazil, 14–21 October 2007; pp. 1–8.
14. Chi, C.; Yoo, H.; Ben-Ezra, M. Multi-Spectral Imaging by Optimized Wide Band Illumination. *Int. J. Comput. Vis.* **2008**, *86*, 140–151. [CrossRef]
15. Han, S.; Sato, I.; Okabe, T.; Sato, Y. Fast Spectral Reflectance Recovery Using DLP Projector. In *Computer Vision. ACCV 2010*; Lecture Notes in Computer Science; Springer: Berlin/Heidelberg, Germany, 2011; pp. 323–335.
16. Zhang, W.; Song, H.; He, X.; Huang, L.; Zhang, X.; Zheng, J.; Shen, W.; Hao, X.; Liu, X. Deeply learned broadband encoding stochastic hyperspectral imaging. *Light Sci. Appl.* **2021**, *10*, 108. [CrossRef] [PubMed]
17. MacKinnon, N.; Stange, U.; Lane, P.; MacAulay, C.; Quatrevalet, M. Spectrally programmable light engine for in vitro or in vivo molecular imaging and spectroscopy. *Appl. Opt.* **2005**, *44*, 2033–2040. [CrossRef] [PubMed]
18. Chuang, C.H.; Lo, Y.L. Digital programmable light spectrum synthesis system using a digital micromirror device. *Appl. Opt.* **2006**, *45*, 8308–8314. [CrossRef] [PubMed]
19. Luo, D.; Taphanel, M.; Längle, T.; Beyerer, J. Programmable light source based on an echellogram of a supercontinuum laser. *Appl. Opt.* **2017**, *56*, 2359–2367. [CrossRef] [PubMed]
20. Hornbeck, L.J.; Rice, J.P.; Douglass, M.R.; Neira, J.E.; Kehoe, M.; Swanson, R. DMD diffraction measurements to support design of projectors for test and evaluation of multispectral and hyperspectral imaging sensors. In Proceedings of the Emerging Digital Micromirror Device Based Systems and Applications, San Jose, CA, USA, 24–29 January 2009.
21. Hirai, K.; Irie, D.; Horiuchi, T. Multi-primary image projector using programmable spectral light source. *J. Soc. Inf. Disp.* **2016**, *24*, 144–153. [CrossRef]
22. Xu, D.; Sun, G.; Zhang, G.; Zhang, Y.; Wang, L.; Liu, S.; Zhong, J.; Liang, S.; Yang, J. Design of a digital tunable stellar spectrum calibration source based on a digital micromirror device. *Measurement* **2022**, *191*, 110651. [CrossRef]
23. Zhang, Y.; Chen, X.; Wu, H.; Mo, J.; Wen, K.; Wang, S.; Gao, Y.; Wang, Y. Multiwavelength switchable fiber laser employing a DMD as tuning element and variable optical attenuator. *Opt. Laser Technol.* **2021**, *142*, 107276. [CrossRef]
24. Wang, X.; Li, Z. A spectrally tunable calibration source using Ebert-Fastie configuration. *Meas. Sci. Technol.* **2018**, *29*, 035903. [CrossRef]
25. Liang, J.; Zhang, G.; Zhang, J.; Xu, D.; Chong, W.; Sun, J. An indoor calibration light source of the transmissometers based on spatial light modulation. *Optoelectron. Lett.* **2022**, *18*, 65–71. [CrossRef]
26. Li, Z.; Wang, X.; Zheng, Y.; Li, F. Absolute detector-based spectrally tunable radiant source using digital micromirror device and supercontinuum fiber laser. *Appl. Opt.* **2017**, *56*, 5073–5079. [CrossRef] [PubMed]
27. Michael, G.; Stephen, B. CVX: Matlab Software for Disciplined Convex Programming, Version 2.0 Beta. Available online: http://cvxr.com/cvx (accessed on 15 July 2021).

28. Grant, M.C.; Boyd, S.P. Graph Implementations for Nonsmooth Convex Programs. In *Recent Advances in Learning and Control*; Lecture Notes in Control and Information Sciences; Springer: London, UK, 2008; Volume 371, pp. 95–110.
29. Fu, Y.; Zou, Y.; Zheng, Y.; Huang, H. Spectral reflectance recovery using optimal illuminations. *Opt. Express* **2019**, *27*, 30502–30516. [CrossRef] [PubMed]
30. Kokaly, R.F.; Clark, R.N.; Swayze, G.A.; Livo, K.E.; Hoefen, T.M.; Pearson, N.C.; Wise, R.A.; Benzel, W.M.; Lowers, H.A.; Driscoll, R.L.; et al. *USGS Spectral Library Version 7 Data: U.S. Geological Survey Data Release*; USGS: Reston, VA, USA, 2017. [CrossRef]

Disclaimer/Publisher's Note: The statements, opinions and data contained in all publications are solely those of the individual author(s) and contributor(s) and not of MDPI and/or the editor(s). MDPI and/or the editor(s) disclaim responsibility for any injury to people or property resulting from any ideas, methods, instructions or products referred to in the content.

sensors

MDPI

Article

Performance-Enhanced Static Modulated Fourier Transform Spectrometer with a Spectral Reconstruction

Ju Yong Cho [1], Seunghoon Lee [2] and Won Kweon Jang [1,*]

[1] Department of Aeronautic Electricity, Hanseo University, 46, Hanseo 1-ro, Seosan-si 31962, Republic of Korea
[2] Satellite Research Directorate, Korea Aerospace Research Institute, 169-82, Gwahak-ro, Daejeon 34133, Republic of Korea
* Correspondence: jwk@hanseo.ac.kr; Tel.: +82-41-660-1324

Abstract: A static modulated Fourier transform spectrometer has been noted to be a compact and fast evaluation tool for spectroscopic inspection, and many novel structures have been reported to support its performance. However, it still suffers from poor spectral resolution due to the limited sampling data points, which marks its intrinsic drawback. In this paper, we outline the enhanced performance of a static modulated Fourier transform spectrometer with a spectral reconstruction method that can compensate for the insufficient data points. An enhanced spectrum can be reconstructed by applying a linear regression method to a measured interferogram. We obtain the transfer function of a spectrometer by analyzing what interferogram can be detected with different values of parameters such as focal length of the Fourier lens, mirror displacement, and wavenumber range, instead of direct measurement of the transfer function. Additionally, the optimal experimental conditions for the narrowest spectral width are investigated. Application of the spectral reconstruction method achieves an improved spectral resolution from 74 cm^{-1} when spectral reconstruction is not applied to 8.9 cm^{-1}, and a narrowed spectral width from 414 cm^{-1} to 371 cm^{-1}, which are close to the values of the spectral reference. In conclusion, the spectral reconstruction method in a compact static modulated Fourier transform spectrometer effectively enhances its performance without any additional optic in the structure.

Keywords: Fourier transform spectrometer; static modulation; spectral reconstruction; performance enhancement; transfer function

Citation: Cho, J.Y.; Lee, S.; Jang, W.K. Performance-Enhanced Static Modulated Fourier Transform Spectrometer with a Spectral Reconstruction. *Sensors* **2023**, *23*, 2603. https://doi.org/10.3390/s23052603

Academic Editors: Qing Yu, Ran Tu, Ting Liu and Lina Li

Received: 21 December 2022
Revised: 20 February 2023
Accepted: 23 February 2023
Published: 27 February 2023

Copyright: © 2023 by the authors. Licensee MDPI, Basel, Switzerland. This article is an open access article distributed under the terms and conditions of the Creative Commons Attribution (CC BY) license (https://creativecommons.org/licenses/by/4.0/).

1. Introduction

Concerning spectrometry, the Fourier transform spectrometer is interesting because it offers a high spectral resolution in the infrared region. Unlike diffractive optics-based spectrometry, it does not have the drawback of a diffraction limit in the infrared region; instead, the spectral resolution improves as the wavelength lengthens. A commercial Fourier transform spectrometer composed of a Michelson interferometer plays a critical role in many applications such as refineries, environmental monitoring [1], medical fields [2], space exploration [3], remote sensing [4], and other scientific areas [5–12]. To obtain a complete interferogram, the optical path difference between two divided beams at the beam splitter in the Michelson interferometer needs to be altered by moving one of the two mirrors back and forth. This is called dynamic modulation. The spectral resolution depends on the maximum optical path difference. Therefore, the mirror displacement needs to be long to achieve a long maximum optical path difference. Despite many advantages of Fourier transform spectrometers, the field application of dynamic modulated commercial Fourier transform spectrometers has been limited by strict operation prerequisites. They should be operated under a nitrogen environment to obtain clear spectral information and kept under high mechanical stability to protect the data acquisition from any vibration during operation.

The demand for fast measurement and immunity to vibration has led to the development of various static modulated Fourier transform spectrometers.

A static modulated Fourier transform spectrometer produces an optical path difference that is spatially distributed, resulting in an interferogram as a function of space. Some of these structures are reported as mirror-based interferometers [7,8,10,12–14] and birefringent prism-based interferometers [5,9,11,15,16]. However, the spectral resolution and spectral range of static modulated Fourier transform spectrometers can be worse than those of dynamic modulated Fourier transform spectrometers due to the limited number of data points, depending on the number of pixels of the detector. Moreover, the maximum optical path difference is also limited by the pixel pitch of the detector, resulting in a poor spectral resolution. Though the static modulated Fourier transform spectrometers have shown superior performance in terms of faster measurement of rapidly changing spectral properties of target substances, the main drawbacks of a worse spectral resolution and limited spectral range remain as major problems to be overcome to surpass dynamic modulated Fourier transform spectrometers.

A retarder array [11] and a stepped-mirror array [12,14] were introduced to extend the maximum optical path difference. A sensor shift method was suggested to enhance the spectral resolution [9]. However, these reports are based on utilizing additional optical components, which brings complications to a spectrometer. In addition, the spectral resolution was not comparable with those of dynamic modulated Fourier transform spectrometers. A signal-padding method was suggested to enhance the spectral resolution but it is only valid in limited conditions [10].

Recently, a static modulated Fourier transform spectrometer with multiple optical switches was reported [17,18]. An optical path difference was obtained by permutating pre-implemented optical switches. However, the data points were still insufficient and limited the spectral resolution and spectral range.

This drawback could be solved by employing a reconstruction method. In this study, a spectral reconstruction method was applied in a static modulated Fourier transform spectrometer composed of a modified Sagnac interferometer. After applying the reconstruction method to our static modulated Fourier transform spectrometer, the spectral resolution could be improved to a level comparable with that of a commercial Fourier transform spectrometer.

Applying the reconstruction method requires the transfer function of a spectrometer, which is directly obtainable through an optical measurement of each wavenumber, performed by a monochromator. However, the spectral resolution depends on the performance of the monochromator. To overcome this dependency, we calculated the transfer function based on the relationship between an interferogram and a spectrum. The conditions for better spectral characteristics were investigated in terms of the ratio of a sampling frequency to the wavenumber of a source, mirror displacement, and the maximum optical path difference.

2. Methodology

A static modulated Fourier transform spectrometer was adopted that was composed of a modified Sagnac interferometer, as shown in Figure 1. SD is a source driver with a temperature controller, S is a radiation source, FL is a Fourier lens, and D is a one-dimensional array detector. A modified Sagnac interferometer consists of two mirrors, M1 and M2, and a beam splitter, BS. The two mirrors, M1 and M2, are tilted by 67.5° from the optical axis, and the beam splitter, BS, is tilted by 45° from the optical axis [8,13]. Radiation from the source splits into two beams at the BS, then they travel along each optical axis. If M2 was placed in a symmetrical position (the dashed line) where the displacement of M1 and M2 from the BS was equal, the two beams would have the same optical path length. In this case, an interferogram would not appear at the detector, D. However, if M2 was displaced by a distance, a, from the symmetrical position, the transmitted and reflected beams would be separated by l. An interferogram would be observed in an overlapped area of the two beams.

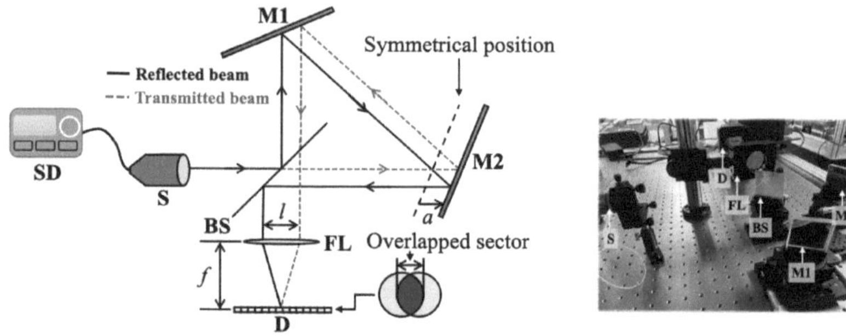

Figure 1. A Fourier transform spectrometer composed of a modified Sagnac interferometer.

The separation distance between a transmitted and a reflected beam is given as follows:

$$l = \sqrt{2}a\,, \tag{1}$$

where l is the separation between a transmitted and a reflected beam, and a is the displacement of mirror M2 from the symmetrical point. Since a larger l provides a longer maximum optical path difference, l should be as long as possible for a better spectral resolution. Since the incident beam size to the detector is defined by the pupil size of the interferometer, and a larger l induces a narrower overlapped area width, the obtainable optical path difference must be limited. The maximum optical path difference and the effective sampling data points are limited by the separation of two beams, l.

In a static modulated Fourier transform spectrometer, a larger maximum optical path difference leads to an insufficient number of sampling data points, resulting in a poor spectral resolution and limited spectral range. Herein, we suggest using the spectral reconstruction method to solve this problem. It is assumed that an interferogram and a spectrum are linearly related as follows [18]:

$$\mathbf{I} = \mathbf{T}\mathbf{B}, \tag{2}$$

where **I** is an interferogram, **T** is the transfer function, and **B** is spectral information. **I** and **B** are column vectors with M and N elements, respectively. **T** is a matrix with $M \times N$ elements. A detector is placed in a focal position to record a spatially distributed interferogram. An interferogram is expressed as the sum of cosine waves multiplied by spectral information associated with wavenumbers. Moreover, a cosine wave depends on parameters of a spectrometer such as the focal length of a lens, separation distance, and pixel pitch. We can assume that the transfer function of a spectrometer is interpreted as a cosine wave. Hence, Equation (2) is rewritten as follows:

$$\begin{pmatrix} I(1y_{\min}) \\ \vdots \\ I(ny_{\min}) \end{pmatrix} = \begin{pmatrix} \cos(2\pi\overline{\nu}_{\min}1y_{\min}\frac{l}{f}) & \cdots & \cos(2\pi\overline{\nu}_{\max}1y_{\min}\frac{l}{f}) \\ \vdots & \ddots & \vdots \\ \cos(2\pi\overline{\nu}_{\min}ny_{\min}\frac{l}{f}) & \cdots & \cos(2\pi\overline{\nu}_{\max}ny_{\min}\frac{l}{f}) \end{pmatrix} \begin{pmatrix} B(\overline{\nu}_{\min}) \\ \vdots \\ B(\overline{\nu}_{\max}) \end{pmatrix}, \tag{3}$$

where $\overline{\nu}_{\min}$ and $\overline{\nu}_{\max}$ are the minimum and the maximum wavenumbers, respectively. n is an integer representing the number of data points in the overlapped area, and $y_{\min}$ is the pixel pitch. Spectral reconstruction can be performed when an inverse matrix of the transfer function of a spectrometer exists.

Figure 2 shows the procedures used when applying the spectral reconstruction method. The transfer function was calculated with optical parameters such as lens focal length, pixel pitch, and mirror displacement. There was no phase information in the transfer function. A recorded interferogram contains phase errors due to thermal and electrical noise, sampling errors, and refraction, resulting in a shifted position of the zero path difference. Before the spectral reconstruction, a phase error should be corrected. We applied Mertz phase correction to the interferogram because it is commonly used in Fourier transform spectroscopy. Figure 2a shows the procedure for obtaining parameters for the transfer function of the spectrometer and phase information. The parameters and the phase information could be obtained from the spectrum. Since pixel pitch, focal length, and mirror displacement depend on the experimental setup, they were easily measured, but the focal length was slightly mispositioned due to refraction. To obtain an effective focal length, the spectrum must be obtained from the Fourier transform of an interferogram. The spectrum also contains a phase error that can be expressed as an angle along a wavenumber. The phase error can be calculated by the ratio of the imaginary part to the real part of the spectrum. Multiplying the calculated angle by the obtained spectrum compensates for the phase error. Since the center wavenumber of the radiation source was known, calibration could be performed and then the focal length calculated. Figure 2b shows the procedure for spectral reconstruction. An interferogram without any phase error could be obtained by inverse Fourier transforming a phase-corrected spectrum. The transfer function was calculated by Equation (3). Then the spectrum was reconstructed by calculating the residual sum of squares after rearranging Equation (2), given as follows [19,20]:

$$\mathrm{RSS}(\mathbf{B}) = \left(\mathbf{T}^{\mathrm{T}}\mathbf{B} - \mathbf{I}\right)^2, \tag{4}$$

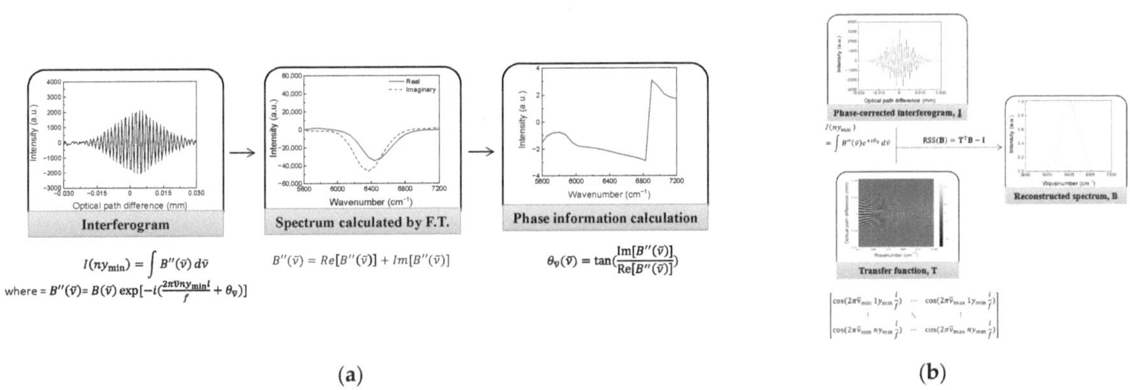

Figure 2. Spectral reconstruction procedure. (**a**) Procedure used to obtain phase information and parameters of the transfer function. (**b**) Reconstruction procedure.

A general standard regression was performed. However, the reconstructed spectrum still contained problems of overfitting and bias. For better reconstruction of the spectrum, an elastic-net method was employed [17,18]. This method takes advantage of both ridge and lasso regression so that overfitting and bias problems can be avoided.

A reconstruction method is valid when an interferogram is properly sampled according to the Nyquist sampling theorem and Shannon sampling criterion. Sampling frequency in a static modulated Fourier transform spectrometer is inversely proportional to the optical path difference corresponding to a single pixel that is given as follows:

$$\overline{\nu}_{\mathrm{s}} = \frac{1}{\Delta_{\min}} = \frac{f}{\sqrt{2}a y_{\min}}, \tag{5}$$

where $\overline{\nu}_s$ is the sampling frequency of a static modulated Fourier transform spectrometer, and Δ_{min} is the optical path difference corresponding to a single pixel. To avoid aliasing, the sampling frequency must be at least twice the maximum wavenumber of the source.

$$\varphi = \frac{\overline{\nu}_s}{\overline{\nu}_{max}} \geq 2, \tag{6}$$

where φ is the ratio of the sampling frequency to the maximum wavenumber of the source, and $\overline{\nu}_{max}$ is the maximum wavenumber of the source. The minimum wavenumber $\overline{\nu}_{min}$ is 0. Associated with Equation (6), the spectral resolution depending on the width of the overlapped area created by mirror displacement is given as follows:

$$\overline{\nu}_R = \frac{2(\overline{\nu}_{max} - \overline{\nu}_{min})}{N_{overlap}} = \frac{\overline{\nu}_s}{N_{overlap}} = \frac{f}{\sqrt{2}a N_{overlap} y_{min}} = \frac{f}{(\frac{2r}{\sqrt{2}a} - 1)2a^2}, \tag{7}$$

where $\overline{\nu}_R$ is the spectral resolution, $N_{overlap}$ is the number of pixels in the overlapped area, and r is the incident beam radius to the detector. $N_{overlap}$ is given by $\frac{A}{y_{min}}$, and A is the width of an overlapped area given by $2r$, which reduces with a.

3. Results and Discussion

The experiments were performed in a static modulated Fourier transform spectrometer composed of a modified Sagnac interferometer. In the experimental setup described in Figure 1, two square mirrors of 2 inches in width are silver-coated. A beam splitter with a transmittance of 50% in the wavenumber range from 3920 to 11,110 cm^{-1} is used. An LED is employed as the incident light source. The center wavenumber and the maximum optical power are 6451 cm^{-1} and 2 mW at a current of 0.55 A, respectively. A parabolic mirror is coupled with the LED to generate a parallel beam. In our study, a thermo-electric module was attached to the LED to control the temperature to 25 °C. At the width of the e^{-2} point from the highest intensity, the beam diameter was measured to be 8.23 mm. Pixel pitch and the number of pixels of the detector were 25 μm and 512, respectively. A 16-bit analog-to-digital converter was employed and the exposure time of the detector set to 4 ms. To compare the experimental data and the reconstructed results with the true spectral information, the reference spectrum was measured by a monochromator.

For successful spectral reconstruction, an interferogram should be recorded with the least distortion possible. Moreover, the ratio of the sampling frequency to the maximum wavenumber of the source needs to be considered. Figure 3 shows the optical path difference corresponding to a single pixel and the ratio of the sampling frequency to the maximum wavenumber φ along mirror displacement a. The focal length of the Fourier lens was 300 mm. As mirror displacement a increases, the optical path difference Δ_{min} linearly. The ratio φ decreases along a and falls below two when a is greater than 6.3 mm. When the ratio φ is below two, an interferogram is insufficiently sampled, leading to aliasing in the spectrum. a may be no longer than 6.3 mm to avoid aliasing.

As a increases, the distance l increases proportionally. As a result, the length of an overlapped area decreases, leading to a decrease in sampling data points. Figure 4 shows the variation of data points in an overlapped area and the spectral resolution along mirror displacement. As the mirror displacement increases, the optical path difference corresponding to single pixel Δ_{min} increases so that the maximum optical path difference increases as the spectral resolution is enhanced. However, as the length of an overlapped area decreases, the maximum optical path difference begins to decrease, resulting in deterioration of spectral resolution. In this experiment, the longest maximum optical path difference was obtained at a mirror displacement of 5 mm. The best spectral resolution was expected to be 74 cm^{-1}. Moreover, since φ was close to two, it was highly possible to obtain an interferogram with the lack of sampling data points.

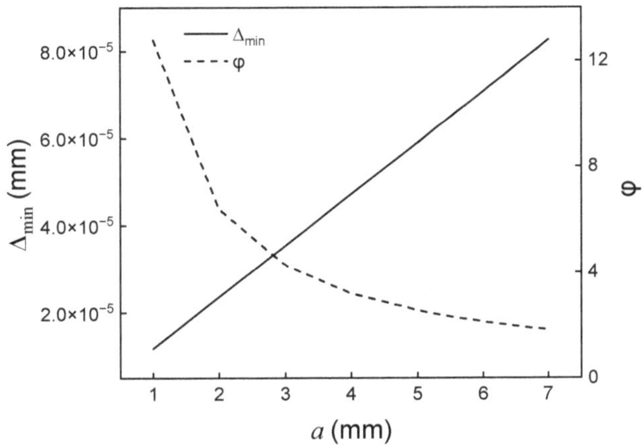

Figure 3. Variation of optical path difference of a single pixel and the ratio of sampling frequency to maximum wavenumber of source radiation along mirror displacement.

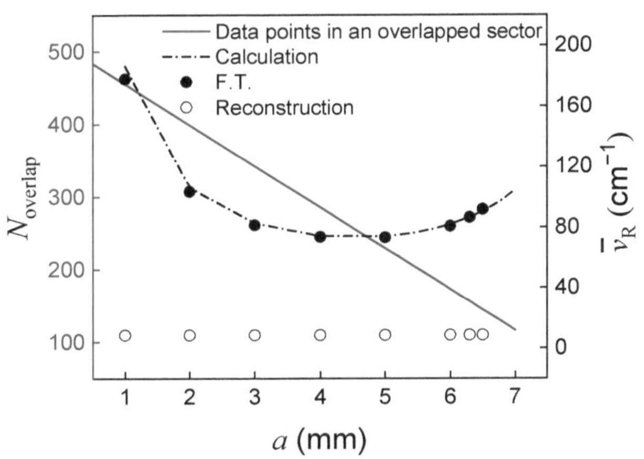

Figure 4. Variation of sampling data points in an overlapped area and spectral resolution along mirror displacement.

As shown in Figure 4, the static modulated Fourier transform spectrometer with a modified Sagnac interferometer has a poor spectral resolution, similar to other reports. However, through a reconstruction method, the spectral resolution of the system can be significantly improved. When the reconstruction method is applied to a measured spectrum, the spectral resolution becomes dependent on the number of wavenumber components in the transfer function. In this experiment, the wavenumber range of the transfer function is from 3584 to 12,472 cm^{-1}, which is coincident with that of the detector, and the number of components is 1024, resulting in a spectral resolution of 8.7 cm^{-1} at any mirror displacement. The number of components should be 1024 at most because the spectrum cannot successfully be reconstructed when the number of components is greater than 1024. This is because the complexity inducing the bias and overfitting problem grows as the number of components increases. In addition, to obtain the narrow spectral width, an interferogram needs to be sampled for a long maximum optical path difference according to the Shannon criterion. The optimal mirror displacement should be decided by careful consideration. Figure 5 shows interferograms and Fourier transformed spectra along mirror displacement. Figure 5a is an interferogram obtained at a mirror displacement of 1 mm. The number of data points in the overlapped area is 457 and the maximum optical path difference is 5.6×10^{-2} mm. The maximum optical path difference Δ_{min} is determine by multiplying $N_{overlap}$. Figure 5b shows the spectrum at a mirror displacement of 1 mm. The monochromator (black solid line) is the reference spectrum. F.T. (black dashed line) is a Fourier transformed spectrum. When Fourier transform is performed, the spectral width is 596 cm^{-1} and the spectral resolution is 178 cm^{-1}. However, after the reconstruction method is applied, the spectral width reduces to 468 cm^{-1} and the spectral resolution improves to 8.9 cm^{-1}. The spectral resolution could be significantly improved but the spectral width was still wider compared to that of the reference spectrum due to the small value of the maximum optical path difference. We could obtain spectra along mirror displacement by performing Fourier transform and reconstruction to find the condition under which the spectral width became the narrowest. The best spectral resolution of 73 cm^{-1} was obtained at a mirror displacement of 5 mm and a spectral width of 428 cm^{-1}. However, in this experiment, the narrowest spectral width was obtained at mirror displacement of 4 mm. The difference might have arisen from sampling errors that broadened the spectral width, with the effects of the errors dominant when φ became closer to the condition in Equation (7). At the mirror displacement of 4 mm, φ was 3.2, which was greater than the 2.6 obtained for a mirror displacement of 5 mm. Therefore, the optimal mirror displacement should be adjusted by considering the optimal value of φ. Figure 5c is an interferogram obtained at a mirror displacement of 4 mm. The number of data points in the overlapped area is 288, and the maximum optical path difference is 13.6×10^{-2} mm. Figure 5d shows the spectrum at a mirror displacement of 4 mm. When Fourier transform is performed, the spectral width and spectral resolution are 414 cm^{-1} and 74 cm^{-1}, respectively. When the reconstruction is performed, the spectral width reduces to 371 cm^{-1} and the spectral resolution improves to 8.9 cm^{-1}. The reconstructed spectrum shows much closer characteristics to the reference spectrum.

Figure 5e is an interferogram obtained at a mirror displacement of 6 mm. Though mirror displacement increases, the spectral width becomes broader than that at a mirror displacement of 4 mm. The number of data points in the overlapped area is 175, and a maximum optical path difference is 12.3×10^{-2} mm. Compared to Figure 5c, the number of data points decreases and the maximum optical path difference shortens. Figure 5f shows a spectrum at a mirror displacement of 6 mm. When Fourier transform is performed, the spectral width is 450 cm^{-1} and the spectral resolution is 81 cm^{-1}. When the reconstruction is performed, the spectral width reduces to 407 cm^{-1} and the spectral resolution improves to 8.9 cm^{-1}.

Figure 5g shows a spectral width and φ along mirror displacement. A black solid rectangle and a hollow rectangle show the spectral width obtained by performing Fourier transform and reconstruction, respectively, along mirror displacement. The hollow circle is φ. The spectral width of the spectrum obtained by the reconstruction method is narrower than that obtained by Fourier transform. Moreover, the reconstructed spectrum is more accurate when the maximum optical path difference is long in an obtained interferogram. The maximum optical path difference is found for the longest mirror displacement of 5 mm; however, the spectral width is not the narrowest. It should be considered that, for the best performance of the spectrometer, not only the maximum optical path difference but also the sampling frequency should be considered. After correction of sampling errors, the optimal φ value of 3.2 was obtained at a mirror displacement of 4 mm.

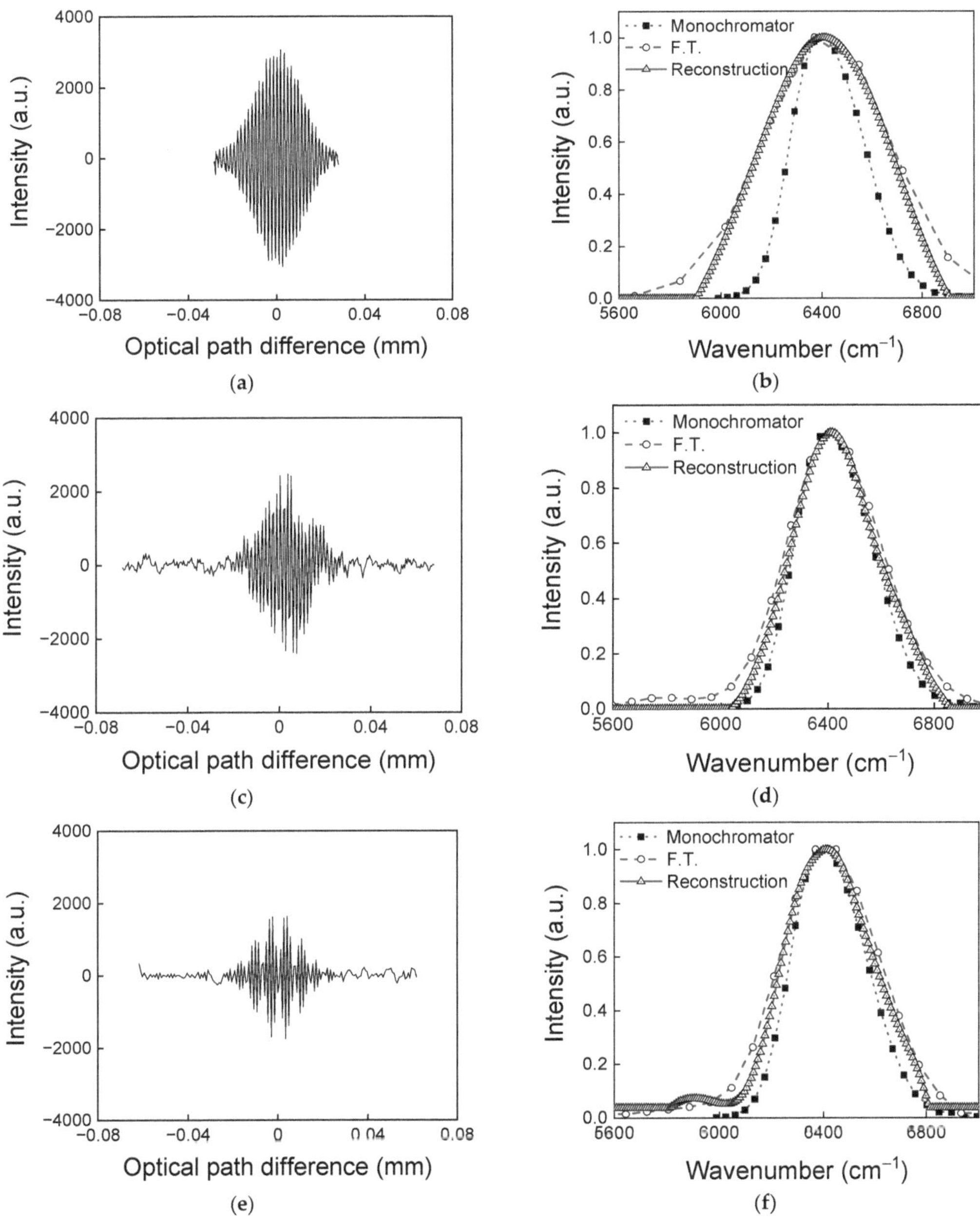

Figure 5. *Cont.*

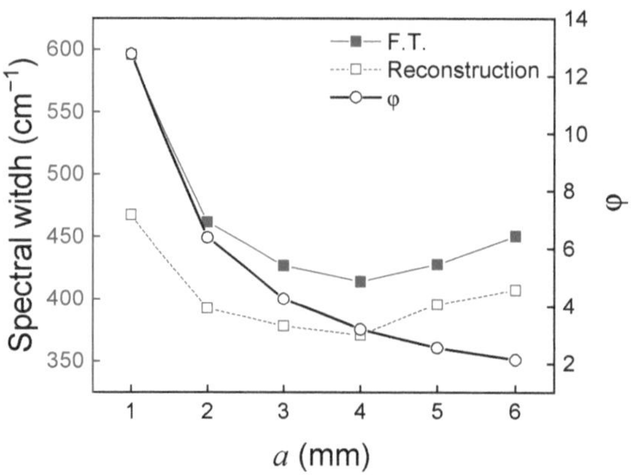

(g)

Figure 5. An interferogram and spectra along with mirror displacement. (**a**) An interferogram obtained at a mirror displacement of 1 mm, (**b**) spectra corresponding to (**a**), (**c**) an interferogram obtained at a mirror displacement of 4 mm, (**d**) spectra corresponding to (**c**), (**e**) an interferogram obtained at a mirror displacement of 6 mm, (**f**) spectra corresponding to (**e**), and (**g**) the spectral width associated with the method and φ, along with mirror displacement..

4. Conclusions

A static modulated Fourier transform spectrometer is composed of a modified Sagnac interferometer. Due to insufficient sampling data points in the overlapped area, the maximum wavenumber is limited, resulting in distortion such as spectral folding. A modified Sagnac interferometer is readily optimized to overcome the distortion, thus improving on other static modulated interferometers such as a single mirror-based interferometer and a birefringent prism-based interferometer.

In this investigation, from the relationship between the interferogram and the spectrum, the transfer function of a spectrometer could be expressed as a cosine function. Important parameters of the transfer function were focal length of the lens, mirror displacement, and pixel pitch. In spectral reconstruction, the spectral resolution depends on the number of wavenumber components of the transfer function. The spectral resolution could be improved to 8.9 cm^{-1} at any mirror displacement when the number of wavenumber components in a transfer function was 1024 in the range from 3584 to 12,472 cm^{-1}. This marked a significant improvement compared to the value of 74.0 cm^{-1} obtained by performing Fourier transform. To obtain better spectral characteristics, a larger maximum optical path difference is required. Thus, the ratio of the sampling frequency of a spectrometer to the maximum wavenumber should be considered. In our experiment, when φ became closer to two, the errors dominantly affected the spectrum, and as a result, the spectral width was broadened. The best condition for reconstruction was at a mirror displacement of 4 mm, where φ became 3.2. Before applying the reconstruction method suggested in this study, the obtained Fourier transformed spectral width was 414 cm^{-1}; then, after the reconstruction method was applied, the spectral width was narrowed to 371 cm^{-1}, which was very close to the reference spectral value. The spectral reconstruction was successful in improving the spectral characteristics in a static modulated Fourier transform spectrometer. The spectral resolution could be greatly improved without requiring any additional optical component to increase the maximum optical path difference. However, the accuracy of the spectrum is still dependent on the maximum optical path difference and sampling frequency. An approach to solving this drawback needs to be discussed. The reconstruction method that we suggest in this paper allows a static modulated Fourier transform spectrometer to remain compact with high performance.

Author Contributions: Conceptualization, W.K.J.; methodology, W.K.J., S.L. and J.Y.C.; software, W.K.J. and J.Y.C.; validation, W.K.J. and J.Y.C.; formal analysis, W.K.J., S.L. and J.Y.C.; investigation, W.K.J., S.L. and J.Y.C.; resources, W.K.J. and J.Y.C.; data curation, W.K.J. and J.Y.C.; writing—original draft preparation, J.Y.C. and W.K.J.; writing—review and editing, J.Y.C. and W.K.J.; visualization, J.Y.C. and W.K.J.; supervision, W.K.J.; project administration, W.K.J.; funding acquisition, W.K.J. All authors have read and agreed to the published version of the manuscript.

Funding: This research was supported by the Basic Science Research Program through the National Research Foundation of Korea (NRF) funded by the Ministry of Education (NRF-2021R1I1A3052065).

Institutional Review Board Statement: Not applicable.

Informed Consent Statement: Not applicable.

Data Availability Statement: Not applicable.

Conflicts of Interest: The authors declare no conflict of interest.

References

1. Pust, N.J.; Shaw, J.A. Wavelength dependence of the degree of polarization in cloud-free skies: Simulations of real environments. *Opt. Express* **2012**, *20*, 15559–15568. [CrossRef] [PubMed]
2. McIntosh, L.M.; Summers, R.; Jackson, M.; Mantsch, H.H.; Mansfield, J.R.; Howlett, M. Towards Non-Invasive Screening of Skin Lesions by Near-Infrared Spectroscopy. *J. Investig. Dermatol.* **2001**, *116*, 175–181. [CrossRef] [PubMed]
3. Moran, T.G.; Davila, J.M. Three-dimensional polarimetric imaging of coronal mass ejections. *Science* **2004**, *305*, 66–70. [CrossRef] [PubMed]
4. Lacan, A.; Bréon, F.-M.; Rosak, A.; Brachet, F.; Roucayrol, L.; Etcheto, P.; Casteras, C.; Salaün, Y. A static Fourier transform spectrometer for atmospheric sounding: Concept and experimental implementation. *Opt. Express* **2010**, *18*, 8311–8331. [CrossRef] [PubMed]
5. Groner, W.; Winkelman, J.W.; Harris, A.G.; Ince, C.; Bouma, G.J.; Messmer, K.; Nadeau, R.G. Orthogonal polarization spectral imaging: A new method for study of the microcirculation. *Nat. Med.* **1999**, *5*, 1209–1212. [CrossRef] [PubMed]
6. Furukawa, H. Real-time multi-channel Fourier transform spectroscopy and its application to non-invasive blood fat measurement. *Sens. Bio-Sens. Res.* **2016**, *8*, 55–58. [CrossRef]
7. Köhler, M.H.; Schardt, M.; Rauscher, M.S.; Koch, A.W. Gas Measurement Using Static Fourier Transform Infrared Spectrometers. *Sensors* **2017**, *17*, 2612. [CrossRef] [PubMed]
8. Okamoto, T.; Kawata, S.; Minami, S. Fourier transform spectrometer with a self-scanning photodiode array. *Appl. Opt.* **1984**, *23*, 269–273. [CrossRef] [PubMed]
9. Watanabe, A.; Furukawa, H. High-resolution and high-throughput multichannel Fourier transform spectrometer with two-dimensional interferogram warping compensation. *Opt. Commun.* **2018**, *413*, 8–13. [CrossRef]
10. Cho, J.Y.; Lee, S.; Jang, W.K. Improvement of spectral resolution by signal padding method in the spatially modulated Fourier transform spectrometer based on a Sagnac interferometer. *Appl. Opt.* **2019**, *58*, 6755–6761. [CrossRef] [PubMed]
11. Li, J.; Qu, C.; Wu, H.; Qi, C. Spectral resolution enhanced static Fourier transform spectrometer based on a birefringent retarder array. *Opt. Express* **2019**, *27*, 15505–15517. [CrossRef] [PubMed]
12. Köhler, M.H.; Schardt, M.; Müller, M.; Kienle, P.; Wang, K.; Dong, X.; Giebeler, C.; Wiesent, B.R.; Jakobi, M.; Koch, A.W. Static Fourier transform mid-infrared spectrometer with increased spectral resolution using a stepped mirror. *OSA Contin.* **2020**, *3*, 2134–2142. [CrossRef]
13. Cho, J.Y.; Lee, S.; Kim, H.; Jang, W.K. Spectral Reconstruction for High Spectral Resolution in a Static Modulated Fourier-transform Spectrometer. *Curr. Opt. Photonics* **2022**, *6*, 244–251.
14. Chen, Y.; Lv, J.; Yue, W.; Zhao, Y.; Qin, Y.; Tao, J.; Chen, C.; Wang, W.; Liang, J. A Snapshot Infrared Imaging Fourier Transform Spectrometer for Dynamic Target Detection. *Remote Sens.* **2022**, *14*, 1543. [CrossRef]
15. Cho, J.Y.; Lee, S.; Jang, W.K. Dependent factors for improving the spectral resolution and signal-to-noise ratio in a spatially modulated Fourier transform spectrometer comprised of a Wollaston prism. *Opt. Eng.* **2021**, *60*, 094109. [CrossRef]
16. Yan, T.; Zhang, C.; Wang, Y.; Chen, Z. The narrowband full polarization spectra reconstruction for static channeled spectropolarimetry. *Optik* **2021**, *247*, 167852. [CrossRef]
17. Kita, D.M.; Miranda, B.; Favela, D.; Bono, D.; Michon, J.; Lin, H.; Hu, T.; Hu, J. High-performance and scalable on-chip digital Fourier transform spectroscopy. *Nat. Commun.* **2018**, *9*, 4405. [CrossRef] [PubMed]
18. Du, J.; Zhang, H.; Wang, X.; Xu, W.; Lu, L.; Chen, J.; Zhou, L. High-resolution on-chip Fourier transform spectrometer based on cascaded optical switches. *Opt. Lett.* **2022**, *47*, 218–221. [CrossRef] [PubMed]

19. Zhang, Z.; Lai, Z.; Xu, Y.; Shao, L.; Wu, J.; Xie, G.-S. Discriminative Elastic-Net Regularized Linear Regression. *IEEE Trans. Image Process.* **2017**, *26*, 1466–1481. [CrossRef] [PubMed]
20. Zou, H.; Hastie, T. Regularization and variable selection via the elastic net. *J. R. Stat. Soc. B Stat. Methodol.* **2005**, *67*, 301–320. [CrossRef]

Disclaimer/Publisher's Note: The statements, opinions and data contained in all publications are solely those of the individual author(s) and contributor(s) and not of MDPI and/or the editor(s). MDPI and/or the editor(s) disclaim responsibility for any injury to people or property resulting from any ideas, methods, instructions or products referred to in the content.

sensors

MDPI

Article

Design and Research of Chromatic Confocal System for Parallel Non-Coaxial Illumination Based on Optical Fiber Bundle

Yali Zhang [1], Qing Yu [1,2,*], Chong Wang [1], Yaozu Zhang [1], Fang Cheng [1], Yin Wang [1,2], Tianliang Lin [1,2], Ting Liu [1,2] and Lin Xi [3,*]

1 College of Mechanical Engineering and Automation, Huaqiao University, Xiamen 361021, China
2 Fujian Key Laboratory of Green Intelligent Drive and Transmission for Mobile Machinery, Huaqiao University, Xiamen 361021, China
3 School of Mechanical Engineering, Anhui Polytechnic University, Wuhu 241000, China
* Correspondence: yuqing@hqu.edu.cn (Q.Y.); eve_q@ahpu.edu.cn (L.X.)

Abstract: Conventional chromatic confocal systems are mostly single-point coaxial illumination systems with a low signal-to-noise ratio, light energy utility and measurement efficiency. To overcome the above shortcomings, we propose a parallel non-coaxial-illumination chromatic-confocal-measurement system based on an optical fiber bundle. Based on the existing single-point non-coaxial-illumination system, the optical fiber bundle is used as the optical beam splitter to achieve parallel measurements. Thus, the system can yield measurements through line scanning, which greatly improves measurement efficiency. To verify the measurement performance of the system, based on the calibration experiment, the system realizes the measurement of the height of the step, the thickness of the transparent specimen and the reconstruction of the three-dimensional topography of the surface of the step and coin. The experimental results show that the measuring range of the system is 200 μm. The measurement accurcy can reach micron level, and the system can realize a good three-dimensional topography reconstruction effect.

Keywords: chromatic confocal system; optical fiber bundle; non-coaxial illumination; transparent specimen; three-dimensional topography reconstruction

Citation: Zhang, Y.; Yu, Q.; Wang, C.; Zhang, Y.; Cheng, F.; Wang, Y.; Lin, T.; Liu, T.; Xi, L. Design and Research of Chromatic Confocal System for Parallel Non-Coaxial Illumination Based on Optical Fiber Bundle. *Sensors* **2022**, *22*, 9596. https://doi.org/10.3390/s22249596

Academic Editor: Lei Huang

Received: 9 November 2022
Accepted: 4 December 2022
Published: 7 December 2022

Publisher's Note: MDPI stays neutral with regard to jurisdictional claims in published maps and institutional affiliations.

Copyright: © 2022 by the authors. Licensee MDPI, Basel, Switzerland. This article is an open access article distributed under the terms and conditions of the Creative Commons Attribution (CC BY) license (https://creativecommons.org/licenses/by/4.0/).

1. Introduction

With the development of science and technology, three-dimensional topography reconstruction plays an important role in production and life. Chromatic confocal technology is derived from traditional laser confocal technology, which is one of the optical measurement techniques widely used at present. It realizes measurement by using the imaging characteristics of axial dispersion. Additionally, compared with other optical measurement techniques, such as the white-light interferometry [1,2], laser triangle method [3,4], grating projection method [5,6], and laser confocal method [7,8], chromatic confocal technology has no axial scanning, and its measuring efficiency and precision are higher. The application fields of chromatic confocal technology mainly include displacement and thickness measurements [9,10], flatness and roughness tests [11,12], defect and flaw detection [13,14], three-dimensional topography reconstruction [15,16], etc.

In recent years, on the one hand, the research of parallel chromatic confocal technology, which is derived from single-point chromatic confocal technology, is gradually emerging. Similar to laser confocal technology, chromatic confocal technology also uses light-beam splitters [17] to realize parallel measurements. Existing light-beam splitters mainly include the Nipkow turntable [18], micropinhole array [19], microlens array [20], optical fiber bundles [21], the digital micromirror device (DMD) [22]. They all turn single-light spots into multi-light spots to reduce scanning time and improve measurement speed. Among them, the microlens array, DMD and optical fiber bundles have been gradually applied to chromatic confocal parallel-measurement techniques. In 1996, Tiziani H [23] realized multipoint measurement using a microlens array and applied this technology to macro- and microsurface measurements and defect analysis in 2000 [24]. In 2013, Hillenbrand [25] designed a three-point chromatic confocal distance sensor to achieve parallel measurements by two-stage spectral multiplexing. Then, in 2015, the authors of [26] proposed two kinds of chromatic confocal matrix sensors for a three-dimensional-object snapshot collection, which can perform spectral evaluations on all transverse channels at the same time. In 2021, Hao Hu [27] designed a line-scanning chromatic confocal optical path using the slit, which can achieve accurate measurements of three-dimensional topography and the thickness of highly reflective materials. In 2020 and 2021, Yu Qing [28] used optical fiber bundles and the DMD as optical beam splitters to achieve full-field three-dimensional topography measurements of the coaxial-illumination chromatic confocal system.

On the other hand, with the gradual expansion of application fields, some coaxial-illumination chromatic confocal systems can no longer meet the measurement requirements of specific specimens. Therefore, a non-coaxial-illumination system has gradually emerged in recent years. Non-coaxial illumination occurs when the reflected light and the camera are not on the same axis. In contrast, coaxial illumination occurs when both are on the same axis. The optical path structure of the chromatic confocal system mentioned above is mainly the vertical optical path, which is also called coaxial illumination. With the development of research, chromatic confocal research about non-coaxial-illumination structures is gradually emerging. In 2020, G. Berkovic [29] designed a commercial chromatic-confocal-displacement sensor suitable for the detection of inclined-incident targets. The sensor is fitted with collimation mirrors and reverse reflectors to improve its performance. In 2021, Yu Qing [30] proposed a non-coaxial-illumination-light-path structure to measure the thickness of transparent specimens with micron accuracy.

Based on previous research about parallel chromatic confocal measurements and non-coaxial-illumination structures, the two methods have been combined to improve the measurement efficiency, signal-to-noise ratio and light energy utilization of the system in our paper. Through the usage of an optical fiber bundle with the line-scanning method, the vertical and lateral information of the measured object was captured by the system. Then, the system could reconstruct the three-dimensional topography of the measured surface by the related information.

2. Principles of Parallel Chromatic Confocal System with Non-Coaxial Illumination

2.1. Principle of Dispersion

The core principle of the chromatic confocal technique is the spectral coding technique, which focuses the light of different wavelengths onto different axial positions through the dispersion system, as shown in Figure 1. The value of λ_1, λ_2 and λ_3 decrease in turn.

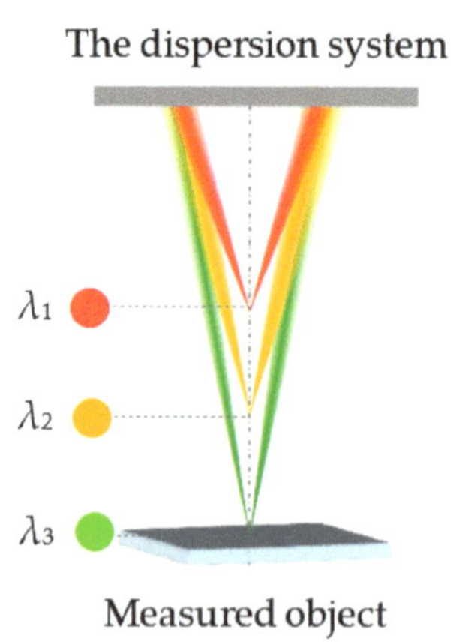

Figure 1. Schematic of the spectral coding technique.

According to the principle of axial dispersion, polychromatic light is dispersed in the direction of the optical axis by the dispersion system. Light of different wavelengths focuses on different axial positions. The wavelength distribution can be expressed as a function of $\lambda(z)$, where z is the axial position and the output of the light intensity in object space is:

$$I(z) = [\frac{\sin(\frac{u}{4})}{\frac{u}{4}}]^4 \quad (1)$$

In this formula, u is the normalized axis coordinate. The normalized axis coordinate u can be expressed as:

$$u = \frac{2\pi}{\lambda(z)}\Delta z \frac{a^2}{f^2} \quad (2)$$

As shown in Equation (2), λ is the wavelength of the incident light, a is the exit pupil radius of the imaging lens and f is the focal length of the lens. Δz is the defocus amount, which is determined by the axial-light-intensity response function of the chromatic-confocal-measurement system. Theoretically, there are a series of intensity peaks on the CCM distribution curve according to $\lambda(z)$. This "multi-peak" feature allows CCM systems to perform axial position measurements without mechanical scanning. Thus, the axial position can be measured by decoding the wavelength information of the reflected light.

2.2. Principle of Single-Point Chromatic Confocal System with Non-Coaxial Illumination

Firstly, different optical path structures are shown in Figure 2. The light from the light source travels to the beam splitter, and the light that arrives at the object is reflected from the surface to the detector. When the incident light arrives at the surface at a vertical angle, the optical path structure is called coaxial illumination, as shown in Figure 2a. On the other hand, when the incident light arrives at the surface at an inclined angle, the optical path structure is called non-coaxial illumination, as shown in Figure 2b.

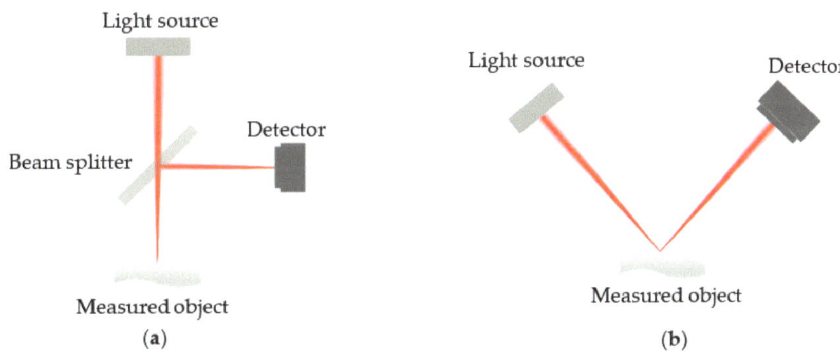

Figure 2. Schematic of the different optical path structures: (**a**) Coaxial illumination; (**b**) Non-coaxial illumination.

Secondly, the schematic of the single-point chromatic confocal system with non-coaxial illumination is shown in Figure 3, where θ represents the angle of the incident light. The light beams with different wavelengths emitted from the white-light source are distributed along the optical axis by the dispersive tube lens and are then focused on different axial locations. In this process, only the light beams focused on the surface of the sample can be reflected and focused by the objective. Finally, the light spots with corresponding color information are captured by the imaging surface of the color camera. Combined with the different color information corresponding to the different heights of the object surface, the three-dimensional topography information can be obtained by the two-dimensional point-scanning method.

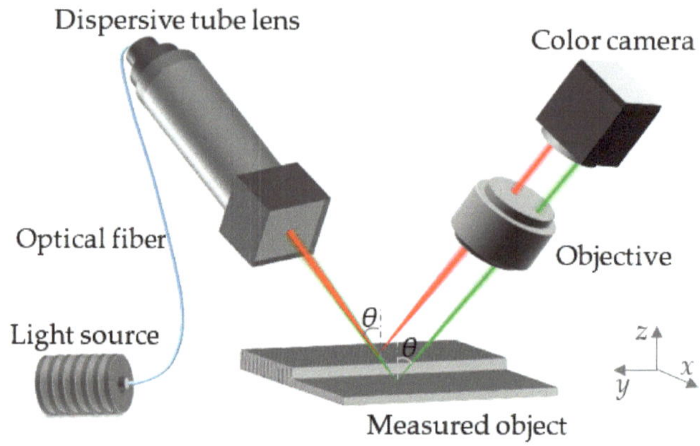

Figure 3. Schematic of the single-point chromatic confocal system with non-coaxial illumination.

The point spread function (PSF) describes the imaging response of the point light source, and it is often used to measure the resolution of reconstructed images.

When the point light source in the single-point system arrives at the surface of the measured object, the PSF function h_0 of the system is:

$$\begin{aligned} &h0(x,y,z) \\ &= \int \delta(x,y,z-\Delta z)\cdot h_1(x,y,z)dz \\ &= \delta(x,y)h_1(x,y,\Delta z) \end{aligned} \tag{3}$$

In Equation (3), Δz represents the longitudinal defocus quantity. Then, the light spot that arrives at the surface is reflected by the measured surface and arrives at the imaging plane. So, the PSF function of the whole single-point system is:

$$\begin{aligned} &h_{conf}(x,y,z) = \delta(x,y)h1(x,y,\Delta z) * h2(x,y,\Delta z) \\ &= \int\int \delta(x',y')h1(x',y',\Delta z)\cdot h2(x-x',y-y',\Delta z)dx'dy' \\ &= h1(0,0,\Delta z)h2(x,y,\Delta z) \end{aligned} \tag{4}$$

2.3. Principle of the Optical Fiber Bundle

The optical fiber bundle, also known as the optical fiber image-transmission bundle, is composed of several optical fibers. It has several array arrangements, including circular, square and linear arrays. In this paper, the light emitted from the optical fiber bundle is linear, as shown in Figure 4. In Figure 4, the number of the optical fibers is represented by M, and the center distance between the adjacent light spots is represented by *d*.

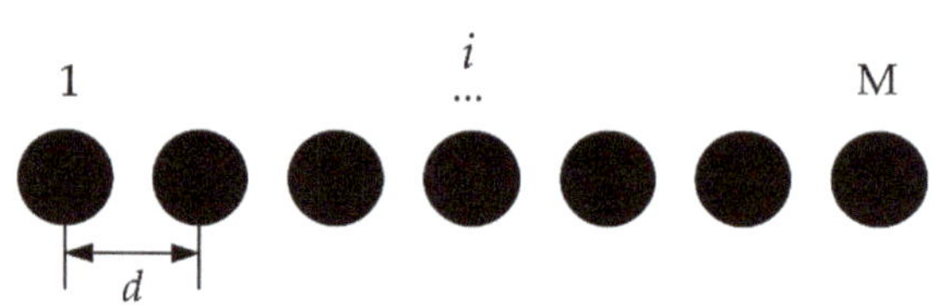

Figure 4. The point source of linear array.

It can be seen from Equation (4) that in parallel measurements, the imaging rule of any point light source satisfies the confocal relation of the impulse response function. Then, the amplitude distribution of the light field of the light source is:

$$G(x0,y0,z0) = \sum_{i=1}^{M} \delta 2(x0-id,y0) \tag{5}$$

In Equation (5), δ_2 represents the two-dimensional δ function and d represents the distance between the adjacent point light sources. Therefore, the PSF function of the whole parallel system is:

$$\begin{aligned} h_{conf} &= \left[\sum_{i=1}^{M} \delta 2(x0 - id, y0) \cdot h1(x - id, y, \Delta z)\right] * h2(x, y, \Delta z) \\ &= \sum_{i=1}^{M} \int\int \delta 2(x' - id, y')h1(x' - id, y', \Delta z) \cdot h2(x - x', y - y', \Delta z)dx'dy' \\ &= \sum_{i=1}^{M} h1(0, 0, \Delta z)h2(x - id, y, \Delta z) \end{aligned} \tag{6}$$

From Equation (6), in the parallel chromatic confocal measurement system, the optical field distribution is a total of each single-point system. It is equivalent to the coefficient of M single-point chromatic confocal subsystems. The above theoretical analysis provides a theoretical basis for the application of optical fiber bundle in the parallel chromatic confocal system.

Under the measurement conditions, the optical fiber bundle applied in this system is shown in Figure 5a. The section of the end coupled with the light source is circular, as shown in Figure 5b. The section of the end coupled with the dispersion tube lens is linear, as shown in Figure 5c. In Figure 5, the number "1" represents the circular end, and the number "2" represents the linear end.

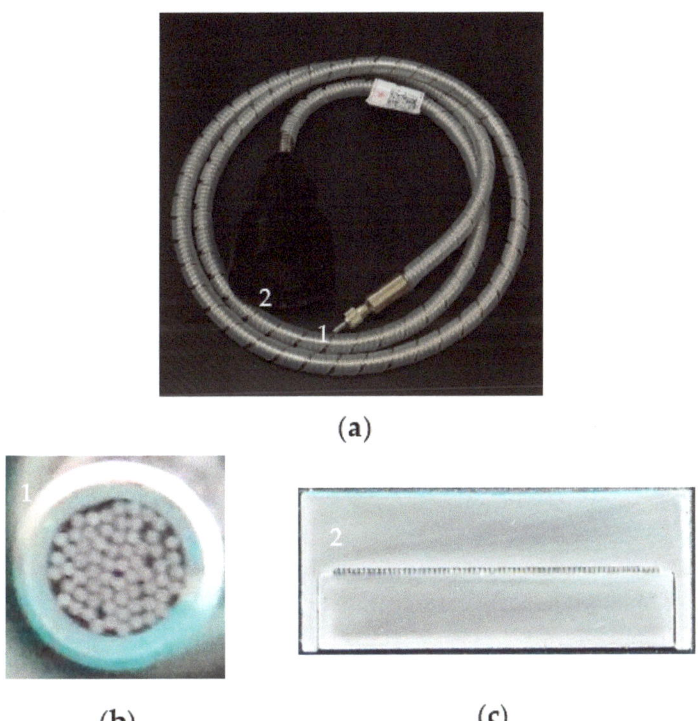

Figure 5. (**a**) The physical picture of the optical fiber bundle; (**b**) Circular end; (**c**) Linear end.

From Figure 5b,c, it can be seen that the optical fiber bundle used in our system is composed of 65 multi-mode fibers, and the material of the fiber core is quartz. The parameter diagram of the optical fiber bundle is shown in Figure 6. The parameters of each optical fiber are as follows: the NA value is 0.22; the diameter is 200 μm; the distance D

between fibers is 120 μm; and the diameter of the circular end is about 2 mm. Due to the limitation of the aperture of the components, only 18 of the 64 optical fibers are applied in our system. Therefore, the actual length of the optical fiber bundle array used in the system is about 5.66 mm.

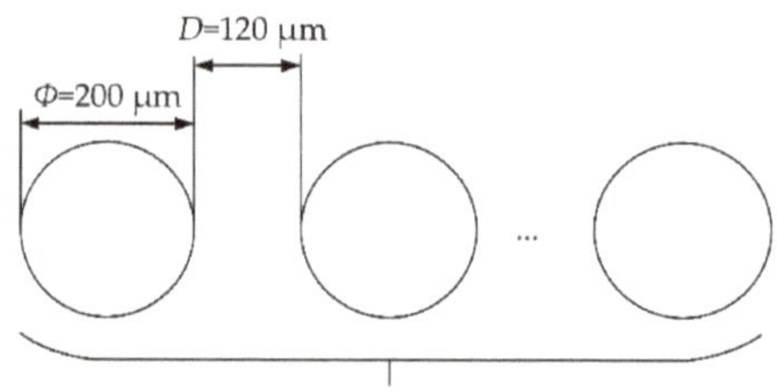

Figure 6. Parameter diagram of the optical fiber bundle.

2.4. Principle of Parallel Chromatic Confocal System with Non-Coaxial Illumination

After introducing the principle of the optical fiber bundle, we apply it to Section 2.2 to obtain the parallel chromatic confocal system with non-coaxial illumination based on the optical fiber bundle, as shown in Figure 7, where θ represents the inclined angle of the incident light. The light generated from the optical fiber bundle is arranged linearly on the measured surface. When the measured object moves along the direction (the x-axis in Figure 7) that is perpendicular to the optical fiber bundle and the z-axis of the system, the measurement can be completed through line scanning. In this process, the three-dimensional topography of the whole measured object can be obtained.

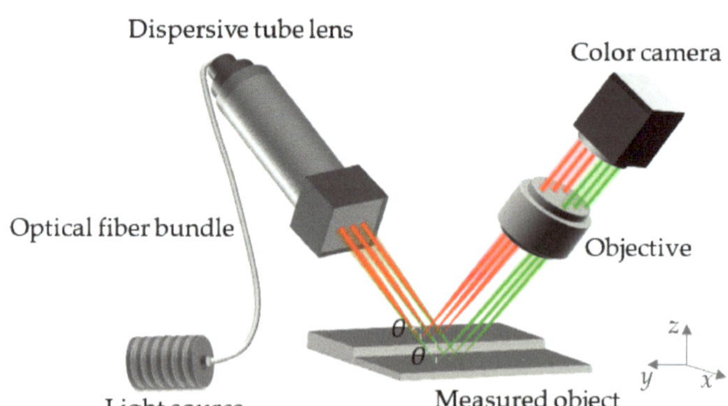

Figure 7. Schematic of the parallel chromatic confocal system with non-coaxial illumination.

2.5. Principle of the Color-Conversion Algorithm

In our study, a color-conversion algorithm is developed, which mainly involves the RGB color space and HSI color space, and the two color-space models are shown in Figure 8.

The HSI color model starts from the human visual system. The three related parameters used to describe colors are H (hue), S (saturation) and I (intensity). The H value, which is a wavelength-dependent parameter, can describe the spectral color and its range is $[0, 2\pi]$. Each spectral color has one angle for itself. For example, the angle of spectral red is 0, the angle of spectral green is $2\pi/3$ and the angle of spectral blue is $4\pi/3$.

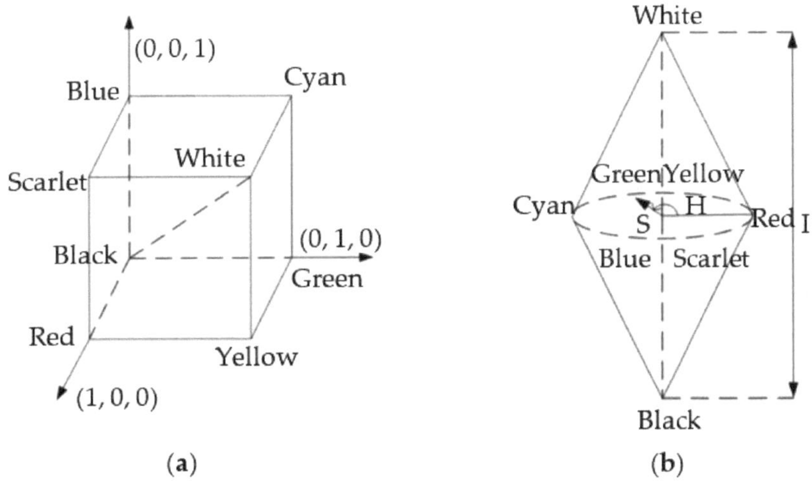

Figure 8. (**a**) Model of the RGB color space; (**b**) Model of the HSI color space.

Because the original data from the color camera is in RGB format, a conversion algorithm for RGB to HSI is needed. The classic geometric derivation conversion formula for the conversion is given as follows:

$$H = \begin{cases} \theta, & G \geq B \\ 2\pi - \theta, & G < B \end{cases} \tag{7}$$

where $\theta = \cos^{-1}\left(\frac{(R-G)+(R-B)}{2\sqrt{(R-G)^2+(R-B)(G-B)}}\right)$.

In the following section, the relevant experiments are discussed to verify the principles discussed above.

3. Experiments

3.1. Construction of the Measurement System

Based on the above theoretical analysis, the parallel chromatic confocal system with a non-coaxial-illumination-measurement experimental platform based on an optical fiber bundle is established, as shown in Figure 9. The components used in the experimental platform are listed in Table 1.

Table 1. List of components used in our system.

Components	Manufacturer	Function
White-light source	Yousheng, (MT G2 Easy White LED)	Produce polychromatic light source
Optical fiber bundle	Yousheng, (Custom-made)	Divide a light beam into several beams
Dispersive tube lens	Self-built	Produce chromatic dispersion
Objective	Motic, (Magnification:10×; N.A value:0.1)	Focus on the light
Platform	Daheng Optics, (GCM-T25MC)	Adjust and provide displacement
Gauge block	WD, (32 pieces of level 0)	As the measured object
Transparent specimen	Sail brand	As the measured object
Inductance micrometer	Tesa, (TT80)	Measure displacement value and true value
Color camera	Basler, (a2A5320-23ucBAS)	Capture color images

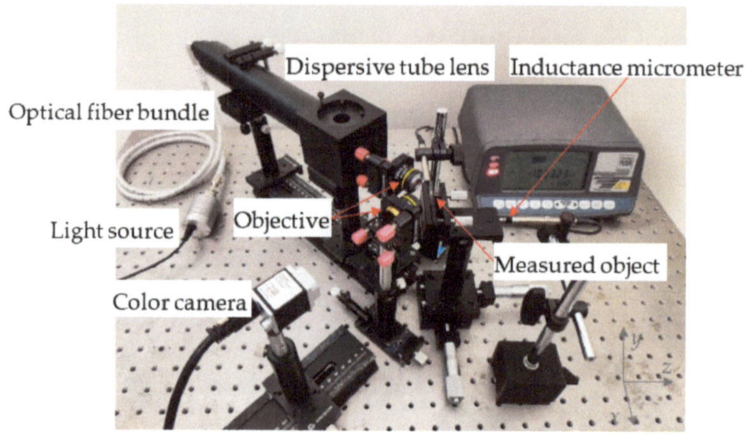

Figure 9. Device diagram of the parallel chromatic confocal system with non-coaxial illumination.

3.2. Calibration Experiment

First of all, the calibration experiment is conducted. After the experimental platform is built, a standard measuring block is selected as the measured object. The platform is controlled by the inductance micrometer to perform line scanning along the z-direction at a fixed step of 50 μm. In the calibration process, the images at different z-axial positions of 50 μm, 350 μm, 650 μm and 950 μm are obtained by the color camera, as shown in Figure 10.

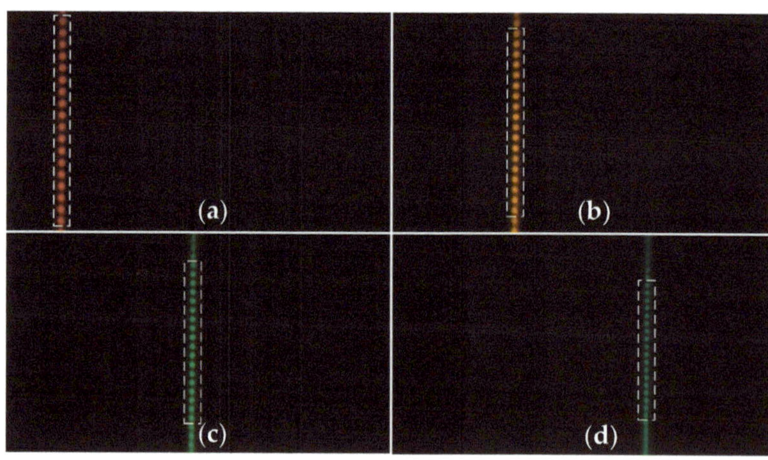

Figure 10. Images at different axial positions obtained by the camera: (**a**): 50 μm; (**b**): 350 μm; (**c**): 650 μm; (**d**): 950 μm.

We use the independently developed color-conversion algorithm mentioned in Section 2.5 to process the obtained color images. According to the color information of each light spot and the corresponding displacement information, the "H value–Axial displacement" calibration curve is obtained by calculating the average H value of 18 optical fibers of spots selected by the dotted box in Figure 10. The calculation results are shown in Table 2.

Table 2. Calibration experimental data.

Number	Axial Displacement (μm)	H Value
1	0	26.93
2	50	28.58
3	100	30.46
4	150	32.88
5	200	35.89
6	250	39.57
7	300	44.24
8	350	49.94
9	400	56.56
10	450	65.47
11	500	76.23
12	550	86.36
13	600	93.12
14	650	97.70
15	700	101.27
16	750	104.63
17	800	107.28
18	850	109.58
19	900	112.03
20	950	115.20

The calibration results are shown in Figure 11a. In the calibration curve, the linear section is the measurement range of the system, which is 200 μm. Through the linear fitting of the calibration data within the linear measurement range, the fitting line is obtained as shown in Figure 11b.

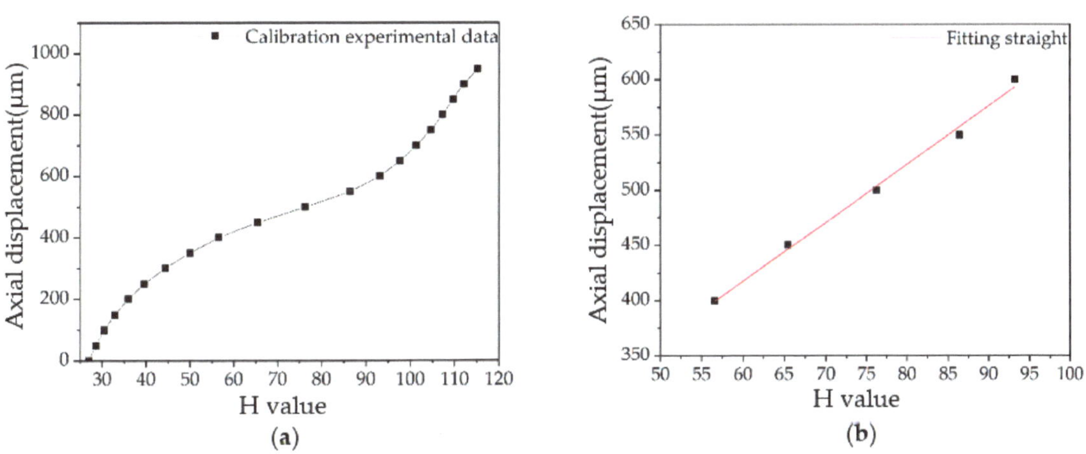

Figure 11. Calibration experiment: (**a**) Experimental data; (**b**) Calibration fitting results.

From Figure 11b, the calibration equation of the system is:

$$y1 = 5.29x + 200.23 \tag{8}$$

where y_1 is the axial displacement value of the measured object, the unit is μm and x represents the corresponding H value. According to the calibration experiment, the measuring range of the system is 200 μm. The linear correlation coefficient of the calibration equation is above 0.99.

3.3. Measurement of Step Height

Based on the calibration experiment, the step measurement experiment is carried out. In the experiment, two standard measuring blocks with a height of 1.03 mm and 1.08 mm are selected. The block of 1.08 mm is Block 1, and the block of 1.03 mm is Block 2. They are ground together on the base by molecular force to form step, and the base block gauge is random. The step is shown in Figure 12a.

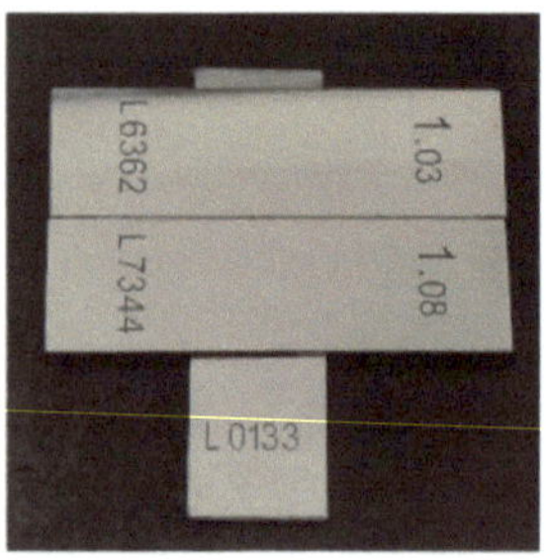

(a)

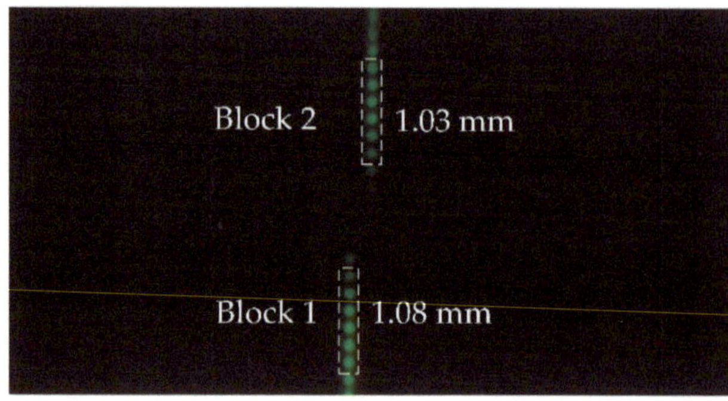

(b)

Figure 12. Measurement of step height: (**a**) Picture of step; (**b**) Schematic diagram of step measurement.

The true height value of the step measured by an inductance micrometer Tesa TT80 is 57.87 μm. The surface of the step is within the linear measurement range of the system and the schematic diagram is shown in Figure 12b.

In Figure 12b, the six light spots fall on the measuring block surface of 1.03 mm and 1.08 mm, respectively. By calculating the average H value of the six light spots of Block 1 and Block 2, the height values and the difference in the height of the step of the two measuring blocks can be obtained. The diagram of measurement results is shown in Figure 13, and the above experimental process was repeated 20 times. The results are shown in Table 3.

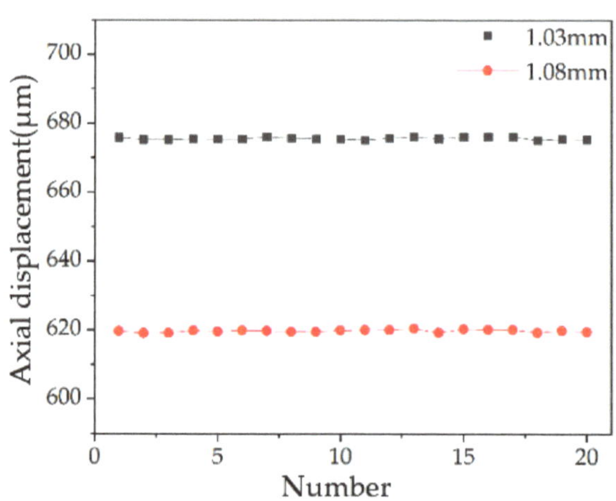

Figure 13. Result diagram of step measurement.

Table 3. Step measurement experiment data.

Number	H Value of Block 1	Displacement of Block 1 (μm)	H Value of Block 2	Displacement of Block 2 (μm)	Height of Difference (μm)
1	89.91	675.85	79.31	619.78	56.07
2	89.80	675.27	79.20	619.20	56.07
3	89.80	675.27	79.21	619.25	56.02
4	89.85	675.54	79.34	619.94	55.60
5	89.84	675.48	79.28	619.62	55.86
6	89.83	675.43	79.35	619.99	55.44
7	89.96	676.12	79.32	619.83	56.29
8	89.89	675.75	79.28	619.62	56.13
9	89.85	675.54	79.26	619.52	56.02
10	89.84	675.48	79.34	619.94	55.54
11	89.79	675.22	79.37	620.10	55.12
12	89.89	675.75	79.37	620.10	55.65
13	89.96	676.12	79.45	620.52	55.60
14	89.87	675.64	79.23	619.36	56.28
15	89.98	676.22	79.43	620.41	55.81
16	89.97	676.17	79.39	620.20	55.97
17	89.98	676.22	79.40	620.26	55.96
18	89.78	675.17	79.24	619.41	55.76
19	89.86	675.59	79.33	619.89	55.70
20	89.83	675.43	79.27	619.57	55.86
		The average value of the step (μm)			55.84
		The difference from the true value (μm)			−2.03
		Relative error			−3.51%
		The standard deviation σ of the step (μm)			0.29

Through the analysis of the above data, it can be seen that the measured height of the step is 55.84 μm, the relative error is −3.51% and the 3σ value is 0.87 μm. Therefore, the measurement accuracy of the system can reach the micron level.

3.4. *Measurement of Transparent Specimen Thickness*

In order to verify the ability of the system to measure the thickness of transparent specimens, a glass slide with the transmittance of 84% (380 nm)–90% (717 nm) is selected as the transparent specimen for measurement. The true value of the glass slide measured by Tesa TT80 is 184.08 μm. The physical image of the glass slide is shown in Figure 14a, and the schematic diagram is shown in Figure 14b. In Figure 14b, the 18 light spots on the left are the H value reflected on the upper surface of the transparent specimen, and the 18 light spots on the right are the H value reflected on the lower surface of the transparent specimen.

Similarly, 20 groups of repetitive measurement experiments are conducted based on the calibration experiment. Additionally, the thickness of the transparent specimen d can be expressed as follows [30]:

$$d = k_1(\mathrm{H_i} - \mathrm{H_0}) \cdot \cos\theta \cdot \frac{(\tan\alpha - \tan\beta)}{(\tan\alpha' - \tan\beta')} \tag{9}$$

The thickness data of transparent specimen obtained is shown in Figure 15 and Table 4.

(a)

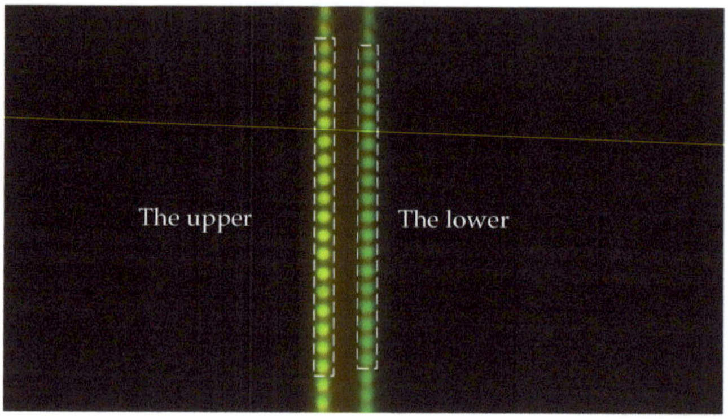

(b)

Figure 14. Measurement of transparent specimen thickness: (**a**) Picture of the transparent specimen; (**b**) Schematic of the measurement of the transparent specimen.

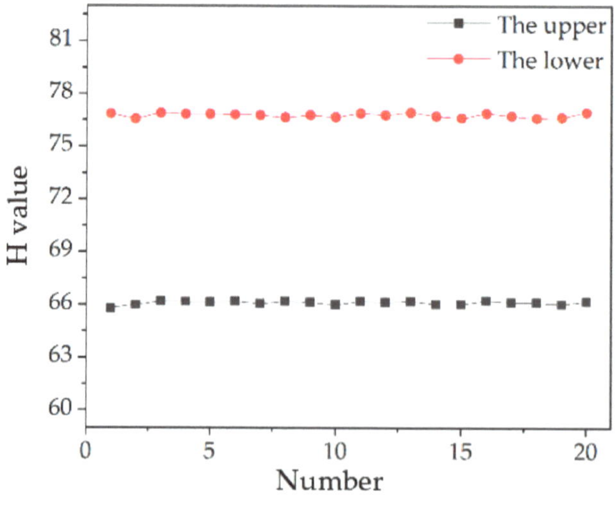

Figure 15. Diagram of measurement results of transparent specimen.

Table 4. Transparent specimen thickness measurement experimental data.

Number	H Value of Upper	H Value of Lower	H Value of the Difference
1	65.80	76.86	11.06
2	65.99	76.57	10.58
3	66.19	76.91	10.72
4	66.17	76.84	10.67
5	66.15	76.84	10.69
6	66.19	76.81	10.62
7	66.06	76.78	10.72
8	66.19	76.65	10.46
9	66.14	76.79	10.65
10	66.00	76.68	10.68
11	66.20	76.90	10.70
12	66.15	76.79	10.64
13	66.21	76.95	10.74
14	66.05	76.74	10.69
15	66.05	76.64	10.59
16	66.23	76.91	10.68
17	66.16	76.75	10.59
18	66.13	76.63	10.50
19	66.05	76.68	10.63
20	66.20	76.97	10.77
The average H value of the difference			10.67
The thickness value of the transparent specimen(μm)			177.58
The difference from the true value(μm)			−6.5
Relative error			−3.53%
The standard deviation σ of the difference(μm)			0.12

The measured thickness of the transparent specimen is 177.58 μm, the relative error is −3.53% and the 3σ value is 0.36 μm. The measurement accuracy of the system can reach the micron level.

3.5. Three-Dimensional Topography by Line-Scanning Method

In order to verify the ability of three-dimensional surface topography, the step in Section 3.3 and the "1" character on the surface of the coin were chosen as measured objects. The physical picture and the schematic diagram of the step and CNY 1 coin are shown in Figure 16.

For the restoration of the step surface, the optical fiber bundle is arranged in the y-axis direction as shown in Figure 16a, and the motor moves in the x-axis direction. In the actual measurement process, the measuring range of the step surface is 3.25 mm × 5.92 mm. The one-dimensional scanning distance is 3.25 mm, and the scanning speed is 50 μm/s. The photography speed of the color camera is 1 piece/s and a total of 66 photos are taken. Each photo has 12 measuring points, so there is a total of 792 measuring points. Through the "H value-displacement" conversion processing of 792 measured points, the three-dimensional topography features of the step surface are obtained, as shown in Figure 17 in which the surface height information of the step surface is about 50 μm.

For the restoration of the "1" character on the surface of the coin, the optical fiber bundle is arranged in the y-axis direction as shown in Figure 16b, and the motor moves in the x-axis direction. In the actual measurement process, the "1" character measuring range is 2.20 mm × 1.02 mm. The one-dimensional scanning distance is 2.20 mm, and the scanning speed is 25 μm/s. The color camera photography speed is 1 piece/s and a total of 89 photos are taken. Each photo has 18 measuring points with a total of 1602 measuring points. By data analysis and the processing of these measuring points, the three-dimensional topography feature of "1" is obtained, as shown in Figure 18 in which the measured surface height is about 50 μm.

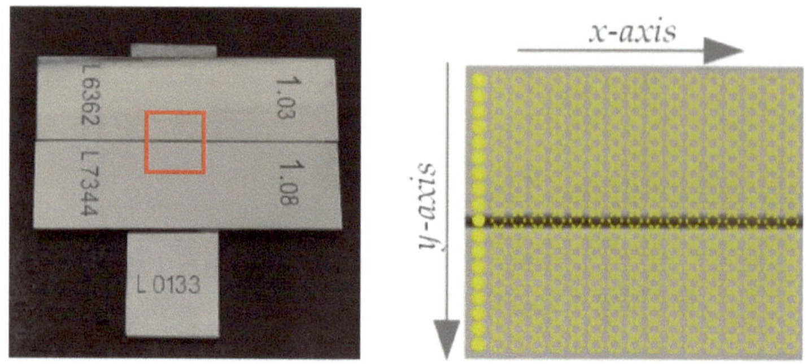

(**a**)

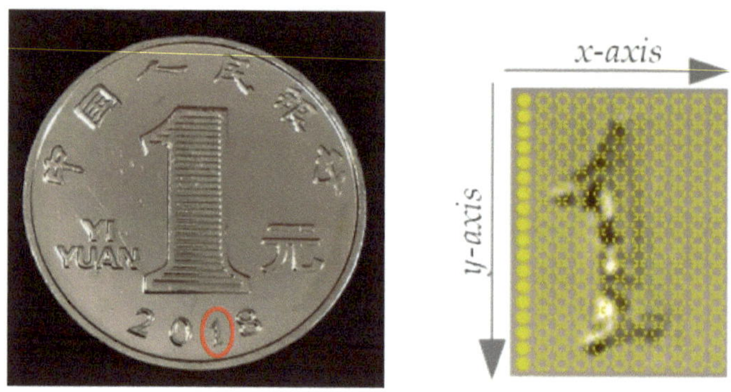

(**b**)

Figure 16. Three-dimensional topography by the line-scanning method: (**a**) Schematic of the step measurement; (**b**) Schematic of the coin measurement.

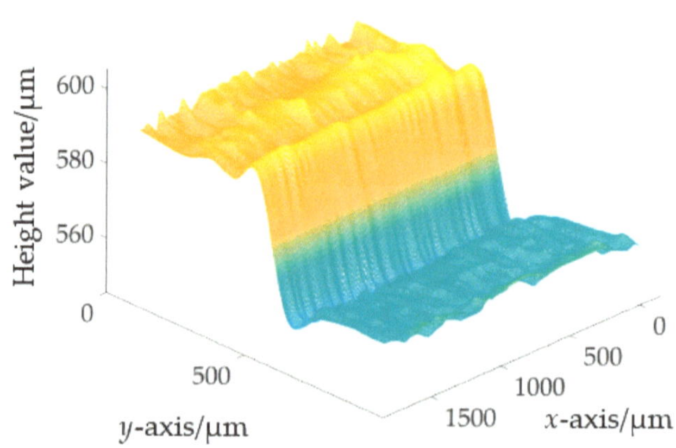

Figure 17. The three-dimensional diagram of the step.

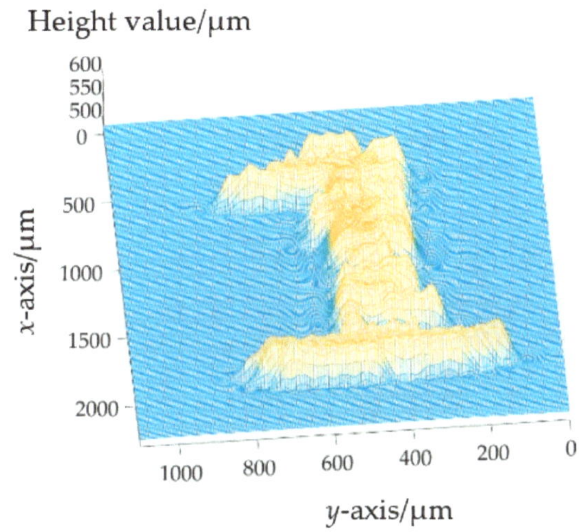

Figure 18. The three-dimensional diagram of "1".

3.6. Contrast Experiment

In order to better verify the improvement of the measurement efficiency of the system in this paper, compared with the single-point non-coaxial-illumination system, the corresponding contrast experiments are conducted.

The experimental platform of the single-point chromatic confocal system with non-coaxial illumination is shown in Figure 19.

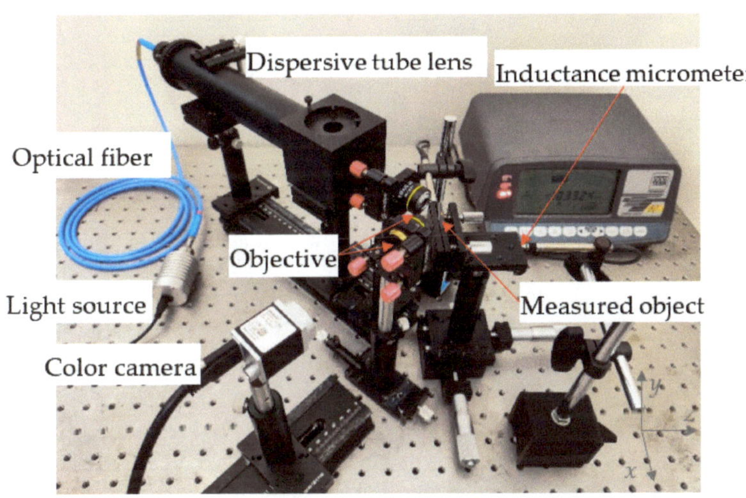

Figure 19. Device diagram of the single-point chromatic confocal system with non-coaxial illumination.

In the above system, under the condition that the other experimental conditions remain unchanged, the optical fiber bundle is replaced with an optical fiber. According to the above-mentioned experimental procedures, the calibration experiment and the transparent specimen-thickness-measurement experiment are carried out. In the calibration process, the images at different z-axial positions of 100 μm, 700 μm, 1300 μm and 1900 μm obtained by the color camera are shown in Figure 20.

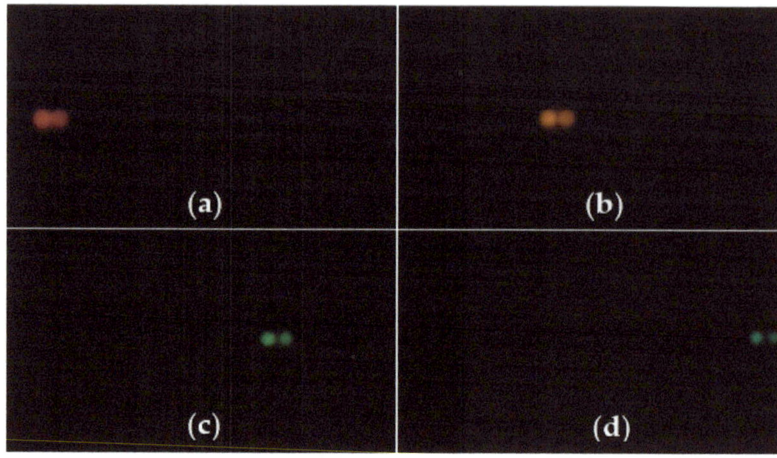

Figure 20. Images at different axial positions obtained by the camera: (**a**): 100 μm; (**b**): 700 μm; (**c**): 1300 μm; (**d**): 1900 μm.

As can be seen from the number of light spots in the calibration results between Figures 10 and 20, the measurement efficiency of the parallel measurement is 18 times that of the single-point measurement. The total experimental results of the parallel measurement are presented in Table 5.

Table 5. Experimental comparison results.

Experiment Type	Experiment Results	Our System	The Comparative System
Calibration experiment	Calibration equation	$y1 = 5.29x + 200.23$	$y2 = 12.70x + 335.12$
	Measuring range (μm)	200	400
	Linear correlation coefficient	>0.99	>0.99
Measurement of transparent specimen thickness	Measured thickness value (μm)	177.58	176.00
	Relative error	−3.53%	−4.39%
	Standard deviation σ (μm)	0.12	0.01

From Table 5, it can be found that the measurement accuracy of the two systems can reach the micron level regardless of whether the optical fiber bundle is used as the beam splitter or not. Because the number of light spots in the optical fiber bundle in the parallel non-coaxial-illumination system is 18 times that in the single-point non-coaxial-illumination system, the measurement efficiency of the parallel non-coaxial-lighting system is significantly improved.

4. Discussion

This paper proposes a chromatic confocal system for parallel non-coaxial illumination based on the optical fiber bundle, and it solves the problem of a low signal-to-noise ratio and low light energy utility which are in the traditional coaxial illumination system. This paper includes the following points:

(1) By the related theoretical analysis of the non-coaxial-illumination system and the properties of the optical fiber bundle, it is proved that the optical fiber bundle can be applied in the chromatic confocal system. Based on this, the existing single-point non-coaxial-illumination system is optimized using the optical fiber bundle as the light-beam splitter to realize parallel measurements, and this optimization can improve the measurement efficiency of the single-point system.

(2) Combined with the color-conversion algorithm, the conclusion of (1) is verified by corresponding step height measurement, transparent specimen thickness measurement and three-dimensional topography restoration. Three-dimensional topography includes the restoration of the step and character "1" of the coin.
(3) The experimental results show that the measuring range of the system is 200 μm, the repeatability is better than 0.87 μm, the relative error is less than ±4% and the measurement accuracy can reach micron level. Additionally, the measurement efficiency of the proposed system is 18 times higher than that of the single-point non-coaxial-illumination system.

5. Conclusions

In this paper, we propose a chromatic confocal system for parallel non-coaxial illumination. Based on the optical fiber bundle and the existing single-point non-coaxial-illumination system, the optical fiber bundle as a light-beam splitter achieves parallel measurements through line scanning, which greatly improves the measurement efficiency of the single-point non-coaxial-illumination system. The experimental results show that the measurement accuracy can reach micron level and the system has a good three-dimensional topography reconstruction effect. In the future, we will further improve the measurement accuracy of the system in three-dimensional topography. Additionally, efforts will also be made to develop the system towards miniaturization and integration.

Author Contributions: Conceptualization, Q.Y. and Y.Z. (Yali Zhang); methodology, Q.Y. and Y.Z. (Yali Zhang); software, C.W. and Y.W.; validation, F.C. and Y.Z. (Yaozu Zhang); formal analysis, T.L. (Tianliang Lin); investigation, L.X. and T.L. (Tianliang Lin); resources, L.X.; data curation, Y.W.; writing—original draft preparation, C.W.; writing—review and editing, Y.Z. (Yali Zhang) and Y.Z. (Yaozu Zhang); visualization, L.X.; supervision, F.C.; project administration, T.L. (Ting Liu); funding acquisition, T.L. (Ting Liu). All authors have read and agreed to the published version of the manuscript.

Funding: This study was supported by the National Natural Science Foundation of China (No. 52075190 and No. 62075067); the Collaborative Innovation Platform of Fuzhou-Xiamen-Quanzhou Independent Innovation Demonstration Area, China (No. 3502ZCQXT202002); the Science and Technology Program of Fujian, China (No. 2019I0013); and the Open Research Fund of Anhui Engineering Technology Research Center of Automotive New Technique (No. QCKJ202006).

Data Availability Statement: Data sharing is not applicable to this article.

Conflicts of Interest: The authors declare no conflict of interest.

References

1. Behrends, G.; Stöbener, D.; Fischer, A. Integrated, Speckle-Based Displacement Measurement for Lateral Scanning White Light Interferometry. *Sensors* **2021**, *21*, 2486. [CrossRef]
2. Im, J.; Kim, H.; Park, W.; Ahn, J.S.; Lee, B.; Choi, S. High-precision white light interferometry based on a color CCD and peak matching algorithm. *J. Korean Phys. Soc.* **2022**, *80*, 599–605. [CrossRef]
3. Pillarz, M.; von Freyberg, A.; Stöbener, D.; Fischer, A. Gear shape measurement potential of laser triangulation and confocal-chromatic distance sensors. *Sensors* **2021**, *21*, 937. [CrossRef] [PubMed]
4. Sun, H.; Wang, S.; Bai, J.; Zhang, J.; Huang, J.; Zhou, X.; Liu, D.; Liu, C. Confocal laser scanning and 3D reconstruction methods for the subsurface damage of polished optics. *Opt. Laser Eng.* **2021**, *136*, 106315. [CrossRef]
5. Ren, H.; Liu, Y.; Wang, Y.; Liu, N.; Yu, X.; Su, X. Uniaxial 3D Measurement with Auto-Synchronous Phase-Shifting and Defocusing Based on a Tilted Grating. *Sensors* **2021**, *21*, 3730. [CrossRef] [PubMed]
6. Iwasa, T.; Ota, K.; Harada, T.; Muramatsu, R. High-resolution surface shape measurement of parabola antenna reflector by using grating projection method with virtual targets. *Acta Astronaut.* **2018**, *153*, 95–108. [CrossRef]
7. Choi, Y.; Yoo, H.; Kang, D. Large-area thickness measurement of transparent multi-layer films based on laser confocal reflection sensor. *Measurement* **2020**, *153*, 107390. [CrossRef]
8. Vasilev, E.; Wang, J.; Knezevic, M. A structure metric for quantitative assessment of fracture surfaces in 3D conceived based on confocal laser scanning microscopy data. *Mater. Charact.* **2022**, *194*, 112369. [CrossRef]
9. Quinten, M. Thickness Determination of Transparent Coatings: Considering a correction factor, chromatic confocal sensors can easily be used to determine the thickness of a transparent workpiece. *PhotonicsViews* **2019**, *16*, 68–71. [CrossRef]

10. Lu, W.; Chen, C.; Wang, J.; Leach, R.; Zhang, C.; Liu, X.; Lei, Z.; Yang, W.; Jiang, X.J. Characterization of the displacement response in chromatic confocal microscopy with a hybrid radial basis function network. *Opt. Express* **2019**, *27*, 22737–22752. [CrossRef]
11. Fu, S.; Kor, W.S.; Cheng, F.; Seah, L.K. In-situ measurement of surface roughness using chromatic confocal sensor. *Procedia CIRP* **2020**, *94*, 780–784. [CrossRef]
12. Claus, D.; Nizami, M.R. Influence of aberrations and roughness on the chromatic confocal signal based on experiments and wave-optical modeling. *Surf. Topogr. Metrol. Prop.* **2020**, *8*, 025031. [CrossRef]
13. Du, H.; Zhang, W.; Ju, B.; Sun, Z.; Sun, A. A new method for detecting surface defects on curved reflective optics using normalized reflectivity. *Rev. Sci. Instrum.* **2020**, *91*, 036103. [CrossRef] [PubMed]
14. Chouhad, H.; El Mansori, M.; Knoblauch, R.; Corleto, C. Smart data driven defect detection method for surface quality control in manufacturing. *Meas. Sci. Technol.* **2021**, *32*, 105403. [CrossRef]
15. Chun, B.S.; Kim, K.; Gweon, D. Three-dimensional surface profile measurement using a beam scanning chromatic confocal microscope. *Rev. Sci. Instrum.* **2009**, *80*, 073706. [CrossRef]
16. Wertjanz, D.; Kern, T.; Csencsics, E.; Stadler, G.; Schitter, G. Compact scanning confocal chromatic sensor enabling precision 3-D measurements. *Appl. Opt.* **2021**, *60*, 7511–7517. [CrossRef]
17. Yu, Q.; Yu, X.F.; Cui, C.; Ye, R. Survey of parallel light source technology in parallel confocal measurement. *Chin. Opt.* **2013**, *6*, 652–659.
18. Grant, D.M.; McGinty, J.; McGhee, E.J.; Bunney, T.D.; Owen, D.M.; Talbot, C.B.; Zhang, W.; Kumar, S.; Munro, I.; Lanigan, P. High speed optically sectioned fluorescence lifetime imaging permits study of live cell signaling events. *Opt. Express* **2007**, *15*, 15656–15673. [CrossRef]
19. Kagawa, K.; Seo, M.; Yasutomi, K.; Terakawa, S.; Kawahito, S. Multi-beam confocal microscopy based on a custom image sensor with focal-plane pinhole array effect. *Opt. Express* **2013**, *21*, 1417–1429. [CrossRef]
20. Tiziani, H.J.; Achi, R.; Krämer, R.N.; Wiegers, L. Theoretical analysis of confocal microscopy with microlenses. *Appl. Opt.* **1996**, *35*, 120–125. [CrossRef]
21. Lane, P.M.; Dlugan, A.L.; Richards-Kortum, R.; MacAulay, C.E. Fiber-optic confocal microscopy using a spatial light modulator. *Opt. Lett.* **2000**, *25*, 1780–1782. [CrossRef] [PubMed]
22. Hou, W.; Zhang, Y. Fast parallel 3D profilometer with DMD technology. In Proceedings of the Seventh International Symposium on Precision Engineering Measurements and Instrumentation, Yunnan, China, 7–11 August 2011; Volume 8321, pp. 155–161.
23. Tiziani, H.J.; Uhde, H. Three-dimensional image sensing by chromatic confocal microscopy. *Appl. Opt.* **1994**, *33*, 1838–1843. [CrossRef] [PubMed]
24. Tiziani, H.J.; Wegner, M.; Steudle, D. Confocal principle for macro-and microscopic surface and defect analysis. *Opt. Eng.* **2000**, *39*, 32–39. [CrossRef]
25. Hillenbrand, M.; Lorenz, L.; Kleindienst, R.; Grewe, A.; Sinzinger, S. Spectrally multiplexed chromatic confocal multipoint sensing. *Opt. Lett.* **2013**, *38*, 4694–4697. [CrossRef]
26. Hillenbrand, M.; Weiss, R.; Endrödy, C.; Grewe, A.; Hoffmann, M.; Sinzinger, S. Chromatic confocal matrix sensor with actuated pinhole arrays. *Appl. Opt.* **2015**, *54*, 4927–4936. [CrossRef] [PubMed]
27. Hu, H.; Mei, S.; Fan, L.; Wang, H. A line-scanning chromatic confocal sensor for three-dimensional profile measurement on highly reflective materials. *Rev. Sci. Instrum.* **2021**, *92*, 053707. [CrossRef]
28. Yu, Q.; Zhang, Y.; Zhang, Y.; Cheng, F.; Shang, W.; Wang, Y. A novel chromatic confocal one-shot 3D measurement system based on DMD. *Measurement* **2021**, *186*, 110140. [CrossRef]
29. Berkovic, G.; Zilberman, S.; Shafir, E.; Rubin, D. Chromatic confocal displacement sensing at oblique incidence angles. *Appl. Opt.* **2020**, *59*, 3183–3186. [CrossRef]
30. Yu, Q.; Zhang, Y.; Shang, W.; Dong, S.; Wang, C.; Wang, Y.; Liu, T.; Cheng, F. Thickness measurement for glass slides based on chromatic confocal microscopy with inclined illumination. *Photonics* **2021**, *8*, 170. [CrossRef]

sensors

MDPI

Article

Characterisation and Quenching Correction for an Al_2O_3:C Optical Fibre Real Time System in Therapeutic Proton, Helium, and Carbon-Charged Beams

Luana de Freitas Nascimento [1,*], Paul Leblans [2], Brent van der Heyden [1], Mark Akselrod [3], Jo Goossens [4,5], Luis Enrique Correa Rocha [6,7], Ana Vaniqui [1] and Dirk Verellen [4,5]

1 Belgian Nuclear Research Centre, SCK CEN, 2400 Mol, Belgium
2 Agfa NV, 2640 Mortsel, Belgium
3 Landauer, Stillwater Crystal Growth Division, Stillwater, OK 74074, USA
4 Faculty of Medicine and Health Sciences, University of Antwerp, 2610 Antwerp, Belgium
5 Iridium Netwerk, University of Antwerp, 2610 Antwerp, Belgium
6 Department of Economics, Ghent University, 9000 Ghent, Belgium
7 Department of Physics and Astronomy, Ghent University, 9000 Ghent, Belgium
* Correspondence: ldfnasci@sckcen.be

Abstract: Real time radioluminescence fibre-based detectors were investigated for application in proton, helium, and carbon therapy dosimetry. The Al_2O_3:C probes are made of one single crystal (1 mm) and two droplets of micro powder in two sizes (38 μm and 4 μm) mixed with a water-equivalent binder. The fibres were irradiated behind different thicknesses of solid slabs, and the Bragg curves presented a quenching effect attributed to the nonlinear response of the radioluminescence (RL) signal as a function of linear energy transfer (LET). Experimental data and Monte Carlo simulations were utilised to acquire a quenching correction method, adapted from Birks' formulation, to restore the linear dose–response for particle therapy beams. The method for quenching correction was applied and yielded the best results for the '4 μm' optical fibre probe, with an agreement at the Bragg peak of 1.4% (160 MeV), and 1.5% (230 MeV) for proton-charged particles; 2.4% (150 MeV/u) for helium-charged particles and of 4.8% (290 MeV/u) and 2.9% (400 MeV/u) for the carbon-charged particles. The most substantial deviations for the '4 μm' optical fibre probe were found at the falloff regions, with ~3% (protons), ~5% (helium) and 6% (carbon).

Keywords: real time dosimetry; hadron therapy; quenching correction

Citation: de Freitas Nascimento, L.; Leblans, P.; van der Heyden, B.; Akselrod, M.; Goossens, J.; Correa Rocha, L.E.; Vaniqui, A.; Verellen, D. Characterisation and Quenching Correction for an Al_2O_3:C Optical Fibre Real Time System in Therapeutic Proton, Helium, and Carbon-Charged Beams. *Sensors* **2022**, 22, 9178. https://doi.org/10.3390/s22239178

Academic Editors: Qing Yu, Ran Tu, Ting Liu and Lina Li

Received: 17 October 2022
Accepted: 23 November 2022
Published: 25 November 2022

Publisher's Note: MDPI stays neutral with regard to jurisdictional claims in published maps and institutional affiliations.

Copyright: © 2022 by the authors. Licensee MDPI, Basel, Switzerland. This article is an open access article distributed under the terms and conditions of the Creative Commons Attribution (CC BY) license (https://creativecommons.org/licenses/by/4.0/).

1. Introduction

Particle therapy has gained popularity as an effective technique for cancer treatment due to its greater precision in dose delivery and less damage to adjacent healthy tissue and organs. The complexity of particle therapy brings several technical challenges for dosimetrists, radiobiologists, and medical physicists. In order to evaluate the performance of particle therapy against conventional radiotherapy, it is essential to test and report the response of various treatment modalities using similar methods across treatment and research centres. International agencies (e.g., the International Atomic Energy Agency-IAEA, the American Association of Physicists in Medicine-AAPM, and the International Commission on Radiation Units and Measurements-ICRU) have been working on global standard protocols to harmonise the reporting of treatments. The standards should cover beam production, dosimetry, relative biological effectiveness (RBE), treatment planning, clinical requirements and protocols [1,2].

According to the report "Dose Reporting in Ion Beam Therapy" proposed by the IAEA (Report No. 1560) [3], to guarantee a cost-effective operation of a therapy facility, the time required for the quality assurance (QA) program must be kept to a minimum while ensuring complete coverage of all critical system parameters. Each particle therapy centre worldwide should thus optimise its QA procedure.

When looking specifically at dosimetry, several authors have published on charged particle QA using either commercial or in-house developed devices [4–7]. Most of the systems in use are based on single ionisation chambers or diodes, or arrays of them, and on radiographic/radiochromic films, which can be considered sufficiently precise but in need of lengthy procedures and, in the case of diodes and films, as having energy-dependent effects with linear energy transfer (LET) [8–13]. The quantity of LET describes the average energy transfer from electronic interactions per unit length travelled by charged primary particles.

One class of promising dosimeters is based on luminescence detectors. One advantage is the diverse (point, 1D, 2D) readout possibilities, both as active (radioluminescence-RL, scintillation) and as passive detectors (thermoluminescence-TL, optically stimulated luminescence-OSL and radiophotoluminescence-RPL) [5,14–20]. One known drawback of luminescence detectors is the existence of the quenching effect [21], which is the dose nonlinearity effect caused by the high ionisation density in particles of high LET, i.e., the higher the particle LET, the lower the light production efficiency from the luminescence detector, resulting in substantially under-estimated doses [22–27].

In this paper, we further explore a novel quenching correction method to restore the linear dose–response for particle therapy beams. Our approach builds on a method previously used to correct for quenching in plastic scintillator detectors using the Birks equation [28,29]. The novelty of our method is to include the contribution from fragments generated along the Bragg curves by proposing a novel general expression for quenching correction factors. In a previous attempt to correct the response to proton and carbon-charged particle beams in 2D, real time RL films, we observed that a better agreement could be reached by including more fragments in our calculations [30].

The aforementioned method [30] is now applied on three types of RL optical fibre probes, composed of Al_2O_3:C, in five therapeutic charged particle beams: 160 and 230 MeV protons, 150 MeV/u helium ions, and 290 and 400 MeV/u carbon ions. For all the optical fibre probe types, the quenching-corrected doses along the Bragg curves resulted in substantial improvement when compared to uncorrected data.

2. Materials and Methods

2.1. Irradiations at HIMAC and SCK CEN

The Heavy Ion Medical Accelerator facility (HIMAC) in Chiba, Japan, consists of three clinical treatment rooms, one biological experiment room (BIO), two large general experimental halls, and a low-energy experimental room. Figure 1 shows the BIO room with a horizontal beam line (indicated by (I)), a pair of wobbler magnets and a scatterer used to produce uniform irradiation fields. The range shifter is used for adjusting the residual range of the heavy ions in the target. A set of binary filters changes the depth in the measurements (II).

For our experiments, we irradiated optical fibres coupled to Al_2O_3:C sensors with 160 and 230 MeV proton, 150 MeV/u helium, and 290 and 400 MeV/u carbon mono-energetic beams. Actual energies, based on reference measured Bragg curves and Monte Carlo simulations, were 154.5 MeV (LETw = 5.33 MeV/cm) and 226.5 MeV (LETw = 4.15 MeV/cm) protons, 142.6 MeV/u (LETw = 22.44 MeV/cm) helium ions, and 273.8 MeV/u (LETw = 132.9 MeV/cm) and 383.2 MeV/u (LETw = 111.1 MeV/cm) carbon ions (Figure 2). The experimental set-up consisted in positioning Al_2O_3:C+fibre probes in front of a Polymethylmethacrylate (PMMA, density = 1.19 g cm^{-3}) binary filter ("III" in Figure 1) with different water-equivalent thicknesses (depth in H_2O.). The filters consist of 9 plates of PMMA of 0.5, 1, 2, 4, 8, 16, 32, 64, and 128 mm thickness, covering a circular 10 cm radiation field.

Figure 1. Biological experiment room (BIO). The area around "I" indicates the horizontal beam line, "II" indicates the set of binary filters, and "III" is the position of the fibre probes.

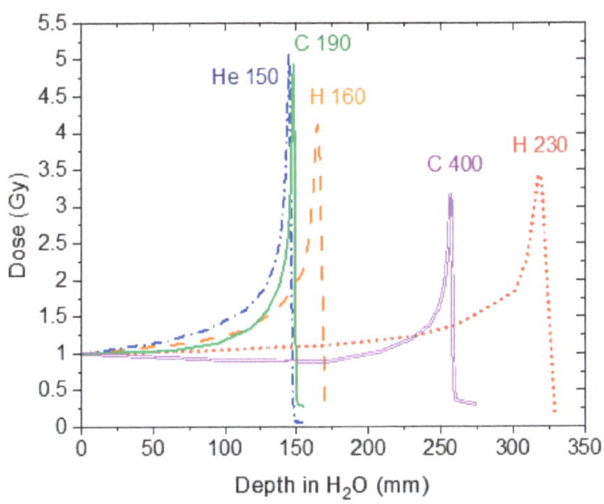

Figure 2. Bragg curves from 160 and 230 MeV proton, 150 MeV/u helium and 290 and 400 MeV/n carbon mono-energetic beams measured with reference ion chambers.

Reference data (absorbed doses and depth-dose profiles) were acquired using a Markus ionisation chamber [31,32] (Figure 2). The given absorbed doses to water and dose rates for each beam/energy type are summarised in Table 1. Bragg curves had different depths in water within a 10 cm × 10 cm lateral field, with flatness better than 3%.

Table 1. Details of each measurement campaign at HIMAC with types and energies of beams, dose rates, and types and quantities of fibre probes.

Beam Type and Energy [MeV/u]	Dose Rate [Gy/min]	Fibber Probes (Quantity)	Dose to Water at Entrance (d = 0 mm H_2O)
H 160	0.08 (a) 0.25 (b)	Single Crystal (1) 38 µm (1) 4 µm (1)	**Dose–response**: from 0.04 to 0.10 Gy (a) from 0.1 to 5.0 Gy (b) **Bragg curve**: 0.05 Gy (a) 0.25 Gy (b)
H 230	0.36 (a) 0.49 (b) 0.65 (c)	Single Crystal (1) 38 µm (2) 4 µm (2)	**Dose–response**: from 0.04 to 0.50 Gy (a), from 0.06 to 2.00 Gy (b) from 0.5 to 2.0 Gy (c) **Bragg curve**: 0.5 and 1.0 Gy (a) 0.05 and 0.50 Gy (b) 0.5 and 1.0 Gy (c)
He 150	4.50	Single Crystal (1) 38 µm (2) 4 µm (2)	**Dose–response**: 0.3 to 10.0 Gy **Bragg curve**: 0.5, 1.0, and 2.0 Gy
C 290	6.90 (a) 7.34 (b) 3.04 (c)	Single Crystal (1) 38 µm (2) 4 µm (2)	**Dose–response**: 0.5 to 10.0 Gy (a) and (c) 5 to 60 Gy (b), **Bragg curve**: 0.5 and 1.0 Gy (a) and (c) 1, 2 and 5 Gy (b)
C 400	6.90	Single Crystal (1) 38 µm (1) 4 µm (1)	**Dose–response**: 0.1 to 60.0 Gy **Bragg curve**: 0.5, 2.0 Gy

Supplementary dose reference data were obtained using ^{60}Co gamma-ray irradiator Theratron 780 at the Belgian Laboratory for Standard Dosimetry Calibrations (LNK, SCK CEN, Belgium) [33]. The fibre probes were exposed with a reference nominal dose rate of 0.5 Gy/min and a total dose of 1.0 Gy (D_{Co}).

These reference irradiations with ^{60}Co gamma beams were used to compare the RL signal with high LET (RL_{LET}) to the signals with low LET from photons (RL_{Co}) and to calculate the luminescence efficiency further. Reference irradiations took place before and after the measurements at HIMAC to account for any changes in material sensitivity.

The relative luminescence efficiency μ used in this paper was previously employed by Sawakuchi et al. [34] and Kalef-Ezra and Horowitz [35]. It is defined as the ratio of the luminescence signal to the radiation field k, and the luminescence signal to a reference radiation field l, for a specific dose D_k or D_l. Equation (1) gives the relative luminescence efficiency μ for heavy charged particles (HCP) and ^{60}Co gamma rays.

$$\mu = \frac{RL_{HCP}/D_{HCP}}{RL_{Co}/D_{Co}} = C \cdot RL_{HCP}/D_{HCP} \quad (1)$$

where RL_{HCP} and RL_{Co} are the measured luminescence signals (RL), and D_{HCP} and D_{Co} are the absorbed doses, respectively, from the irradiations with heavy charged particles (HCP) and ^{60}Co gamma rays. The quantity RL_{Co}/D_{Co} is a constant (C) for each fibre type.

2.2. Optical Fibres and Detectors

We tested several fibre probes with Al_2O_3:C crystals and powder grains produced by Landauer, Stillwater, OK, USA: one 'Single Crystal'-type, with one Al_2O_3:C crystal ($2 \times 1 \times 1$ mm^3); two '38 µm'-types, with droplets containing Al_2O_3:C with average micro-crystal (or grain) size of 38 µm [36,37] (r = 0.5 mm and l = 200 µm); and two '4 µm'-types, with droplets containing Al_2O_3:C with an average crystalline grain size of 4 µm (r = 0.5 mm and l = 200 µm) [38]. All PMMA optical fibres were 15 m long, with a 1 mm diameter. We irradiated the detectors before the experiments in HIMAC to fill deep traps to saturation [39]. A bi-alkali photomultiplier tube (PMT) P30USB (Sens-Tech^{TM}) reads the RL signal from the probes, while two 2 mm 425 nm Hard Coated Broadband Bandpass Interference Filters

(Edmund Optics, Nether Poppleton, York, United Kingdom) allow only the slow 420 nm component from Al_2O_3:C to pass. For all measurements, we sampled at 200 points per second using a NI USB 6341 DAQ card (National Instruments, Austin, TX 78759, USA) for data acquisition and control via an in-house developed LabVIEW software. More details of the RL prototype can be found in previous publications [38,40,41].

The evaluation of the relative efficiency μ by Equation (1) requires the doses D_{HCP} and D_{Co} to be in the linear range of the dose–response. In order to check if the RL_{HCP} is linearly proportional to D_{HCP}, the fibres were irradiated with nominal doses D_{HCP} (in water) in the ranges presented in Table 1 for each beam type, energy, dose rate, and type and quantity of fibre probes.

2.3. Quenching Correction

To correct the fibre's measured luminescence for quenching along the central axis of the particle beams, we used the Birks law, further adapting the method proposed by Robertson et al. and Almurayshid et al. [29,42,43] by combining the contribution of fragments along the Bragg curve and the relative luminescence efficiency (μ) described in Section 2.1.

The Birks model describes the RL light emission in terms of the stopping power of the phosphor for the particle beam, according to Equation (2) below.

$$\frac{\mathrm{dRL}}{\mathrm{dx}} = \mathrm{RL_0} \cdot \frac{\mathrm{dE/dx}}{1 + \mathrm{kB} \cdot \mathrm{dE/dx}} \tag{2}$$

where RL is the luminescence intensity, dE/dx is the specific energy deposited by the particles per unit of path length x in the medium, kB is the Birks constant (μg MeV^{-1} cm^{-2}), which depends on the charged particle type and the material and RL_0 is the relative luminescence efficiency of the medium. We rewrite Equation (2) in terms of finite voxels to describe a more realistic therapeutic charged particle beam, where we replace the stopping power term of the Birks equation with LET [44] as follows (Equation (3)):

$$\mathrm{RL_v} = \left(\frac{\mathrm{RL_0} \cdot \mathrm{LET_v}}{1 + \mathrm{kB} \cdot \mathrm{LET_v}} \right) \cdot \phi_\mathrm{v} \tag{3}$$

where RL_v is the light emitted from a voxel of volume "v" (Al_2O_3:C droplets or crystal described in Section 2.2), LET_v is the fluence-averaged LET within the voxel, and ϕ_v is the particle fluence in the voxel. The fluence and fluence averaged-LET (LET_f) from the nuclear fragments of the primary beams were generated via the "TOol for PArticle Simulation" (TOPAS) Monte Carlo code [45] Monte Carlo calculations (Section 2.4).

The finite size of the active volume in the Al_2O_3:C probes caused an averaging of the dose gradients along the Bragg curves. The dose and LET are scored in 0.1 mm volumes in TOPAS, while the probes have different volumes, as described in Section 2.2. The deviation between the dose and LET scored in such voxels compared to the same quantities scored in 0.1 mm wide voxels is taken into account and corrected.

In Equation (3), $RL_v = RL_{HCP}$, gives a direct link between the measurements with the fibre probes ('RL signal'), the nominal given doses measured with the reference Markus chamber (D_{HCP}), and the Birks law for quenching.

In order to correct the measured dose for quenching, a correction factor η is required. This factor takes the form of Equation (4), where the ratio of deposited energy (E_v) to the emitted RL_v light in the voxel "v" can be expressed as:

$$\eta_v = \left(\frac{E_v}{RL_v} \right) = \left(\frac{\phi_v \cdot LET_v}{RL_v} \right) = \frac{1 + kB \cdot LET_v}{RL_o} \tag{4}$$

Our proposed quenching correction factor η [30] is the sum of the fluence-weighted quenching corrections η_i for each particle type (primary and fragment) for specific LET ranges in water.

$$\eta = \sum_{d=0}^{n} \sum_i \left(f_d^i \cdot \eta_i\right) = \sum_{d=0}^{n} \sum_i \left[f_d^i \cdot \left(\frac{1 + kB_i \cdot LET_d^i}{RL_{o_i}}\right)\right] \quad (5)$$

where f_d^i is the percentage contribution in fluence of particle "i" at position "d", multiplied by the correction factor corresponding to the LET at depth d. Each particle (primary and fragment) has its own Birks constant kB_i and multiplication factor $1/RL_{o_i}$.

We determined the Birks constant, kB and the relative luminescence efficiency RL_0 for each fibre type (single crystal, 38 and 4 µm powder) by plotting the normalised 'RL signal' $= [(\mu \cdot D_{HCP_i})/DHCP_{entrance}]$ (i = depth in H_2O) versus LET_f (TOPAS) and then fitting the curves by using Equation (3) in the nonlinear curve fit option in the "fitting" routine in Origin(Pro) (Version 2020b, OriginLab Corporation, Northampton, MA, USA). The parameter RL_o is a scaling factor dependent on the detector geometry and the fluence in the Monte Carlo calculation, while the Birks constant unit is $mg \cdot MeV^{-1} cm^{-2}$.

By applying the quenching correction factors to all 'RL signal' along the Bragg curves, weighted by the relative luminescence efficiency (that correlates 'RL signal' to reference D_{HCP}), one obtains a corrected dose distribution for each fibre probe type.

2.4. Monte Carlo Simulations

The LET values used to correct the quenching from the optical fibre probes are based on fluence-averaged LET (LET_f). Fluence-based (LET_f) and dose-based (LET_D) LET values can vary considerably with depth [46] according to the choice of step limit. This effect strongly affects the LET_D for small step sizes (<500 µm) because Monte Carlo codes usually only consider collisions where the kinetic energy imparted to secondary electrons is below a given threshold, restricting the quantity to shorter-range electrons and giving better characterisation when one wants to correlate the radiation effects to RBE or microdosimetry [47]. This step-limiting effect was studied by Guan et al. [46,48] and further addressed and used by other authors [18,29,49–51]. The agreement is that the step limit effect is negligible for LET_f although it strongly affects LET_D results [52,53]. Since the size of the detectors used in our study is not at the cellular scale (µm), we decided to show only the results related to LET_f.

The "TOol for PArticle Simulation" (TOPAS) Monte Carlo code [45] was used to simulate the fluence and LET_f from the primary beams and their nuclear fragments. The proton, helium, and carbon ion simulations were performed respectively with $10{*}10^6$, $20{*}10^6$, and $25{*}10^5$ histories. The error statistics in output results (fluence) were (a) <0.01% along the 160 MeV proton beam up to the Bragg peak (0.04% at the 80% distal falloff depth), (b) <0.02% along the 230 MeV proton beam up to the Bragg peak (0.04% at the 80% distal falloff depth), (c) <0.01% along the 150 MeV/u helium beam up to the Bragg peak (0.07% at the 80% distal falloff depth), (d) <0.05% along the 290 MeV/u carbon beam up to the Bragg peak (0.55% at the 80% distal falloff depth), and (e) <0.1% along the 400 MeV/u carbon beam up to the Bragg peak (0.2% at the 80% distal falloff depth). In TOPAS, a particle fluence scorer and a fluence-averaged LET scorer were attached to the simulated water volume in function of beam penetration depth. A dedicated filter was assigned to both active scorers to separate the scored fluence and LET signals for the primary beam (^1H, ^4He or ^{12}C) and a list of nuclear fragments (^1H, ^4He, ^6Li, ^7Be, ^{10}B, ^{14}N and ^{16}O). In post-processing software, written in Matlab R2020b (The Mathworks Inc., Natick, MA, USA), the output of the energy deposit scorer was divided by the output of the fluence scorer, multiplied by the voxel volume to obtain LET in MeV/mm units. The final LET values were converted to MeV/cm or keV/µm. Fluence and fluence LET were scored with the resolution of 0.1 mm, so that the entrance position for TOPAS simulations is defined as within the first 0.1 mm in water.

2.5. Overview of Tests

The dose–response curves of proton, helium, and carbon-charged particles were evaluated by placing the fibre probes at entrance depth, where each dose relates to the 'RL signal' in the beam's isocenter, as defined in Section 2.1 and Figure 1. The RL emission from Al_2O_3:C is proportional to the dose rate [38,40]. To correlate 'RL signal' to the given dose, the 'ΔRL' is calculated as the sum of the 'RL signal' from the start of irradiation (t = 0) until the end (t = T), corrected for the averaged background ($\overline{Bkg}$) for each independent irradiation (Equation (6)). Each $\overline{Bkg}$ was acquired by averaging 25 measured points prior and 25 measured points post-irradiation (m = 50).

$$'\text{RL signal}' = \Delta RL = \sum_{t=0}^{t=T} RL_t - \left[\frac{1}{m}\cdot\sum_{n=1}^{m} Bkg_n\right] = \sum_{t=0}^{t=T} RL_t - \overline{Bkg} \quad (6)$$

The linearity of the dose response was evaluated by calculating the linearity index of the measurements. The linearity index describes the departure of the detector's response from linearity at a chosen calibration dose D_0. The sensitivity of the fibre probe at dose D_i for the i-th radiation is related to the observed RL signal ('RL signal'$_i$/D_i) and is further normalised by the chosen calibration dose D_0 = 1 Gy so that the linearity index is: ('RL signal'$_i$/D_i)/('RL signal'$_{1\,Gy}$/$D_{1\,Gy}$). Fitting curves were derived to correct further the 'RL signal' for the dose ranges where deviations from linearity were observed (Section 3.1).

The Al_2O_3:C optical fibre probes' dependence on radiation LET was assessed by measuring the Bragg curves for proton, helium, and carbon-charged particles. The measurements in the beam's isocenter were rescaled according to the relative luminescence efficiencies at entrance doses (d in H_2O = 0 mm) and compared with the curves assessed with a Markus ionisation chamber in terms of peak-to-plateau ratio signals (Section 3.2). The Monte Carlo simulated fluence and fluence-averaged LET (Section 3.3) was used to correct the depth-dose curves for quenching using the quenching model for proton, helium and carbon-charged particles determined in Section 2.3. The model was first implemented assuming only the primary proton-charged particles and later expanded considering a combination of the primary beam plus fragments for the helium (^{1}H fragment) and carbon- (^{1}H and ^{4}He fragments) charged particles (Sections 3.3 and 3.5).

3. Results

In this Section, we present the results and analysis of the measurement campaigns, starting with the dose responses and Bragg curves for all probe types, followed by the determination of calibration curves for the and $\eta_{LET,Co}$ compared to LET. We introduce a method to determine the unknown doses and LET of particle therapy fields using a combination of two or more probes. We used the LET dependence of the RL from different Al_2O_3:C probes to establish fluence-LET (LET_f) calibration curves. Our fundamental assumption was that the RL signal does not depend on beam type/energy, as well as dose–rate and absorbed dose, and thus the RL signal can describe averaged LET values. Our results offer a proof of concept of the proposed method. Limitations on applying this method in practical applications will be discussed at the end of this session.

3.1. Fibres Dose Response

Figure 3a–e show the dose calculated from the ΔRL for 'Single Crystal', '38 μm' and '4 μm' fibre probes irradiated with 160 MeV proton, 230 MeV proton, 150 MeV/u helium, 290 MeV/u carbon and 400 MeV/u carbon, respectively. In all figures, each point is the average of independent irradiations (Table 1), and the standard deviations (1 SD), not plotted in the graphs, are below 1% for 38 and 4 μm fibres and 5% for the 'Single Crystal' probe.

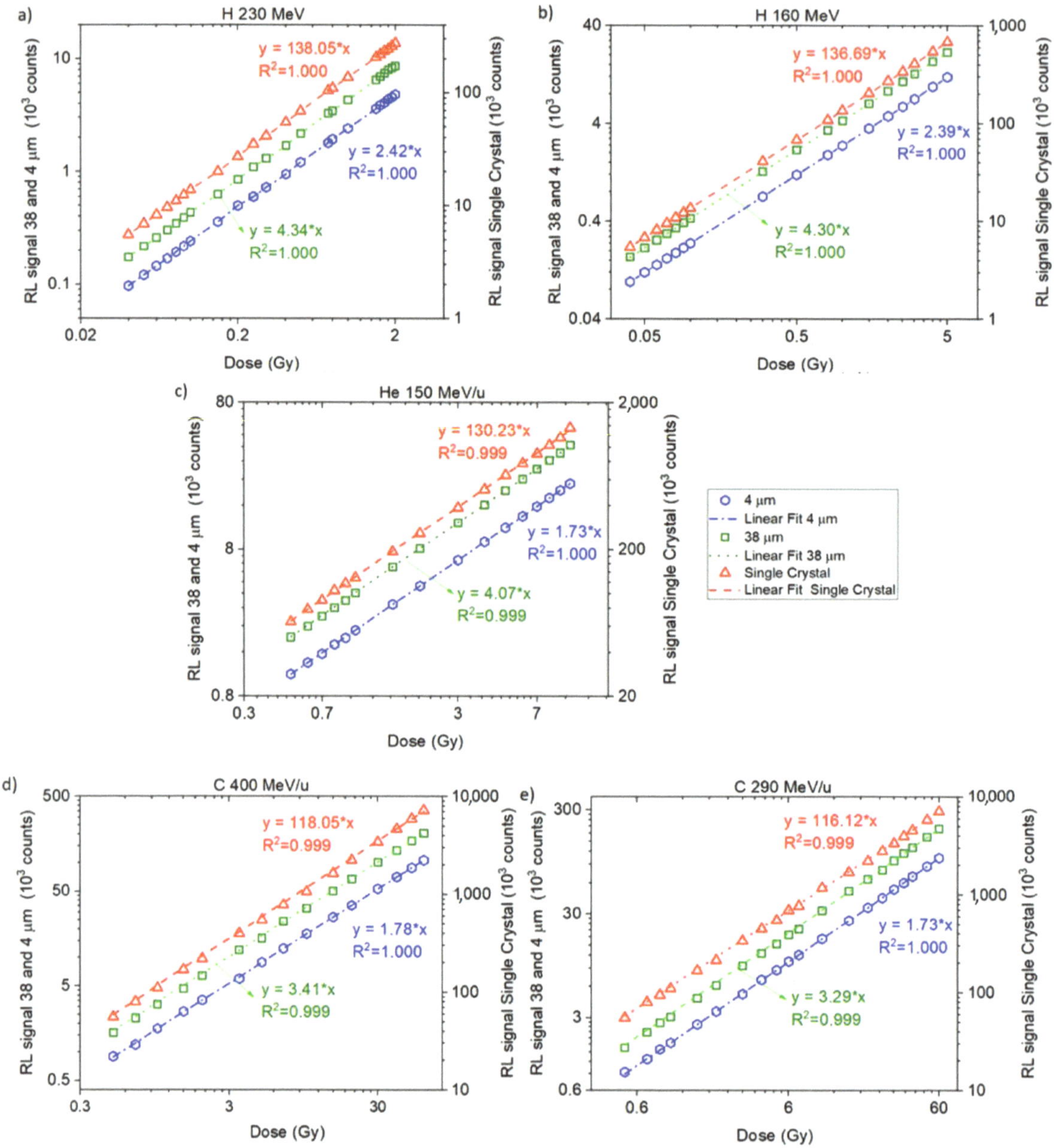

Figure 3. Dose response 'Single Crystal', '38 µm' and '4 µm' from (**a**) 0.04 to 2.00 Gy (0.08 Gy/min) and from 0.1 to 0.5 Gy (0.25 Gy/min) irradiated with 160 MeV protons; (**b**) 0.03 to 0.5 Gy (0.36 Gy/min) and from 0.5 to 4.0 Gy (0.65 Gy/min) irradiated with 230 MeV protons; (**c**) 0.3 to 10.0 Gy (4.5 Gy/min) irradiated with 150 MeV/u helium ions; (**d**) 0.5 to 10.0 Gy (3.04 and 7.34 Gy/min) irradiated with 290 MeV/u carbon ions; (**e**) 0.1 to 5.0 Gy (6.90 Gy/min) irradiated with 400 MeV/u carbon ions.

The proton curves are very similar for all the same fibre types, with slopes (s) of 2.42 and 2.39 ('4 µm'), 4.34 and 4.30 ('38 µm'), and 138.05 and 136.69 ('Single Crystal') for 230 and 160 MeV, respectively, resulting in $\bar{s}$ = 2.405 ± 0.015 (4 µm), $\bar{s}$ = 4.32 ± 0.02 (38 µm), and $\bar{s}$ = 137.37 ± 0.68 ('Single Crystal'). The higher the beam LET, the flatter the curves.

Before measuring the RL signal along the Bragg curve, we tested in which dose ranges the optical probes responded linearly. The available dose rates for irradiations with both 160 and 230 MeV protons were considerably lower than those available for helium and carbon (Table 1). Hence, the dose ranges in Figure 3 are different.

The linearity index is depicted in Figure 4 for the helium and carbon beams. A supralinear behaviour is observed for both '38 μm' and 'Single Crystal' starting from 2 Gy, with maximum overresponses of 7.5% ('38 μm') and 17.5% ('Single Crystal') at 60 Gy for the carbon-heavy charged particles. The linearity index for the '4 μm' fibre probe did not show supralinearity for doses below 60 Gy, and no correction was needed for the subsequent results.

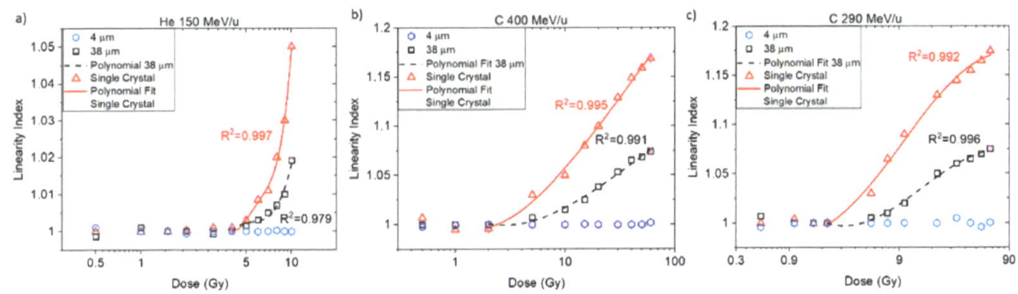

Figure 4. Linearity index relative to 1 Gy for '4 μm', '38 μm' and 'single crystal' optical fibre probes for (**a**) He 150 MeV/u, (**b**) C 400 MeV/u, and (**c**) C 290 MeV/u charged particles.

The nonlinear response in the '38 μm' and 'Single Crystal' curves were corrected by fitting the linearity index as a function of dose (D). A linearity correction factor (LCF) was defined for each '38 μm' and 'Single Crystal' curve as shown in Equation (7) below and was applied in subsequent sections to correct for the doses measured along the Bragg curves.

$$\left\{\begin{array}{l} LCF_{SingleCrystal}^{He\ 150} = 8.11\cdot D^4 - 24.38\cdot D^3 + 27.46\cdot D^2 - 13.66\cdot D + 3.53 \\ LCF_{38\mu m}^{He\ 150} = 0.82\cdot D^3 - 1.83\cdot D^2 - 1.36\cdot D + 0.66 \\ LCF_{SingleCrystal}^{C\ 400} = -0.03\cdot D^3 + 0.12\cdot D^2 - 0.03\cdot D + 0.99 \\ LCF_{38\mu m}^{C\ 400} = -0.02\cdot D^3 + 0.09\cdot D^2 - 0.07\cdot D + 1.01 \\ LCF_{SingleCrystal}^{C\ 290} = -0.06\cdot D^3 + 0.18\cdot D^2 - 0.02\cdot D + 0.99 \\ LCF_{38\mu m}^{C\ 290} = -0.05\cdot D^3 + 0.19\cdot D^2 - 0.17\cdot D + 1.03 \end{array}\right\} \quad (7)$$

3.2. Bragg Curves

To assess the Bragg curves, we chose specific entrance doses in the linear range for each fibre type and energy (Figure 5 and Table 1) or, when necessary, used corrections according to each fibre type-dose response curve (Equation (7)).

The relative luminescence efficiencies (μ) of 'Single Crystal', '38 μm' and '4 μm' were calculated for the RL signals measured at the entrance doses (d = 0.0 mm in H_2O) using Equation (1). Table 2 shows the calculated μ for each probe type and beam energy, corresponding to the average of different dose rate measurements (Table 1), and the error corresponds to one standard deviation (1 SD).

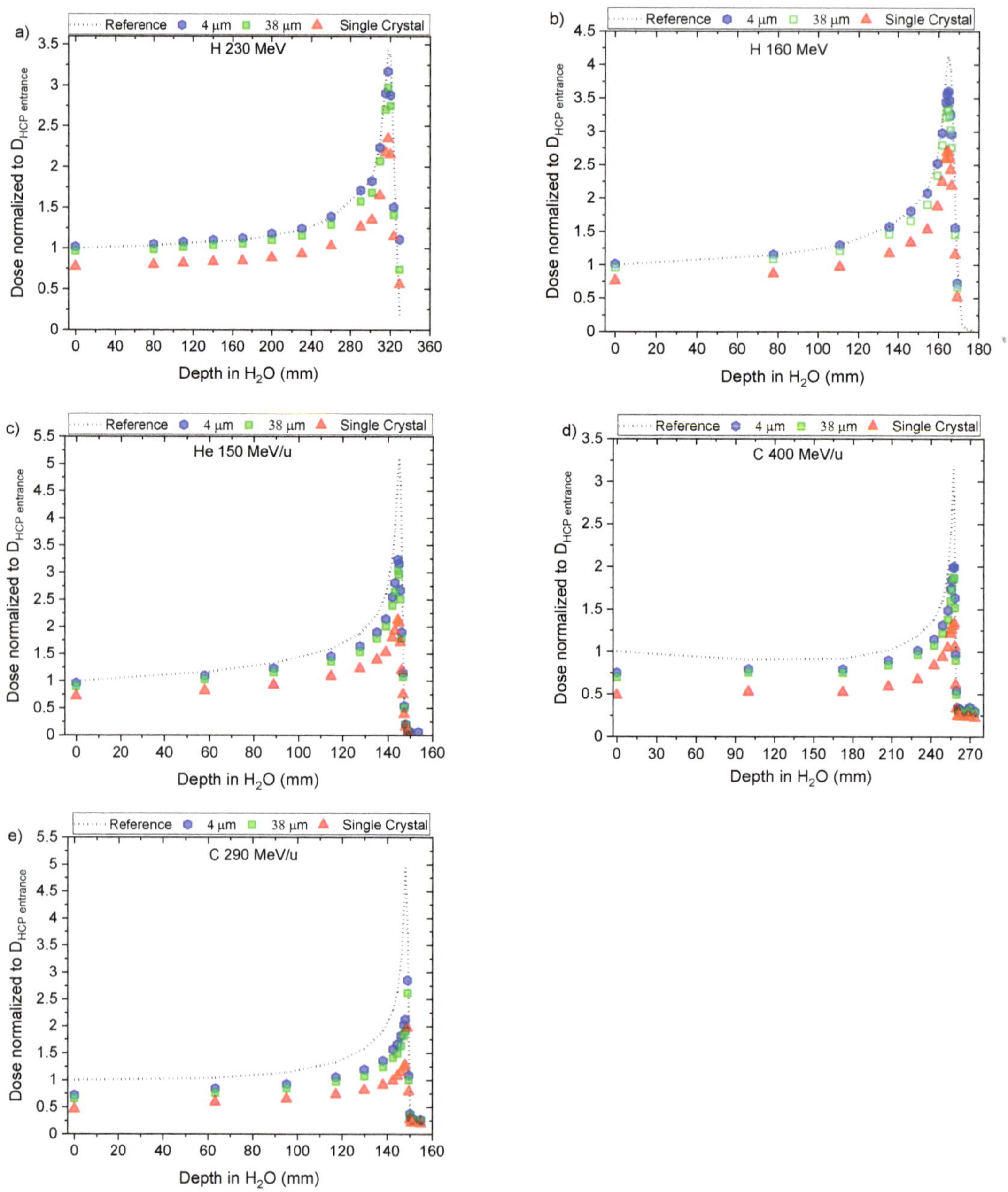

Figure 5. Central axis peak-to-plateau ratios (dose normalised to D_{HCP} entrance) profiles along the Bragg curves for 'Single Crystal', '38 μm' and '4 μm' probes for (**a**) H 160 MeV; (**b**) H 230 MeV; (**c**) He 150 MeV/u; (**d**) for C 400 MeV/u, and (**e**) C 290 MeV/u.

Table 2. Relative luminescence efficiency of fibre probes ('Single Crystal', "38 μm" and "4 μm") at entrance depth in H_2O for H 230, H 160, He 150, C 400, and C 290.

Energy (MeV/u)	LET (keV/μm)	*Relative Luminescence Efficiency (μ) at Entrance Depth in H_2O*		
		4 μm	**38 μm**	**Single Crystal**
H 230	0.41	1.02 ± 0.03	0.97 ± 0.03	0.77 ± 0.06
H 160	0.54	1.01 ± 0.02	0.96 ± 0.03	0.76 ± 0.07
He 150	2.25	0.95 ± 0.02	0.90 ± 0.02	0.71 ± 0.07
C 400	11.22	0.75 ± 0.04	0.71 ± 0.03	0.56 ± 0.08
C 290	13.50	0.73 ± 0.03	0.69 ± 0.02	0.55 ± 0.08

Using the calculated μ at entrance depth from Table 2, we rescaled the doses measured with the optical fibre probes along the Bragg curve for each beam type/energy and fibre type. Figure 5a–e shows the rescaled Bragg curves (quenched) and the ion chamber reference for the 160 and 230 MeV protons, 150 MeV/u helium ions, and 290 and 400 MeV/u carbon ions, respectively.

A comparative analysis of the calculated μ for the three types of optical probes exposed to protons, helium and carbon-charged particles indicates a larger statistical error for 'Single Crystal'. This result is partially due to the spatial non-uniformity of an ion beam at the crystal target area (1 mm) compared to the droplet probes (0.2 mm).

3.3. Fluence and Fluence Averaged-Let

The contribution from the primary beam and its fragments along the Bragg curve concerning fluence and LET_f, is presented in Figure 6 (protons), Figure 7 (helium ions) and Figure 8 (carbon ions).

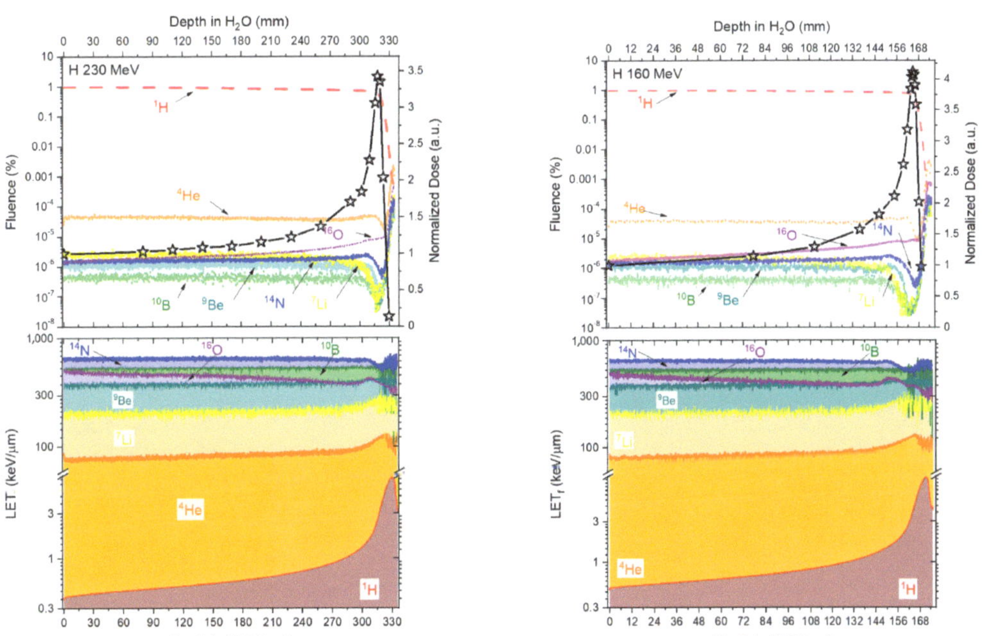

(a) (b)

Figure 6. Fluence (upper-left Y-axis) and LET_f (bottom-left Y-axis) contribution in depth in water from the primary 230 (**a**) and 160 (**b**) MeV proton-charged particles and their fragments (H, He, B, Li, Be, N and O) simulated in TOPAS. Upper-right Y-axis presents the reference measured doses (Markus ion chamber) normalised to the entrance dose ($D_{HCP}/D_{HCP\ entrance}$).

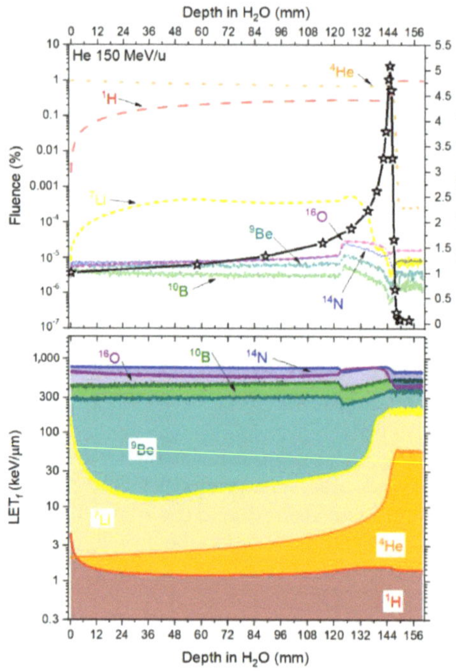

Figure 7. Fluence (upper-left Y-axis) and LET_f (bottom-left Y-axis) contribution in depth in water from the primary 150 MeV/u helium-charged particles and its fragments (H, He, B, Li, Be, N and O) simulated in TOPAS. Upper-right Y-axis presents the reference measured doses (Markus ion chamber) normalised to the entrance dose ($D_{HCP}/D_{HCP\ entrance}$).

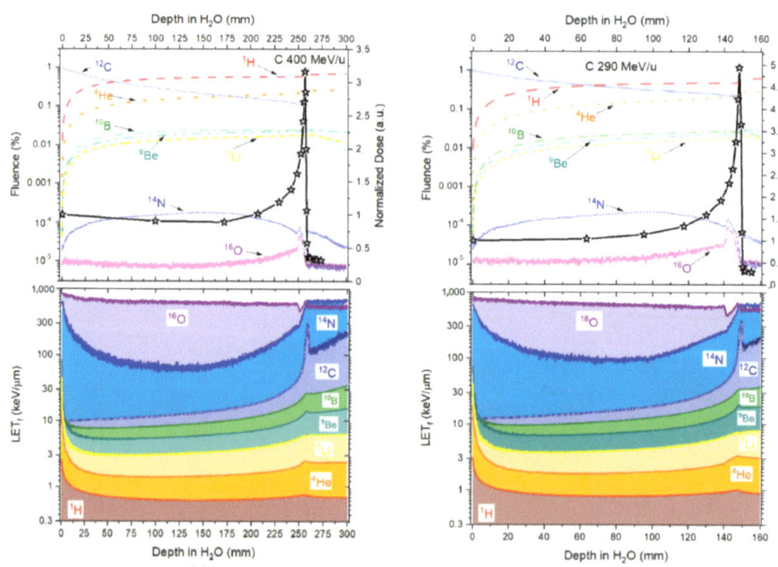

Figure 8. Fluence (upper-left Y-axis) and LET_f (bottom-left Y-axis) contribution in depth in water from the primary 400 (**a**) and 290 (**b**) MeV/u carbon-charged particles and their fragments (H, He, B, Li, Be, N and O) simulated in TOPAS. Upper-right Y-axis presents the reference measured doses (Markus ion chamber) normalised to the entrance dose ($D_{HCP}/D_{HCP\ entrance}$).

Figure 6 and Table 3 show that the contribution, in fluence, from the primary beam (^1H) consists of >99% along the Bragg curves for both 230 and 160 MeV. As such, we assumed that the quenching correction factors (Equation (5)) take the form of Equation (8) below and that, to correct the doses along the Bragg curves, only the primary beam Birks factors were necessary.

$$\eta = \eta_H\left(f_d^H,\ LET_d^H\right) = \left(\frac{1 + kB_{1H} \cdot LET_d^{1H}}{RLo_{1H}}\right) \tag{8}$$

Table 3. Fluence and LET_f were simulated for the 160 MeV and 230 MeV primary beam (^1H) and two fragments (^4He and ^7Li) for four depths in water: entrance, end of the f plateau, Bragg peak, and falloff.

Depth in H_2O (mm)	Fluence (%)	LET_f (keV μm^{-1})	Fluence (%)	LET_f (keV μm^{-1})	Fluence (%)	LET_f (keV μm^{-1})
	^1H		^4He		^7Li	
			230 MeV			
00.10	99.999	0.381	3.6×10^{-3}	83.62	4.5×10^{-5}	502.442
170.42	99.994	0.567	4.2×10^{-3}	82.92	4.1×10^{-5}	490.233
317.74	99.994	3.01	4.1×10^{-3}	122.07	3.3×10^{-6}	519.452
329.4	99.981	6.987	12.8×10^{-3}	115.18	2.3×10^{-4}	575.503
			160 MeV			
00.10	99.995	0.496	3.4×10^{-3}	84.60	3.9×10^{-5}	536.542
110.8	99.995	0.842	3.8×10^{-3}	87.27	4.6×10^{-5}	528.111
164.81	99.996	3.820	2.7×10^{-3}	127.22	2.7×10^{-5}	555.512
168.8	99.997	7.93	1.7×10^{-3}	104.29	7.7×10^{-5}	583.401

Figure 7 and Table 4 show that the fluence coming from the primary beam decreases by around 29% at the Bragg peak (d = 30.68 mm), with 29.4% of the fluence coming from ^1H and 0.004% coming from the other fragments. When looking at the falloff region, the contribution to the total fluence from ^1H amounts to >98%. As such, the quenching correction factor for the Helium curves takes the form of Equation (9), where η is mostly affected by the primary beam (^4He) and the ^1H fragment.

$$\eta = \eta_{He}\left(f_d^{He}, LET_d^{He}\right) + \eta_H\left(f_d^H,\ LET_d^H\right) = \left[f_d^H \cdot \left(\frac{1 + kB_H \cdot LET_d^H}{RL_{oH}}\right) + f_d^{He} \cdot \left(\frac{1 + kB_{He} \cdot LET_d^{He}}{RL_{oHe}}\right)\right] \tag{9}$$

Table 4. Fluence and LET_f simulated for the 150 MeV/u primary beam (^4He) and two fragments (^1H and ^7Li) for four depths in water: entrance, end of plateau, Bragg peak, and falloff.

Depth in H_2O (mm)	Fluence (%)	LET_f (keV μm^{-1})	Fluence (%)	LET_f (keV μm^{-1})	Fluence (%)	LET_f (keV μm^{-1})
	^4He		^1H		^7Li	
			150 MeV/u			
00.10	99.636	2.077	0.360	3.94	4.00×10^{-4}	423.292
88.91	74.705	3.221	25.255	1.241	3.30×10^{-4}	27.360
144.91	70.525	22.659	29.471	1.562	6.29×10^{-5}	205.123
148.1	1.227	56.400	98.768	1.464	1.49×10^{-5}	330.995

Figure 8 and Table 5, the primary carbon-charged particles are responsible for ~98% in fluence contribution (d = 0.01 mm), with a rapid increase in fragments contribution with deeper depths, reaching a contribution in fluence of ~61% (^1H) and ~24% (^4He) at the Bragg peak (d = 256.9 mm) for 400 MeV/u and for ~51% (^1H) and ~23% (^4He) at the Bragg peak (d = 147.92 mm) for 290 MeV/u. At the falloff region, primary carbon-charged particles are almost negligible. The quenching correction factor for the carbon-charged particles takes the form of Equation (10), where the quenching correction factor has the contribution from the primary beam (^{12}C) and the ^{1}H and ^{4}He fragments.

$$\eta = \eta_C(f_d^C,\ LET_d^C)\ +\eta_{He}(f_d^{He}, LET_d^{He}) + \eta_H(f_d^H,\ LET_d^H) = \left[f_d^H \cdot \left(\frac{1+kB_H \cdot LET_d^H}{RL_{o_H}}\right) + f_d^{He} \cdot \left(\frac{1+kB_{He} \cdot LET_d^{He}}{RL_{o_{He}}}\right) + f_d^C \cdot \left(\frac{1+kB_C \cdot LET_d^C}{RL_{o_C}}\right)\right] \tag{10}$$

Table 5. Fluence and LETf simulated for the 400 MeV/u and 290 MeV/u primary beam (^{12}C) and two fragments (^{1}H and ^{4}He) for four depths in water: entrance, end of plateau, Bragg peak and falloff.

Depth in H_2O (mm)	Fluence (%)	LET_f (keV μm^{-1})	Fluence (%)	LET_f (keV μm^{-1})	Fluence (%)	LET_f (keV μm^{-1})
	^{12}C		^{1}H		^{4}He	
			400 MeV/u			
00.10	98.196	9.883	1.646	2.510	0.117	16.517
172.24	18.948	14.972	56.712	0.636	18.215	1.494
256.9	8.031	167.429	61.438	0.763	23.790	2.455
259.21	0.054	291.223	66.900	0.740	25.929	2.411
			290 MeV/u			
00.10	98.600	11.985	1.237	2.790	0.110	15.473
95.03	32.500	18.089	45.839	0.836	15.946	1.875
147.92	18.900	182.151	50.947	1.036	23.109	3.366
150.15	0.0001	170.56	63.010	0.981	28.714	3.278

3.4. Relative Luminescence Efficiency Curves (μ)

Figure 9 shows the calculated relative luminescence efficiency (μ) of all the optical probes at entrance depth (d = 0.01 mm) compared with data provided by Yukihara et al. using Al_2O_3:C OSL crystals mixed with a binder to form detectors with a diameter of 7 mm and thickness of 0.3 mm [54]. These detectors were read out, such as the LET dependence was acquired by combining the two known OSL emissions (called UV and blue). One can observe that the Al_2O_3:C RL and OSL relative luminescence efficiencies do follow the same decay trend, with the '4 µm' showing the closest agreement with the Al_2O_3:C OSL.

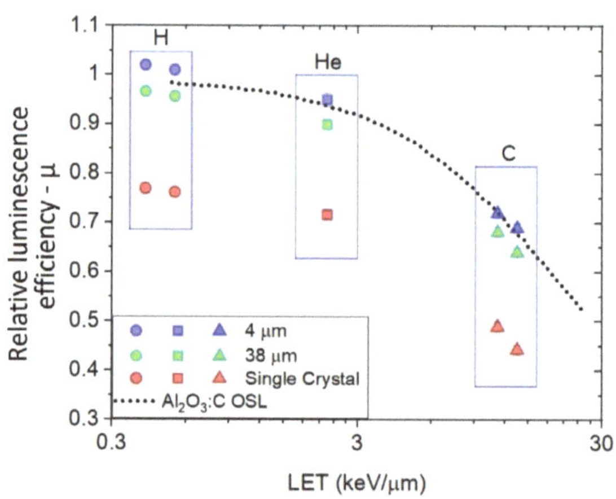

Figure 9. Relative luminescence efficiencies (µ) of fibre probes ('Single Crystal', "38 µm" and "4 µm") at depth in H_2O = 0 mm compared to Al_2O_3:C OSL results.

As a next step, we calculated the relative luminescence efficiencies along the Bragg curves and plotted the results against the simulated primary LET_f at each depth position. In Figure 10, the µ curve from the '4 µm' fibre clearly follows the same trend as observed for the Al_2O_3:C OSL for the μ calculated using the primary LET_f from 160 (orange hexagons) and 230 (red circles) MeV proton-charged particles. At deeper depths, the μ calculated using the primary LET_f from 150 MeV/u helium-charged particles (green triangles) decreased to 30 keV/μm, where the curve started increasing again. For the data calculated using the primary LET_f from the 400 (purple squares) and 290 (blue rhombi) MeV/u carbon-charged particles, one observes first a jump from μ calculated at the entrance, with the following points following a similar trend as observed for the primary helium-charged particles.

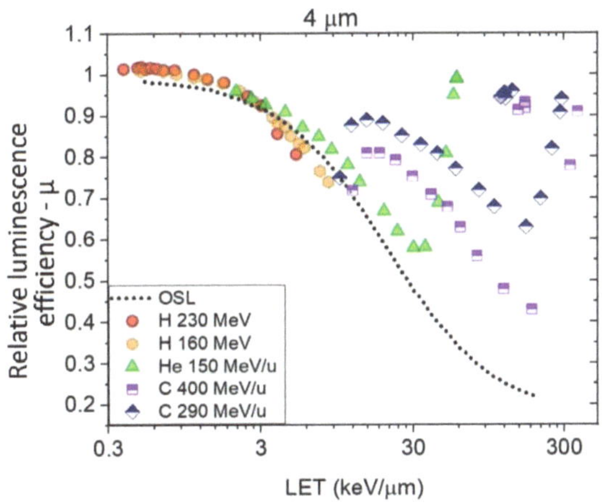

Figure 10. Combination of all the '4 μm' Relative luminescence efficiencies (μ) acquired from different beam types and energies vs. primary LET_f.

Similar results were observed for the ´38 μm´ and ´Single Crystal´ fibre probes when plotting the relative luminescence efficiencies along the Bragg curves vs. simulated primary LET_f, with the difference that the calculated μ are shifted to lower values when compared to the '4 μm' (as observed in Figure 9).

Suppose one uses the fluence weighted contribution of LET_f from the primary and each fragment (Equation (11)) instead of using the primary charged particles to plot the relative luminescence efficiencies. In that case, the points in Figure 10 are rearranged in the form presented in Figure 11.

$$LET_f = \sum_i LET_{f,i} \times fluence(\%)_i \; i = {}^{1}H,\ {}^{4}He,\ {}^{7}Li,\ {}^{9}Be,\ {}^{10}B,\ {}^{12}C,\ {}^{14}N,\ {}^{16}O \quad (11)$$

Figure 11a–c show the relative luminescence efficiencies (μ) and fluence weighted LET_f (primary + fragments) for '4 μm', '38 μm' and ´Single Crystal´, respectively. In all the plots, the curves of both proton beams (160 MeV and 230 MeV) superimpose (i.e., similar μ for similar averaged LET_f), smoothly connecting to the curves generated by the helium-charged particles (150 MeV/u) and to the two carbon-charged particles (290 MeV/u and 400 MeV/u). The combination of all curves reveals a trend consistent with previous results for OSL Al_2O_3:C [54] (shown in Figure 11d).

A fitting exponential curve can describe the combination of the calculated μ vs. LET_f (primary + fragments), as shown in Figure 11a–c by the full black lines. The calculated coefficients of the determination indicate a good correlation, with R^2 = 0.994 ('4 μm'), R^2 = 0.991 ('38 μm') and R^2 = 0.989 ('Single Crystal'). Based on the results, we observed that (i) each probe presents a unique curve μ and (ii) that μ is independent of beam quality (i.e., only depends on the averaged LET_f).

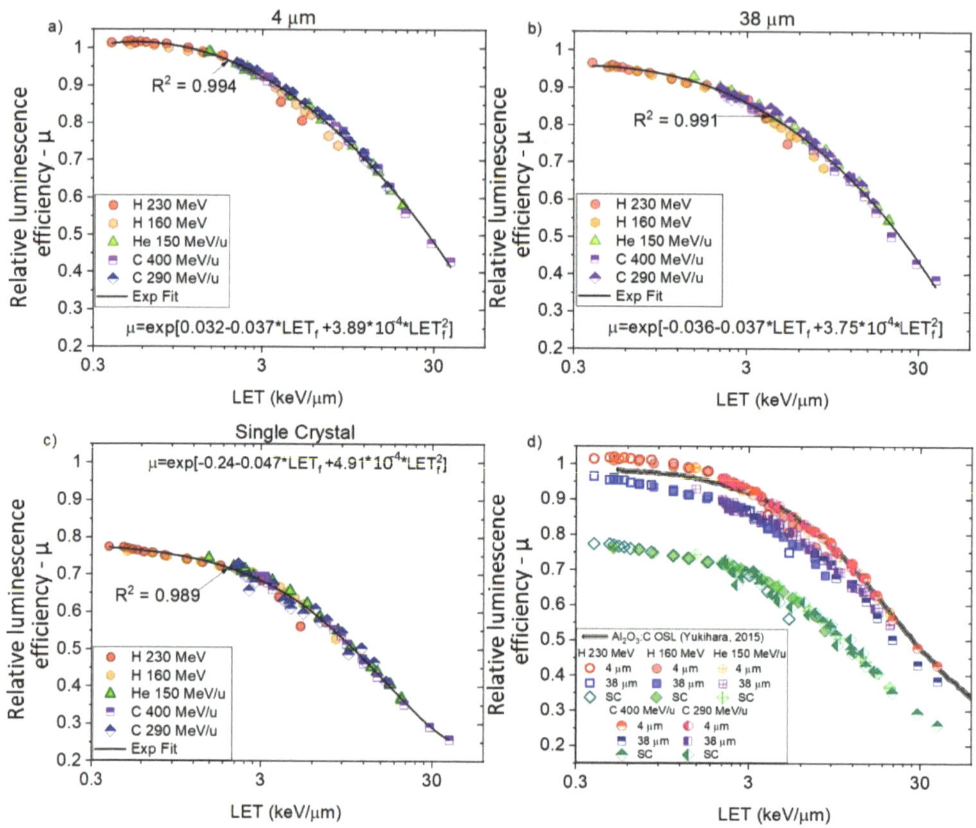

Figure 11. Relative luminescence efficiency (μ) curves and exponential fitting (ExpFit) for (**a**) '4 μm' (R^2 = 0.994), (**b**) '38 μm' (R^2 = 0.991) and (**c**) "Single Crystal" (R^2 = 0.989) acquired from different beam types and energies vs. LET_f and (**d**) the combination of all curves plus A_2O_3:C OSL (data provided by Dr. Yukihara).

3.5. Determination of the Birks Factors and Quenching Corrected Curves

The values of RL_0 and kB (Table 6) were determined for the ´4 μm´, ´38 μm´ and 'Single Crystal' fibre probes irradiated with protons, individually, according to the fitting curves (Equation (3)) presented in Figure 12, as described in Section 2.3. These values are valid for the LET_f range from 3 to 45 MeV/cm and used further to correct for quenching using the correction factor derived for proton-charged particles (Equation (8)).

Table 6. Comparison of the determined kB and RL_o parameters, as in Equation (3), for '4 μm', '38 μm' and 'Single Crystal' for proton-charged particles.

Beam Type/Energy (MeV/u)	LET_f (MeV cm^{-1})	kB (μg MeV^{-1} cm^{-2})			RL_0		
		4 μm	38 μm	Single Crystal	4 μm	38 μm	Single Crystal
H 230 H 160	3–45	(300 ± 20)	(300 ± 30)	(250 ± 30)	1.13	1.09	0.856

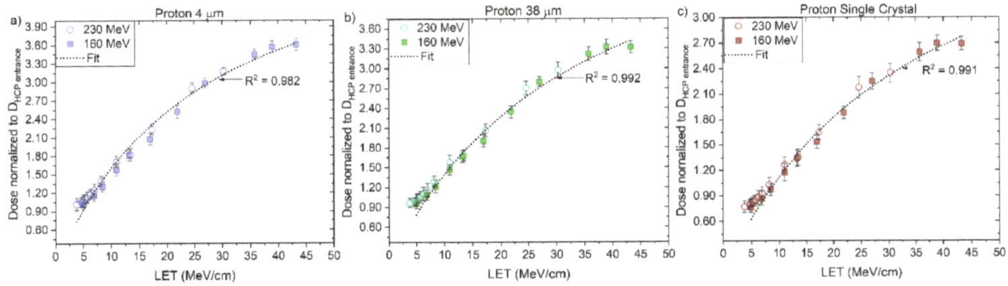

Figure 12. The central axis peak-to-plateau ratios (normalised RL, Figure 5a,b) for the (**a**) '4 μm', (**b**) '38 μm' and (**c**) ´Single Crystal´ fibre probes are shown as a function of averaged electronic stopping power (LET_f) for mono-energetic proton beams with nominal energies 160 and 230 MeV. The RL_o and kB parameters are derived from a nonlinear fit (full line, Equation (3)).

The corrected RL measurements for proton beams, using Equation (8), agreed closely with the reference measurements as shown in Figures 13 and 14a–c, for '4 μm', '38 μm' and ´Single Crystal, respectively. A better overall agreement is again observed for '4 μm' compared to the other two probes. The calculated and corrected Bragg peak heights agreed within 3% ('4 μm'), 4% ('38 μm') and 5% ('Single Crystal') for both proton beams.

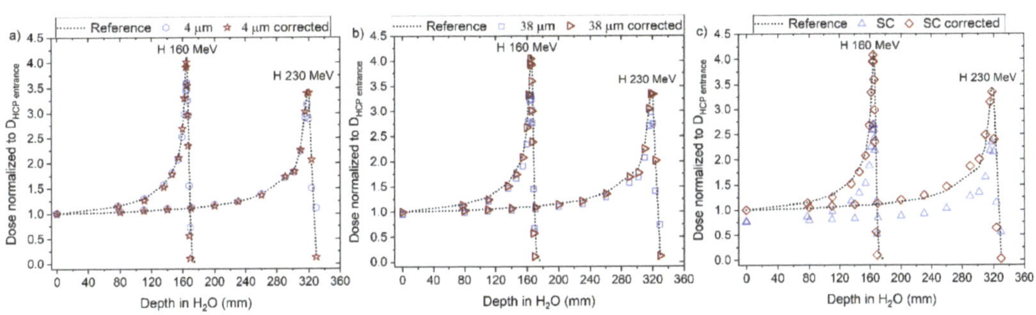

Figure 13. The RL signal converted to dose and normalised to the entrance dose of the reference (D_{HCP} entrance) from the (**a**) 4 μm, (**b**) 38 μm and (**c**) Single Crystal (SC) RL fibre probe before and after the correction factor was applied for 160 and 230 MeV protons beams. The normalised dose from the reference is shown for comparison (dotted line).

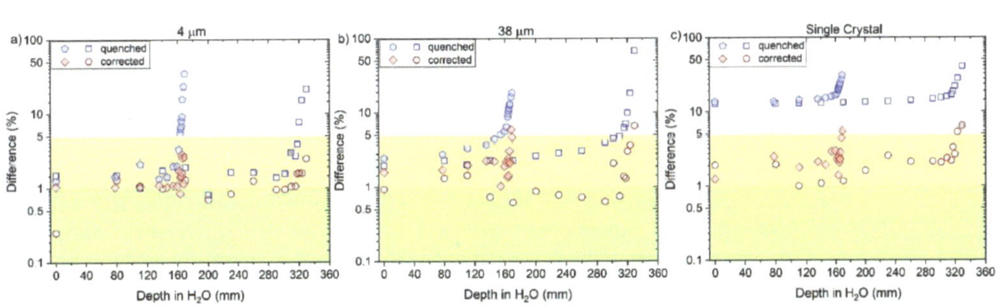

Figure 14. The plots are the difference between the uncorrected (quenched) and corrected RL signal with respect to the reference (ion chamber) from the (**a**) 4 μm, (**b**) 38 μm and (**c**) Single Crystal RL fibre probe before and after the correction factor was applied for 160 and 230 MeV protons beams. The green area in the curve represents the differences < 1%, and the yellow area represents values between 1 and 5%.

As described in Section 3.3, a good approximation for the correction factor function to be applied to the helium-charged particles is based on Equation (9), where the contribution from the primary beam (^{4}He) and the fragment ^{1}H account for most of the beam's fluence.

The Birks factor (kB) is characteristic of the material and can have different values for the same material in different measurements and data treatment conditions. In our study, the kB values were obtained by fitting data for particles of one kind and in some specific energy/LET range. We assumed, as such, that the kB_H and RL_{o_H} from Equation (9) are the same derived from the fitting in Figure 12 and described in Table 6 for each fibre probe type.

The values of RL_0 and kB (Table 7) were determined for the '4 µm', '38 µm' and 'Single Crystal' fibre probes irradiated with 150 MeV/u helium-charged particles, according to the fitting curves (Equation (3)) presented in Figure 15, as described in Section 2.3. These values are valid for the primary LET_f range from 20 to 240 MeV/cm from ^{4}He. The parameters from Tables 6 and 7 are combined to correct for quenching using the correction factor derived for helium-charged particles (Equation (9)).

Table 7. Comparison of the determined kB and RL_o parameters, as in Equation (3), for 4 µm, 38 µm and Single Crystal for helium-charged particles.

Beam Type/Energy (MeV/u)	LET_f (MeV cm^{-1})	kB (µg MeV^{-1} cm^{-2})			RL_0		
		4 µm	38 µm	Single Crystal	4 µm	38 µm	Single Crystal
He 150	20–240	(120 ± 30)	(110 ± 35)	(100 ± 42)	1.3	1.2	0.78

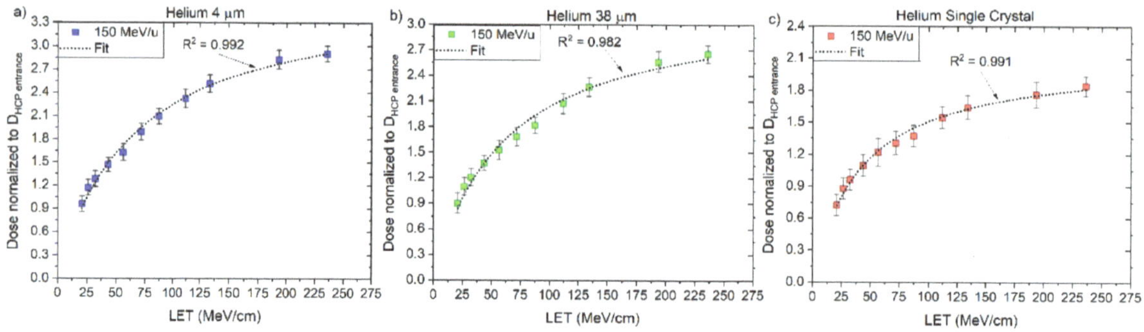

Figure 15. The central axis peak-to-plateau ratios (normalised RL, Figure 5c) for the (**a**) 4 µm, (**b**) 38 µm and (**c**) 'Single Crystal' fibre probes are shown as a function of averaged electronic stopping power (LET_f) for mono-energetic proton beams with nominal energies 160 and 230 MeV. A nonlinear fit (full line, Equation (3)) determines the RL_o and kB.

The corrected RL measurements for the 150 helium beam, using Equation (9), agreed closely with the reference measurements as shown in Figures 16 and 17a–c, for '4 µm', '38 µm' and 'Single Crystal, respectively. The corrected curves for quenching present a clear improvement in the dose–response, especially for points close to the Bragg peak. For example, the difference between the '4 µm' fibre probe and reference improved from ~30% to ~5% at the Bragg peak (144.91 mm). The same type of improvement is also observed for the other two probes.

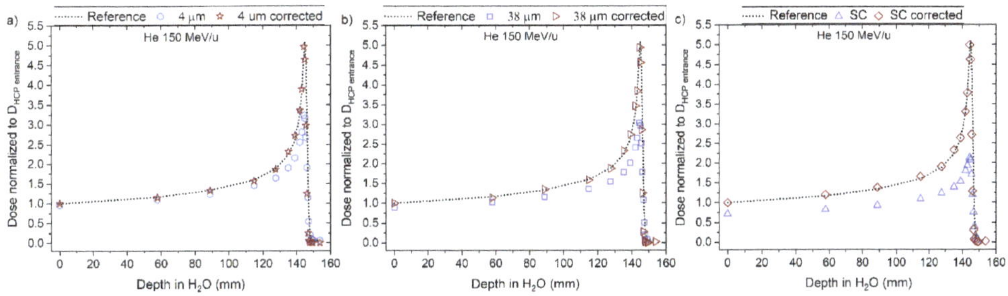

Figure 16. The RL signal converted to dose and normalised to the entrance dose of the reference (D_{HCP} entrance) from the (**a**) 4 μm, (**b**) 38 μm and (**c**) Single Crystal (SC) RL fibre probe before and after the correction factor was applied for 150 MeV/u helium beams. The normalised dose from the reference is shown for comparison (dotted line).

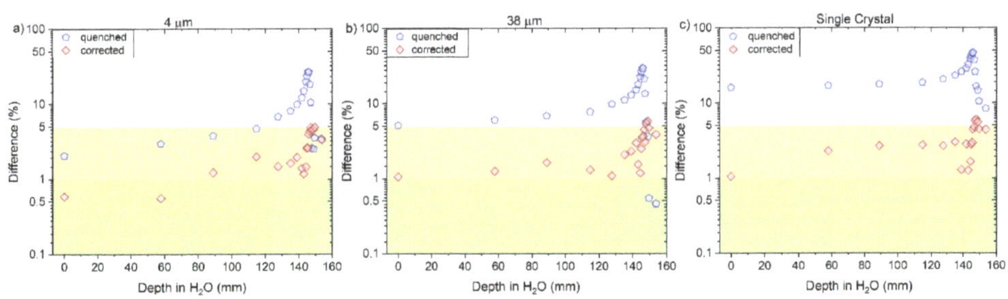

Figure 17. The plots are the difference between the uncorrected (quenched) and corrected RL signal concerning the reference (ion chamber) from the (**a**) 4 μm, (**b**) 38 μm and (**c**) Single Crystal RL fibre probe before and after the correction factor was applied for 150 MeV/u helium beams. The green area in the curve represents the differences < 1%, and the yellow area represents values between 1 and 5%.

Moving forward to correct the quenched curves measured in the carbon-charged particles, Equation (10) is the sum of the contribution from the primary beam (^{12}C) and the fragments ^{1}H and ^{4}He. We assume that the Birks factors (kB) from the fragments are already defined in Table 6 for kB_H and RL_{o_H}, and Table 7 for kB_{He} and $RL_{o_{He}}$ for each fibre probe type.

The values of RL_0 and kB were determined for the '4 μm', '38 μm' and 'Single Crystal' fibre probes irradiated with 290 and 400 MeV/u carbon-charged particles, according to the fitting curves (Equation (3)) presented in Figure 18 and Table 8, as described in Section 2.3.

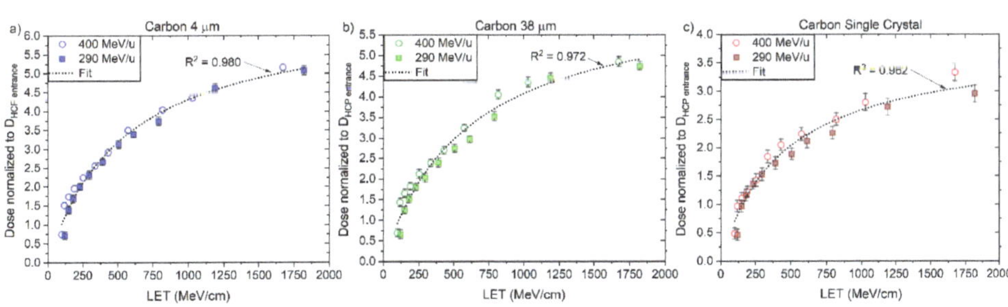

Figure 18. The central axis peak-to-plateau ratios (normalised RL, Figure 5d,e) for the (**a**) 4 μm, (**b**) 38 μm and (**c**) 'Single Crystal' fibre probes are shown as a function of averaged electronic stopping power (LET_f) for mono-energetic carbon beams with nominal energies 290 and 400 MeV/u. The RLo and kB parameters are derived from a nonlinear fit (full line, Equation (3)).

Table 8. Comparison of the determined kB and RL_o parameters, as in Equation (3), for 4 µm, 38 µm and Single Crystal for carbon-charged particles.

Beam Type/Energy (MeV/u)	LET (MeV cm^{-1})	kB (µg MeV^{-1} cm^{-2})			RL_0		
		4 µm	38 µm	Single Crystal	4 µm	38 µm	Single Crystal
C 400 C 290	90–1700	(1.8 ± 0.2)	(1.7 ± 0.3)	(2.5 ± 0.4)	0.2	0.19	0.06

The corrected RL measurements for both carbon beams, using Equation (10), resulted in a significant improvement in the difference values concerning the reference dose measurements, as shown in Figures 19 and 20a–c, for '4 µm', '38 µm' and 'Single Crystal', respectively. The corrected curves for quenching presented differences with respect to the reference between 5 and 8% for doses around the Bragg peak and in the falloff regions.

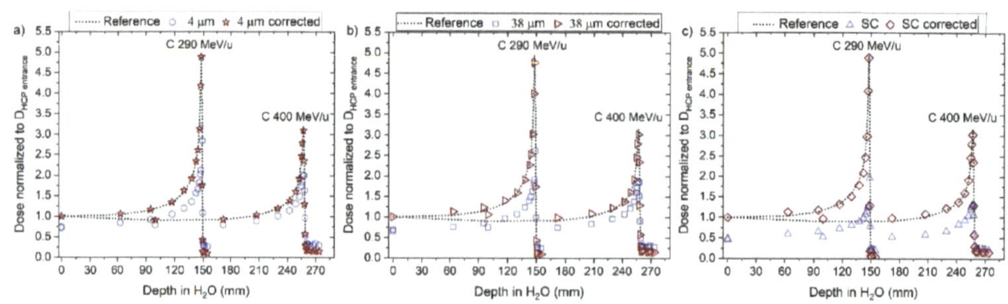

Figure 19. The RL signal converted to dose and normalised to the entrance dose of the reference (D_{HCP} entrance) from the (**a**) 4 µm, (**b**) 38 µm and (**c**) Single Crystal (SC) RL fibre probe before and after the correction factor was applied for 290 and 400 MeV/u carbons beams. The normalised dose from the reference is shown for comparison (dotted line).

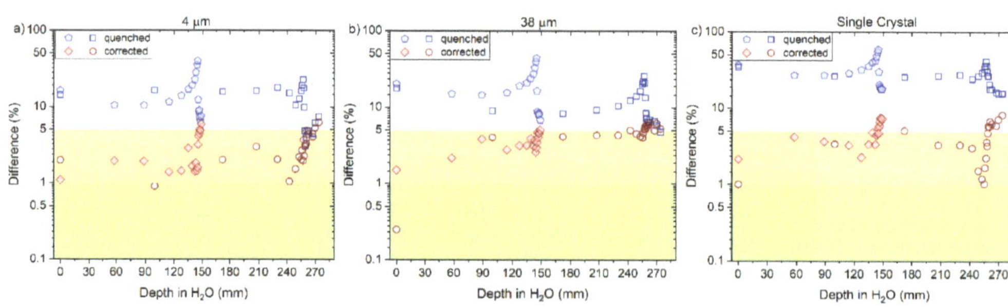

Figure 20. The plots are the difference between the uncorrected (quenched) and corrected RL signal with respect to the reference (ion chamber) from the (**a**) 4 µm, (**b**) 38 µm and (**c**) Single Crystal RL fibre probe before and after the correction factor was applied for 290 and 400 MeV/u carbon beams. The green area in the curve represents the differences < 1%, and the yellow area represents values between 1 and 5%.

4. Discussion

In this paper, we studied a method to correct dose quenching in Al_2O_3:C RL detectors. The Birks formulation was adapted to account for the contribution of fragments generated along the Bragg curves. Our method has been previously applied to correct for quenching in Al_2O_3:C,Mg two-dimensional films irradiated with three different proton and one therapeutic carbon beam. Here, we used our method in three different optical fibre probes and five charged therapeutic beams (protons, helium, and carbon).

For the dose–response test (Figure 3), measured at entrance depth, all the probes presented a linear response for doses up to 2 Gy and at higher doses, supralinearity, with higher deviations observed for the 'Single Crystal' probe followed by the '38 µm'. The results from the '4 µm' fibre probes agreed most with linear dose–response (Figure 4).

The sensitivity of Al_2O_3:C detectors changes with irradiation due to the filling of deep electron and hole traps, competing with dosimetric traps during irradiation and readout [55], a phenomenon generally linked to supralinearity. This effect was previously observed in other studies [38,41,56]. Figure 3 suggests a dependency on crystal size and the deposition of energy, similar to those observed for irradiations with other beam types, such as 6 MV photons [38] and heavily charged particles [30,41]. Although identifying the exact mechanisms explaining the differences between crystal sizes needs further analysis, we believe this effect comes from the competition between the immediate recombination of charge carriers and charge trapping. Pre-irradiated fibres stabilised the RL signal from Al_2O_3:C [39] and Al_2O_3:C, Mg [57]. However, the pre-dosing likely only fills up the charge from the dosimetry traps [58] and does not fill all the deep traps.

If one considers each optical fibre probe as a large cavity, such as the electrons stopping entirely in the RL material, we would not expect differences in quenching observed in probes made with different grain sizes. However, we have observed that the smaller the grain size, the larger the number of ionisations happening in the water-equivalent binder surrounding the grain. In large powder grains and crystals (such as '38 µm' and 'Single Crystal'), there is a higher absorption of the electrons inside the Al_2O_3:C, resulting in a higher ionisation density that causes quenching. This effect with detector size has also been discussed by previous authors using "cavity theory" in OSL/TL passive detectors [59] and scintillators [60] in X-rays.

For measurements along the Bragg curves, we observed quenching for all fibre probes, with a closer agreement to the reference for the '4 µm' fibre, followed by '38 µm' and the 'Single Crystal'. The same trend was observed for all beams and energies. There is also a better agreement for lower LET beams (i.e., 230 MeV protons) than for the higher LET beams (290 MeV/u). The link between quenching and crystal size was observed previously [30,41], where probes with '38 µm' were compared with 'Single Crystal'. We further studied crystal size dependence with LET by adding an extra (smaller) crystal size ('4 µm') and four extra beams. Although the difference between the rescaled Bragg curves from '4 µm' and '38 µm' was smaller than the difference between '38 µm' and 'Single Crystal', we did not find a linear correlation with crystal size.

Quenching was previously observed for Al_2O_3:C when used as both passive (OSL) and active (RL) detectors. Andersen et al. studied the Al_2O_3:C RL vs. absorbed dose–rate during 175 MeV proton radiotherapy [61]. They observed that in the low 0–0.3 Gy range, the RL signal closely resembles that observed for a clinical 6 MV X-ray beam without any LET-dependent correction factors. In contrast, the relative luminescence efficiency decreased to about 60% for higher doses. Klein et al. tested a thin layer of Al_2O_3:C to resolve the steep gradients of the ion depth-dose curves in 142.66 MeV proton and 270.55 MeV/u carbon ion beams and observed a relative luminescence efficiency dropping for higher LET values [62]. Measurements with helium, carbon, neon and iron ions demonstrated that the Al_2O_3:C OSL signal is also strongly LET-dependent [34,41].

The energy deposition along the Bragg curves can explain the quenching dependence with LET. Near the Bragg Peak and in the falloff region, primary proton, helium, and carbon-charged particles experience a rapid increase in their LET values, nearing the end of their ranges. Fragmentation (most prominent for carbon beams) generates secondary particles with very high LET (such as alpha particles and heavy ion recoils) [63] that will create regions of highly high local dose in the close vicinity of the ion track, saturating RL centres and causing luminescence quenching [64]. As scintillators are used in several applications where heavy particles are present, from medical applications to dark matter studies, many approaches for the calculation of quenching factors have been proposed [43,65–67]. However, there is no standard theory to predict and describe measured quenched response curves. According to the Birks model, two ions with the same LET but a different atomic number (Z) will result in the same ionisation quenching and, consequently, the same kB. Many experiments, however, contradict such a statement [67–69], showing that the Birks factor (kB) is characteristic of the material and can have different values for the same material in various measurements and data treatment conditions.

Here, the kB values were obtained by fitting data for particles of one kind and in some specific energy/LET range. The fitting curves derived from the measured '4 μm' RL signals and simulated LET_f values are $R^2 > 0.982$ for protons, $R^2 > 0.992$ for helium and $R^2 > 0.991$ for carbon-charged particles (Figures 12a, 15a and 18a), from the measured '38 μm' RL signals and simulated LET_f values, are $R^2 > 0.992$ for protons, $R^2 > 0.982$ for helium and $R^2 > 0.991$ for carbon-charged particles (Figures 12b, 15b and 18b) and from the measured 'Single Crystal' RL signals and simulated LET_f values are $R^2 > 0.980$ for protons, $R^2 > 0.972$ for helium and $R^2 > 0.962$ for carbon-charged particles (Figures 12c, 15c and 18c).

Our method shows promising results when applied in the plateau and peak region of the Bragg curves for the five beam types studied. The method is less accurate for the points measured at falloff (see Section 3.5). We believe this is due to the low doses measured and the worse statistics from the Monte Carlo simulations in this region. To improve these errors, we recommend a dedicated measurement campaign, using much higher doses to improve signal-to-noise ratios and new Monte Carlo simulations with a more significant number of events.

5. Conclusions

We found that the response of Al_2O_3:C RL detectors is LET_f-dependent, a general phenomenon observed in solid-state dosimeters. Because of the decrease in relative luminescence efficiencies with LET, doses in heavily charged particle beams cannot be calculated directly from real time RL measurements unless in regions and energies where the relative luminescence efficiencies are flat.

The RL response from three types of Al_2O_3:C optical fibre probes to radiotherapy proton, helium, and carbon-charged particles have been investigated and compared with ionisation chamber measurements in the same conditions. The observed LET-related quenching under response along the Bragg curve was corrected using a new method based on RL measurements and Monte Carlo simulated fluence averaged-LET values. This method demonstrated the linear dose response of all the optical fibre probes.

Author Contributions: L.d.F.N.—Conceptualization, methodology, investigation, Writing—Original Draft, Visualization; J.G.—investigation, Writing—Review and Editing; P.L.—validation, resources, Writing—Review and Editing; B.v.d.H.—resources, Writing—Review and Editing; M.A.—resources, Writing—Review and Editing, Visualization; D.V.—validation, resources, Writing—Review and Editing, Visualization; L.E.C.R.—investigation, Writing—Review and Editing, Visualization; A.V.—Review and Editing. All authors have read and agreed to the published version of the manuscript.

Funding: Luana de Freitas Nascimento has received two travel grants from the Fonds voor Wetenschappelijk Onderzoek (FWO), Belgium. File numbers: K205820N (2019) and K201618N (2017).

Institutional Review Board Statement: Not applicable.

Informed Consent Statement: Not applicable.

Data Availability Statement: Data will be made available on request.

Acknowledgments: The authors thank Satoshi Kodaira (NIRS, Japan) for collaborating during measurements at HIMAC and Eduardo Yukihara; for providing data related to relative luminescence efficiency using OSL materials.

Conflicts of Interest: The authors declare that they have no known competing financial interests or personal relationships that could have appeared to influence the work reported in this paper.

References

1. Amaldi, U.; Braccini, S. Present challenges in hadrontherapy techniques. *Eur. Phys. J. Plus* **2011**, *126*, 70. [CrossRef]
2. Solov'yov, A.V. *Nanoscale Insights into Ion-Beam Cancer Therapy*; Springer: Berlin/Heidelberg, Germany, 2016.
3. Agency, I.A.E. *Dose Reporting in Ion Beam Therapy*; IAEA TECDOC Series; International Atomic Energy Agency: Vienna, Austria, 2007.
4. Patera, V.; Sarti, A. Recent advances in detector technologies for particle therapy beam monitoring and dosimetry. *IEEE Trans. Radiat. Plasma Med. Sci.* **2019**, *4*, 133–146. [CrossRef]
5. Beaulieu, L.; Beddar, S. Review of plastic and liquid scintillation dosimetry for photon, electron, and proton therapy. *Phys. Med. Biol.* **2016**, *61*, R305. [CrossRef] [PubMed]
6. Russo, S.; Mirandola, A.; Molinelli, S.; Mastella, E.; Vai, A.; Magro, G.; Mairani, A.; Boi, D.; Donetti, M.; Ciocca, M. Characterization of a commercial scintillation detector for 2-D dosimetry in scanned proton and carbon ion beams. *Phys. Med.* **2017**, *34*, 48–54. [CrossRef]
7. Marafini, M.; Attili, A.; Battistoni, G.; Belcari, N.; Bisogni, M.G.; Camarlinghi, N.; Cappucci, F.; Cecchetti, M.; Cerello, P.; Ciciriello, F.; et al. The INSIDE project: Innovative solutions for in-beam dosimetry in hadrontherapy. *Acta Phys. Pol. A* **2015**, *127*, 1465. [CrossRef]
8. Grusell, E.; Medin, J. General characteristics of the use of silicon diode detectors for clinical dosimetry in proton beams. *Phys. Med. Biol.* **2000**, *45*, 2573. [CrossRef]
9. Darafsheh, A.; León-Marroquín, E.Y.; Mulrow, D.J.; Baradaran-Ghahfarokhi, M.; Zhao, T.; Khan, R. On the spectral characterization of radiochromic films irradiated with clinical proton beams. *Phys. Med. Biol.* **2019**, *64*, 135016. [CrossRef]
10. Matsufuji, N. Miniature spherical silicon diode as a dosimeter during carbon-ion beam therapy. *Nucl. Instrum. Methods Phys. Res. Sect. A Accel. Spectrom. Detect. Assoc. Equip.* **2019**, *935*, 156–160. [CrossRef]
11. Deng, Y.; Chen, Z.; Wang, Q.; Yepes, P.; Shen, Z.; Chen, H.; Li, J.; Moyers, M.F. Comparison of pencil beam and Monte Carlo calculations with ion chamber array measurements for patient-specific quality assurance. *Radiat. Med. Prot.* **2022**, *3*, 115–122. [CrossRef]
12. Arjomandy, B.; Sahoo, N.; Ding, X.; Gillin, M. Use of a two-dimensional ionization chamber array for proton therapy beam quality assurance. *Med. Phys.* **2008**, *35*, 3889–3894. [CrossRef]
13. Jensen, S.V.; Valdetaro, L.B.; Poulsen, P.R.; Balling, P.; Petersen, J.B.; Muren, L.P. Dose-response of deformable radiochromic dosimeters for spot scanning proton therapy. *Phys. Imaging Radiat. Oncol.* **2020**, *16*, 134–137. [CrossRef] [PubMed]
14. Veronese, I.; Cantone, M.C.; Chiodini, N.; Coray, A.; Fasoli, M.; Lomax, A.; Mones, E.; Moretti, F.; Vedda, A. Feasibility study for the use of cerium-doped silica fibres in proton therapy. *Radiat. Meas.* **2010**, *45*, 635–639. [CrossRef]
15. De Saint-Hubert, M.; Castellano, F.; Leblans, P.; Sterckx, P.; Kodaira, S.; Swakoń, J.; de Freitas Nascimento, L. Characterization of 2D Al_2O_3: C, Mg radiophotoluminescence films in charged particle beams. *Radiat. Meas.* **2021**, *141*, 106518. [CrossRef]
16. Mijnheer, B.; Beddar, S.; Izewska, J.; Reft, C. In vivo dosimetry in external beam radiotherapy. *Med. Phys.* **2013**, *40*, 070903. [CrossRef]
17. Klosowski, M.; Czopyk, L.; Olko, P.; Rebisz, M.; Voss, B.; Waligórski, M.P.R. TL efficiency of LiF: Mg, Cu, P (MCP-N) 2-D thermoluminescence detectors to raster-scanned carbon ion beams. *Radiat. Meas.* **2008**, *43*, 994–997. [CrossRef]
18. Christensen, J.B.; Togno, M.; Bossin, L.; Pakari, O.V.; Safai, S.; Yukihara, E.G. Improved simultaneous LET and dose measurements in proton therapy. *Sci. Rep.* **2022**, *12*, 8262. [CrossRef]
19. Christensen, J.B.; Togno, M.; Nesteruk, K.P.; Psoroulas, S.; Meer, D.; Weber, D.C.; Lomax, T.; Yukihara, E.; Safai, S. Al_2O_3: C optically stimulated luminescence dosimeters (OSLDs) for ultra-high dose rate proton dosimetry. *Phys. Med. Biol.* **2021**, *66*, 085003. [CrossRef]
20. Piccinini, M.; Nichelatti, E.; Ampollini, A.; Picardi, L.; Ronsivalle, C.; Bonfigli, F.; Libera, S.; Vincenti, M.A.; Montereali, R.M. Proton beam dose-mapping via color centers in LiF thin-film detectors by fluorescence microscopy. *EPL (Europhys. Lett.)* **2017**, *117*, 37004. [CrossRef]
21. Beddar, S.; Beaulieu, L. *Scintillation Dosimetry*; CRC Press: Boca Raton, FL, USA, 2016.
22. Zubenko, D.A.; Noginov, M.A.; Smirnov, V.A.; Shcherbakov, I.A. Different mechanisms of nonlinear quenching of luminescence. *Phys. Rev. B* **1997**, *55*, 8881. [CrossRef]
23. Akselrod, M.S.; Agersnap Larsen, N.; Whitley, V.; McKeever, S.W.S. Thermal quenching of F-center luminescence in Al_2O_3: C. *J. Appl. Phys.* **1998**, *84*, 3364–3373. [CrossRef]
24. Ogiegło, J.M.; Katelnikovas, A.; Zych, A.; Justel, T.; Meijerink, A.; Ronda, C.R. Luminescence and luminescence quenching in Gd_3 (Ga, $Al)_5O_{12}$ scintillators doped with Ce^{3+}. *J. Phys. Chem. A* **2013**, *117*, 2479–2484. [CrossRef] [PubMed]
25. Janulevicius, M.; Marmokas, P.; Misevicius, M.; Grigorjevaite, J.; Mikoliunaite, L.; Sakirzanovas, S.; Katelnikovas, A. Luminescence and luminescence quenching of highly efficient $Y_2Mo_4O_{15}$: Eu^{3+} phosphors and ceramics. *Sci. Rep.* **2016**, *6*, 26098. [CrossRef] [PubMed]
26. Zhao, Y.; Zhang, H.; Jin, Q.; Jia, D.; Liu, T. Ratiometric Optical Fiber Dissolved Oxygen Sensor Based on Fluorescence Quenching Principle. *Sensors* **2022**, *22*, 4811. [CrossRef] [PubMed]
27. Pratte, J.-F.; Nolet, F.; Parent, S.; Vachon, F.; Roy, N.; Rossignol, T.; Deslandes, K.; Dautel, H.; Fontaine, R.; Charlebois, S.A. 3D photon-to-digital converter for radiation instrumentation: Motivation and future works. *Sensors* **2021**, *21*, 598. [CrossRef]

28. Wang LL, W.; Perles, L.A.; Archambault, L.; Sahoo, N.; Mirkovic, D.; Beddar, S. Determination of the quenching correction factors for plastic scintillation detectors in therapeutic high-energy proton beams. *Phys. Med. Biol.* **2012**, *57*, 7767. [CrossRef]
29. Robertson, D.; Mirkovic, D.; Sahoo, N.; Beddar, S. Quenching correction for volumetric scintillation dosimetry of proton beams. *Phys. Med. Biol.* **2012**, *58*, 261. [CrossRef]
30. Nascimento, L.F.; Leblans, P.; van der Heyden, B.; Akselrod, M.; Goossens, J.; Verellen, D.; Kodaira, S. Characterization and quenching correction for a 2D real time radioluminescent system in therapeutic proton and carbon charged beams. *Sens. Actuators A Phys.* **2022**, *345*, 113781. [CrossRef]
31. Kanai, T.; Fukumura, A.; Kusano, Y.; Shimbo, M.; Nishio, T. Cross-calibration of ionization chambers in proton and carbon beams. *Phys. Med. Biol.* **2004**, *49*, 771. [CrossRef]
32. Murakami, T.; Tsujii, H.; Furusawa, Y.; Ando, K.; Kanai, T.; Yamada, S.; Kawachi, K. Medical and other applications of high-energy heavy-ion beams from HIMAC. *J. Nucl. Mater.* **1997**, *248*, 360–368. [CrossRef]
33. Barish, R.J.; Fleischman, R.C.; Pipman, Y.M. Teletherapy beam characteristics: The first second. *Med. Phys.* **1987**, *14*, 657–661. [CrossRef]
34. Sawakuchi, G.O.; Yukihara, E.G.; McKeever SW, S.; Benton, E.R.; Gaza, R.; Uchihori, Y.; Yasuda, N.; Kitamura, H. Relative optically stimulated luminescence and thermoluminescence efficiencies of Al_2O_3: C dosimeters to heavy charged particles with energies relevant to space and radiotherapy dosimetry. *J. Appl. Phys.* **2008**, *104*, 124903. [CrossRef]
35. Kalef-Ezra, J.; Horowitz, Y. Heavy charged particle thermoluminescence dosimetry: Track structure theory and experiments. *Int. J. Appl. Radiat. Isot.* **1982**, *33*, 1085–1100. [CrossRef]
36. de FNascimento, L.; Vanhavere, F.; Souza, R.; Verellen, D. Al_2O_3: C optically stimulated luminescence droplets: Characterization and applications in medical beams. *Radiat. Meas.* **2016**, *94*, 41–48. [CrossRef]
37. Nascimento, L.F.; Saldarriaga, C.V.; Vanhavere, F.; D'Agostino, E.; Defraene, G.; De Deene, Y. Characterization of OSL Al_2O_3:C droplets for medical dosimetry. *Radiat. Meas.* **2013**, *56*, 200–204. [CrossRef]
38. Nascimento, L.F.; Veronese, I.; Loi, G.; Mones, E.; Vanhavere, F.; Verellen, D. Radioluminescence results from an Al_2O_3: C fiber prototype: 6 MV medical beam. *Sens. Actuators A Phys.* **2018**, *274*, 1–9. [CrossRef]
39. Andersen, C.E.; Damkjær SM, S.; Kertzscher, G.; Greilich, S.; Aznar, M.C. Fiber-coupled radioluminescence dosimetry with saturated Al_2O_3:C crystals: Characterization in 6 and 18 MV photon beams. *Radiat. Meas.* **2011**, *46*, 1090–1098. [CrossRef]
40. Nascimento, L.F.; Vanhavere, F.; Boogers, E.; Vandecasteele, J.; De Deene, Y. Medical Dosimetry Using a RL/OSL Prototype. *Radiat. Meas.* **2014**, *71*, 359–363. [CrossRef]
41. Nascimento, L.F.; Vanhavere, F.; Kodaira, S.; Kitamura, H.; Verellen, D.; De Deene, Y. Application of Al_2O_3: C+ fibre dosimeters for 290MeV/n carbon therapeutic beam dosimetry. *Radiat. Phys. Chem.* **2015**, *115*, 75–80. [CrossRef]
42. Almurayshid, M.; Helo, Y.; Kacperek, A.; Griffiths, J.; Hebden, J.; Gibson, A. Quality assurance in proton beam therapy using a plastic scintillator and a commercially available digital camera. *J. Appl. Clin. Med. Phys.* **2017**, *18*, 210–219. [CrossRef]
43. Birks, J.B. Scintillations from organic crystals: Specific fluorescence and relative response to different radiations. *Proc. Phys. Soc. Sect. A* **1951**, *64*, 874. [CrossRef]
44. Berger, M.J. *Penetration of Proton Beams through Water 1. Depth-Dose Distribution, Spectra and LET Distribution*; US Department of Commerce, National Institute of Standards and Technology: Gaithersburg, MD, USA, 1993.
45. Perl, J.; Shin, J.; Schümann, J.; Faddegon, B.; Paganetti, H. TOPAS: An innovative proton Monte Carlo platform for research and clinical applications. *Med. Phys.* **2012**, *39*, 6818–6837. [CrossRef] [PubMed]
46. Guan, F.; Peeler, C.; Bronk, L.; Geng, C.; Taleei, R.; Randeniya, S.; Ge, S.; Mirkovic, D.; Grosshans, D.; Mohan, R.; et al. Analysis of the track-and dose-averaged LET and LET spectra in proton therapy using the geant4 Monte Carlo code. *Med. Phys.* **2015**, *42*, 6234–6247. [CrossRef] [PubMed]
47. Grimes, D.R.; Warren, D.R.; Partridge, M.J.S.r. An approximate analytical solution of the Bethe equation for charged particles in the radiotherapeutic energy range. *Sci. Rep.* **2017**, *7*, 9781. [CrossRef]
48. Guan, F.; Peeler, C.; Bronk, L.; Geng, C.; Taleei, R.; Randeniya, S.; Ge, S.; Mirkovic, D.; Grosshans, D.; Mohan, R.; et al. Erratum: "Analysis of the track- and dose-averaged LET and LET spectra in proton therapy using the geant4 Monte Carlo code" [Med. Phys. 42 (11), page range 6234–6247(2015)]. *Med. Phys.* **2018**, *45*, 1302. [CrossRef] [PubMed]
49. Mohan, R.; Peeler, C.R.; Guan, F.; Bronk, L.; Cao, W.; Grosshans, D.R. Radiobiological issues in proton therapy. *Acta Oncol.* **2017**, *56*, 1367–1373. [CrossRef]
50. Dahle, T.J.; Rykkelid, A.M.; Stokkevåg, C.H.; Mairani, A.; Görgen, A.; Edin, N.J.; Rørvik, E.; Fjæra, L.F.; Malinen, E.; Kristian, S. Ytre-Hauge Monte Carlo simulations of a low energy proton beamline for radiobiological experiments. *Acta Oncol.* **2017**, *56*, 779–786. [CrossRef]
51. Christensen, J.B.; Almhagen, E.; Stolarczyk, L.; Vestergaard, A.; Bassler, N.; Andersen, C.E. Ionization quenching in scintillators used for dosimetry of mixed particle fields. *Phys. Med. Biol.* **2019**, *64*, 095018. [CrossRef]
52. Paganetti, H. Relative biological effectiveness (RBE) values for proton beam therapy. Variations as a function of biological endpoint, dose, and linear energy transfer. *Phys. Med. Biol.* **2014**, *59*, R419. [CrossRef]
53. Grassberger, C.; Paganetti, H. Elevated LET components in clinical proton beams. *Phys. Med. Biol.* **2011**, *56*, 6677. [CrossRef]
54. Yukihara, E.G.; Doull, B.A.; Ahmed, M.; Brons, S.; Tessonnier, T.; Jäkel, O.; Greilich, S. Time-resolved optically stimulated luminescence of Al_2O_3: C for ion beam therapy dosimetry. *Phys. Med. Biol.* **2015**, *60*, 6613. [CrossRef]

55. Yukihara, E.G.; Gaza, R.; McKeever SW, S.; Soares, C.G. Optically stimulated luminescence and thermoluminescence efficiencies for high-energy heavy charged particle irradiation in Al_2O_3: C. *Radiat. Meas.* **2004**, *38*, 59–70. [CrossRef] [PubMed]
56. Damkjær, S.M.S.; Andersen, C.E.; Aznar, M. Improved real-time dosimetry using the radioluminescence signal from Al_2O_3: C. *Radiat. Meas.* **2008**, *43*, 893–897. [CrossRef]
57. Kalita, J.M.; Chithambo, M.L. On the sensitivity of thermally and optically stimulated luminescence of α-Al_2O_3: C and α-Al_2O_3: C, Mg. *Radiat. Meas.* **2017**, *99*, 18–24. [CrossRef]
58. Pagonis, V.; Lawless, J.; Chen, R.; Andersen, C. Radioluminescence in Al_2O_3: C–analytical and numerical simulation results. *J. Phys. D Appl. Phys.* **2009**, *42*, 175107. [CrossRef]
59. Scarboro, S.; Kry, S.F. Characterisation of energy response of Al_2O_3: C optically stimulated luminescent dosemeters (OSLDs) using cavity theory. *Radiat. Prot. Dosim.* **2013**, *153*, 23–31. [CrossRef]
60. Santurio, G.V.; Pinto, M.; Andersen, C.E. Evaluation of the ionization quenching effect in an organic plastic scintillator using Kv X-rays and a modified Birks model with explicit account of secondary electrons. *Radiat. Meas.* **2020**, *131*, 106222. [CrossRef]
61. Andersen, C.E.; Edmund, J.M.; Medin, J.; Grusell, E.; Jain, M.; Mattsson, S. Medical proton dosimetry using radioluminescence from aluminium oxide crystals attached to optical-fiber cables. *Nucl. Instrum. Methods Phys. Res. Sect. A Accel. Spectrom. Detect. Assoc. Equip.* **2007**, *580*, 466–468. [CrossRef]
62. Klein, F.A.; Greilich, S.; Andersen, C.E.; Lindvold, L.R.; Jäkel, O. A thin layer fiber-coupled luminescence dosimeter based on Al_2O_3:C. *Radiat. Meas.* **2011**, *46*, 1607–1609. [CrossRef]
63. Rosenfeld, A.B.; Bradley, P.D.; Cornelius, I.; Kaplan, G.I.; Allen, B.J.; Flanz, J.B.; Goitein, M.; Van Meerbeeck, A.; Schubert, J.; Bailey, J.; et al. A new silicon detector for microdosimetry applications in proton therapy. *IEEE Trans. Nucl. Sci.* **2000**, *47*, 1386–1394. [CrossRef]
64. Benton, E.R.; Johnson, C.E.; DeWitt, J.; Yasuda, N.; Benton, E.V.; Moyers, M.H.; Frank, A.L. Observations of short-range, high-LET recoil tracks in CR-39 plastic nuclear track detector by visible light microscopy. *Radiat. Meas.* **2011**, *46*, 527–532. [CrossRef]
65. Tretyak, V. Semi-empirical calculation of quenching factors for ions in scintillators. *Astropart. Phys.* **2010**, *33*, 40–53. [CrossRef]
66. Murray, R.; Meyer, A. Scintillation response of activated inorganic crystals to various charged particles. *Phys. Rev.* **1961**, *122*, 815. [CrossRef]
67. Tretyak, V.I. Semi-empirical calculation of quenching factors for scintillators: New results. In *EPJ Web of Conferences*; EDP Sciences: Les Ulis, France, 2014; Volume 65, p. 02002.
68. Newman, E.; Steigert, F. Response of NaI (Tl) to energetic heavy ions. *Phys. Rev.* **1960**, *118*, 1575. [CrossRef]
69. Boivin, J.; Beddar, S.; Bonde, C.; Schmidt, D.; Culberson, W.; Guillemette, M.; Beaulieu, L. A systematic characterization of the low-energy photon response of plastic scintillation detectors. *Phys. Med. Biol.* **2016**, *61*, 5569. [CrossRef]

sensors

MDPI

Article

Location of Latent Forensic Traces Using Multispectral Bands

Samuel Miralles-Mosquera [1], Bernardo Alarcos [2] and Alfredo Gardel [2,*]

[1] Police—Specialist of Forensic Image of General Headquarters of Forensics Police, National Police, c/Julián González Segador s/n, 28043 Madrid, Spain

[2] Polytechnic School, University of Alcala, 28805 Alcala de Henares, Spain

* Correspondence: alfredo.gardel@uah.es; Tel.: +34-918-856-585

Abstract: In this paper, a conventional camera modified to capture multispectral images, has been used to locate latent forensic traces with a smart combination of wavelength filters, capturing angle, and illumination sources. There are commercial multispectral capture devices adapted to the specific tasks of the police, but due to their high cost and operation not well adapted to the field work in a crime scene, they are not currently used by forensic units. In our work, we have used a digital SLR camera modified to obtain a nominal sensitivity beyond the visible spectrum. The goal is to obtain forensic evidences from a crime scene using the multispectral camera by an expert in the field knowing which wavelength filters and correct illumination sources should be used, making visible latent evidences hidden from the human-eye. In this paper, we show a procedure to retrieve from latent forensic traces, showing the validity of the system in different real cases (blood stains, hidden/erased tattoos, unlocking patterns on mobile devices). This work opens the possibility of applying multispectral inspections in the forensic field specially for operational units for the location of latent through non-invasive optical procedures.

Keywords: forensic science; photography; multispectral; latent traces; blood; ultraviolet; infrared

Citation: Miralles-Mosquera, S.; Alarcos, B.; Gardel, A. Location of Latent Forensic Traces Using Multispectral Bands. *Sensors* **2022**, *22*, 9142. https://doi.org/10.3390/s22239142

Academic Editors: Qing Yu, Ran Tu, Ting Liu and Lina Li

Received: 16 October 2022
Accepted: 23 November 2022
Published: 25 November 2022

Publisher's Note: MDPI stays neutral with regard to jurisdictional claims in published maps and institutional affiliations.

Copyright: © 2022 by the authors. Licensee MDPI, Basel, Switzerland. This article is an open access article distributed under the terms and conditions of the Creative Commons Attribution (CC BY) license (https://creativecommons.org/licenses/by/4.0/).

1. Introduction

If there is an axiom that has gone along with photography since its creation, it is the advertising slogan from the *Ermanox* camera, (the camera used by the pioneering photojournalists of the early 20th century) promising the capture of all kinds of indiscreet snapshots, even in impossible lighting conditions: *"with Ermanox, my eye becomes my exposure meter, if I can see it, I can photograph it!"* [1]. By its nature and intended purpose, photography has always been traditionally related to the capture of images within the visible light spectrum, trying to imitate with it the strict patterns of perception of the human eye. However, the demands of a discipline, such as forensic photography, have made it necessary to progressively incorporate new techniques, in many cases far away from the so-called conventional photography. The advances in tools and methods have taken forensic photography far beyond the limited human perception, being especially significant multispectral technology that, despite the associated problems of aliasing [2,3] and artifacts derived from image conditions [4] as a non-invasive procedure for locating vestiges, it has allowed new fields of research to open [5] and minimizing the risks of deterioration or accidental deletion of latent traces, a common problem of classical inspection procedures [6]. There are multiple related fields where multispectral image analysis has been used, for example, to detect certain chemicals in food [7] to detect dangerous gas emissions [8] and, in forensic sciences to retrieve shooting distances [9]. Remote sensing is the main research field where multiple bands of information is considered, thus several advances in hyperspectral images related to reduce band interference and select the best band has been proposed [10]. In our paper, we focus the research work on the forensic inspection to retrieve latent vestiges, considering the advantage of the method because the risks of deterioration or accidental deletion of traces is reduced compared with traditional inspection procedures [11].

The continuous evolution of optical image capture devices, in any of its variants, has contributed, not without some initial scepticism, to the definitive incorporation of this technology into the field of Forensic Sciences, being today an essential tool for any forensic researcher. This progress, especially in the last 45 years with the emergence of electronics and computing in the photography sector, has not only increased the performance of this technology but also improving the sensitivity, precision and definition of current optical devices [12,13]. In addition, the appearance of new techniques and capture image devices has led to greater and faster access to the latent information to be obtained, regardless of whether or not they are initially visible.

Every investigator looking into a crime scene to locate those vestiges relevant to the clarification of a case, must use, during the forensic inspection of the scene, each and every one of the available detection procedures and tools. Therefore, from the definitive incorporation of photography into the field of forensic sciences, one of the main search tools has been the "illumination" of the scene with visible light (VL), which represents only a small fraction of the electromagnetic spectrum contained in the 380–740 nm wavelength band, corresponding to a frequency band of about 405–790 THz. Radiations with wavelengths bordering the limits of the VL, especially ultraviolet (UV) and infrared (IR) radiation, are also used in criminalistics for the detection and photographing of vestiges, both visible and latent, thanks to the incorporation of forensic lights, named as alternate lighting sources (ALS) which use tunable wavelengths for illumination of the scene [14]. A large part of the ALS radiated spectrum is not perceptible by the human eye, being outside the spectral curve of luminous efficiency of its photoreceptors, but it is captured by certain photosensitive emulsions used in old argentic/analogue photography or, currently, multispectral/hyperspectral capture devices. These optical localization procedures provide the researcher with several advantages [15]. In addition to providing more information about the scene than narrow-band visible sensors [16]. On the one hand, as a non-contact inspection method there is no risk of deterioration of traces nor cross-contamination of the scene itself. On the other hand, the inspection results can be evaluated instantaneously, reducing the processing times and permanence of the researchers at the scene, improving the performance and moving the searching process in the right direction. Finally, it is worth noting that the optical inspection can be complemented with other traditional procedures to locate latent vestiges by means of physical developers or chemical components, such as blood chemo locators, Luminol, Bluestar, powdery developers of sweat and sebaceous secretions, black magnetic, wax or fluorescent powder.

Just as the resolution of a digital image is determined by the number of pixels of its sensor, the visual perception that the human being has of its environment is given, at first, by the density of photoreceptors located in its retina, called cones and rods for the characteristic shape of its external segments. Its main function is to absorb light and, thanks to the visual pigments they contain—proteins such as opsin and rhodopsin—transform it into electricity, which is the only form of energy that can be processed by the brain. The cones are responsible for the so-called photopic or diurnal vision, while the rods on the other hand are more sensitive (scotopic or nocturnal vision conditions) but their response to light is more moderate, which translates into lower visual acuity and reduced discrimination of colour. Figure 1 shows a comparison of the spectral curve of luminous efficiency of the cones and rods of the human retina, whether under conditions of photopic or scotopic vision, with those of a conventional digital camera with a CMOS silicon sensor. It can be observed that the nominal spectrum sensitivity of the camera is considerably higher than the human-eye photoreceptors, specifically from about 380 to 740 nm versus 300 to 1150 nm, both at short wavelengths (UV radiation) and at long wavelengths (IR radiation).

To avoid certain optical aberrations associated with the radiations of the non-visible spectrum, that is, those that are below 380 and above 740 nm wavelengths, the manufacturers of cameras place in front of the silicon sensor a colour filter array (CFA) and a bandpass filter, generically called hot mirror, whose mission is to block all UV and IR radiation, allowing only VL wavelengths to reach it. The function of the *hot mirror* is to limit the spectral sensitivity of the sensor to adapt it to human perception, eliminating possible optical aberrations associated with these specific regions of the spectrum, so that the images captured by the camera are the same as those perceived by the photographer. On the other hand, Forensic Photography requires that the optical devices capture regions of the spectrum outside the strict human perception, since it is often there where the reactions of the electromagnetic radiation with the relevant forensic vestiges occur and their observation is critical for the subsequent clarification of the fact to be investigated.

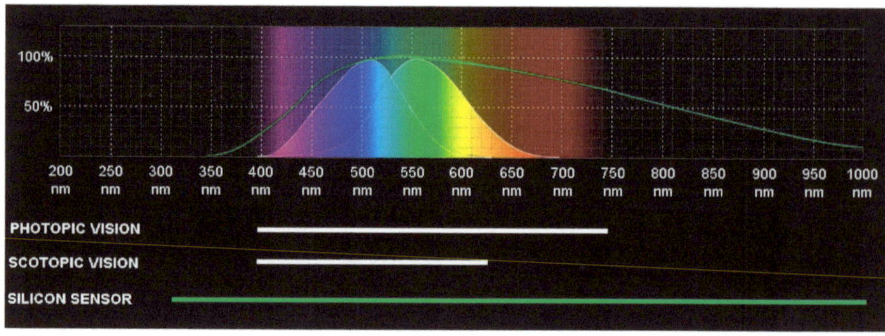

Figure 1. Normalized spectrum sensitivity for photopic and scotopic vision of the cones and rods of a human eye compared with a CMOS sensor.

Generically, multispectral imaging refers to the capture of images using more than one spectral band, regardless of whether they are contiguous and whether they are visible to the human eye in which the result is processed as an individual image. Therefore, multispectral imaging uses a subset of specific wavelengths within a defined spectral range. The concept of multispectral imaging has its origins in the mapping of the Earth's surface using artificial satellites, such as the pioneer LandSat-1 launched in 1972 and equipped with a quad-band multispectral scanner system. As this technology became more affordable, it was taking more and more presence, expanding its uses to the digitization of cultural goods in the 1980s, and in recent years, to the location of vestiges in the field of Criminalistics [17].

Currently, the multispectral capture devices available in the market, incorporate different types of sensors, depending on the model and the specific nominal sensitivity to the spectrum that is necessary according to the demands of the addressed work.

On the one hand, there would be the sensors InGaAs, indium gallium arsenide, used in infrared cameras and which have, depending on the manufacturer, a spectral curve of luminous efficiency wavelengths of 700 to 1.700 nm. These devices are commonly used in the inspection of industrial phenomena but, from a strictly forensic point of view, their sensors, although they are useful in the detection of certain latent evidence that reacts to long wavelengths. In the cases of blood, alterations in documents or shooting residues, the captured images do not have sufficient quality to provide the vestige with the necessary identifying value, with a general lack of sharpness and contrast that reduces its effectiveness [18].

Another type of sensors would be the so-called Vidicom tubes which, in essence, are video cameras with a lead oxysulfide sensor. They are capture devices commonly used in examinations of artworks, in particular paintings on canvas. Its spectral curve of luminous efficiency is variable, being able to reach up to 1.900 nm wavelength. Its application in Criminalistics has several drawbacks, derived largely from its low resolution, which ultimately translates into a difficulty in reproducing elements or details of size less than 2 mm. In addition, the captured images present large geometric aberrations that make it impossible to obtain measurements of elements within the image. Finally, the reduced field of view forces to work with image mosaics so their use during a police technical inspection is very scarce [19].

Thermal imaging sensors based on uncooled microbolometers, or lead selenide or indium antimonide, offer a nominal spectral sensitivity of 1.500 nm and higher wavelengths [20]. The main limitation presented by this type of sensors for forensic use is that the main vestiges of criminalistic relevance do not emit any thermal reaction, beyond the heat generated by the action of cadaveric fauna.

In the same way that there has been a natural evolution of the traditional systems of capturing images in monochrome bands to those others in colour bands, already assuming a considerable improvement in the quality and usefulness of them, also the capture devices have evolved from capturing only the visible spectrum to those capable of capturing images in multispectral bands not visible to human-eye. The incorporation of these devices into the field of criminalistics has had a direct utility as a tool to detect elements not perceived with the naked eye, through a direct comparison with the object to be inspected or with the visible image of it, visually analysing the differences between the two. In forensic sciences, the usefulness of these multispectral devices lies in the fact that they allow to visualize and document the reactions that, in the face of electromagnetic radiation of a specific wavelength, present certain components of both the vestiges to be located and the surface where they settle.

The multispectral capture systems used today in criminalistics for the search for latent vestiges, offer the possibility of locating and photographing evidence in multiple scenarios and work circumstances. These tools are presented in the form of portable tablets with limited functionality, such as the ForenScope tool, intended for fieldwork during the police technical inspection at the scene or as independent workstations, as is the case of foster and Freeman's DCS 5, oriented exclusively to the work of localization and documentation of pieces of evidence in a forensic laboratory [21]. Both proposals incorporate autonomous filtering and lighting systems in the form of LED rings located around the optics that emit, according to the vestige to be located, different wavelengths, but without the possibility of varying their angle of incidence on the search surface. However, reasons, such as their high cost or a functionality often poorly adapted to fieldwork, mean that they are currently not used by many criminalistics sections. These are the reasons why the Scientific Police in Spain use a low-budget multispectral modified digital camera, without reducing its operability and effectiveness when it comes to locating latent vestiges in a crime scene.

The modified camera, together with a quartz or fluorite lens and an adequate longpass or shortpass filter, would allow to locate and graphically document certain latent traces/vestiges, such as latent blood stains, tattoos partially erased or covered under new tattoos and unlock patterns on mobile phones, among others. Currently, such vestiges are located by physical developers and chemical reagents for forensic use, such as Luminol, Bluestar, Hemascein or Benzidine in the case of latent blood; or magnetic, wax, fluorescent powder, cyanoacrylate or diazafluorenone plus flavin, in the case of traces of sebaceous and sudoriparous secretions.

These classic localization procedures have a direct impact in the scene as they require direct contact with the surface on which the vestige might be present. This fact means that, when applied, either by means of sprayers or fiberglass brushes, they do not allow the analysis of the trace's morphology nor the subsequent search for other latent vestiges, such as fingerprints. This problem is solved using a non-contact localization procedure, such as the modified multispectral modular system presented in this paper.

The rest of the paper is organized as follows. Next section presents the materials and methodology used in the research, to show the correct procedure to retrieve latent forensic traces. Section 3 presents the conducted experiments, resume tables and detection results for real use cases (blood stains, hidden/erased tattoos, unlocking patterns on mobile devices, etc.) showing the validity of the proposal. Finally, Section 4 resume the main conclusions of the research work.

2. Materials and Methods

Electromagnetic radiation can interact with the matter through an exchange of energy, being able to classify this phenomenon according to several criteria, such as nature of the matter involved, specific incident spectral band or the possible reactions of the interaction. Starting from the fact that the laws of spectrophotometry were enunciated based on monochromatic electromagnetic radiation acting on a homogeneous system, when electromagnetic radiation affects a certain material, we understand that, for the purposes of this study, a vestige or the surface where it sits, the beam will produce different reactions depending on the energy levels in its atoms, it may be absorbed, transmitted, dispersed, reflected or it may induce photoluminescence [22].

When the atoms of a certain material are irradiated with electromagnetic radiation, part of that form of energy can be absorbed by the atoms of the material which, as a consequence, will pass from a lower energy state (basal state or E_1) to a state of higher energy (excited state or E_2). For this absorbance reaction to occur, the energy of the photons of the incident beam ($h \cdot v$), must be equal to the energy difference between E_1 of the atoms of the material and posterior energy E_2.

$$\Delta E = E_2 - E_1 = h \cdot v \quad (1)$$

The atoms of the material can return to their E_1 by converting the energy of their E_2 either into heat, through a luminescent reaction or through a photochemical reaction, which can be documented, by means of the appropriate capture devices, evidenced by a darkening of the irradiated material for the duration of the induced reaction. The transmission of electromagnetic radiation assumes that the beam that is not absorbed or reflected, will pass through the sample without suffering perceptible or energetic changes. As the wavelength of the incident radiant flux increases, the absorption of radiation by the sample decreases, increasing the transmittance, so that the energy absorbed and reflected is less than that transmitted, which will be evidenced in a partial disappearance of the element in the captured image. Dispersion occurs when the incident beam is absorbed and immediately emitted uniformly in all directions, without any energetic change. A reaction that occurs when radiation hits particles in the sample that are smaller than the wavelength of the incident beam itself, so they polarize and oscillate at the same frequency as the radiation, acting as a source that propagates in all directions. Photoluminescence involves an excitation of the particles that make up the sample by absorbing the incident radiation, emitting fluorescence, as happens in scattering, in all directions.

The relevance of these phenomena, for the purposes of this study, is that a large part of the relevant forensic vestiges for the possible clarification of facts of police interest and on which the subsequent expert and technical reports in each of the forensic disciplines will be based, present distinct reactions, visible or not to the human eye, depending on the wavelengths of spectrum used to induce them and the specific region of the spectrum where they occur. This feature leads to a characteristic behaviour that can be used, as a spectral fingerprint, to achieve its location and subsequent graphic documentation [23].

In this paper for the development of the multispectral camera it has been used a Nikon camera D3500-24.2 megapixels, reflex CMOS APS-C 23.5 × 15.6 mm, commonly available in the Scientific Police units in Spain and that meets the needs of the fieldwork of a crime scene investigator: ease of handling the different modules of the multispectral system, ability to view the scene before capturing, interchangeable bayonet to mount quartz optics e.g., Nikon UV. Reusing the camera currently owned by operational units significantly reduce the cost, making available the multispectral technology. The multispectral techniques shown in the paper has also been tested with other cameras/sensors capturing multispectral images such as smartphones and action cameras, but its operation, focus range and resolution is reduced when compared with digital SLR cameras. Figure 2 shows the modified Nikon D3500 camera. The bandpass *hot mirror* filter—CFA colour filter matrix—has been extracted and replaced by a full spectrum filter, to increase its nominal sensitivity [24]. The sensor has been conveniently shifted closer to the nodal point of the lens, recalibrating it in order to correct the image focus, important when it is necessary to locate latent traces using infrared illumination. Likewise, for those images in which UV radiation is used, the camera lens has been replaced by a quartz or fluorite lenses, both permeable to UV radiation, because the Crown glass used by commercial photographic lenses blocks short wavelengths. This conversion allows the capture of regions of the UV and IR spectrum, losing colour information.

Figure 2. Multispectral capture device modified with quartz optics and Baader UV/IR-cut coupled.

For the analysis of pieces of evidence presented in the next section, different bandpass and shortpass filters have been used when filtering the specific wavelengths of the lighting sources and, thus, making visible the reactions of the latent traces. We have carried out a field study on the effectiveness of multiple long-pass and shortpass filters from different manufacturers to detect latent traces in multiple situations. The working band of the used filters (Wratten 87, Schott RG9 and Baader UV/IR-cut) are shown in Figure 3 where it can be shown their spectral response to the electromagnetic radiation used to locate the latent vestiges.

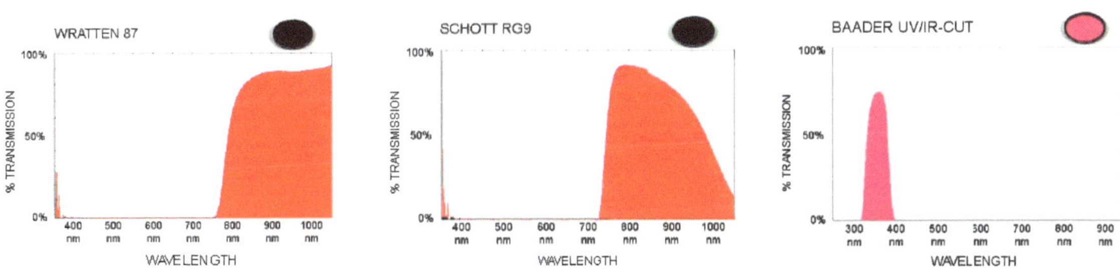

(**a**) (**b**) (**c**)

Figure 3. (**a**) IR Longpass filter Wratten 87; (**b**) IR Longpass filter Schott RG 9; (**c**) UV Shortpass filter Baader UV/IR-cut.

We have named the modular multispectral inspection system as Invespector being composed by the modified camera, optical lenses, band filter and the illumination source. The right combination of the different elements leads to an effective retrieval of latent traces.

3. Results and Discussion

This section presents the results obtained using the Invespector capturing device for the localization of multiple and diverse pieces of evidence: latent blood stains, erased tattoos and unlocking pattern of mobile devices. The source of data comes from real cases analysed by the Spanish Scientific Police.

3.1. Localization of Spots and Latent Blood Patterns

One of the applications in which the effectiveness of the modified multispectral camera has been tested is in the location and photographing of latent blood. Blood stains and patterns are one of the most common vestiges present at any scene of a violent crime. Whether during the police technical inspection conducted at the scene or later in the forensic laboratory, it always acquires special relevance to locate and document this type of biological vestiges because, in addition to being a source of genetic and serological material, they might provide significant information for blood pattern analysts, allowing them to sometimes reconstruct the very commission of the fact, by studying the morphology, size and location of blood stains in the scene or on the objects located in it.

When the blood is recent, has not been manipulated and is settled on surfaces that contrast with the sample itself, its location and photography is a simple task. Unfortunately, it is common for blood stains to be difficult to detect, either because they are settled on dark surfaces that mask them, or because they have been intentionally cleaned or partially erased as a result of the passage of time. It will be then when the researcher uses presumptive methods of locating latent blood, both with chemical and optical means [25]. Blood, human or animal, contains certain components such as hemoglobin iron, lipids and proteins that react through intense absorbance when illuminated with wavelengths typical of ALS forensic lights in its spectrum and near IR wave radiation [26]. This reaction makes the blood vestige latent even in low dilutions (1:200) [27] or small amounts darken, which facilitates its detection and subsequent photographing, provided that the surface where it sits reacts by reflecting or transmitting those specific wavelengths used to induce the reaction [28].

Figure 4 shows the detection results for a latent blood settled in a black cotton shirt. The classical visible images were compared with the image captured using a multispectral Invespector device with a source of IR radiation of 850 nm, a capture angle of 90° and attaching to the camera lens an IR longpass filter Schott RG9. As it can be seen, it is possible not only to locate and document latent blood settled in a black cotton shirt, but to respect the morphology of the stain itself, in this case a footstep. Additionally, the cotton shirt is treated with the Bluestar chemo locator which can react with blood in dilutions of up to 1:10,000.

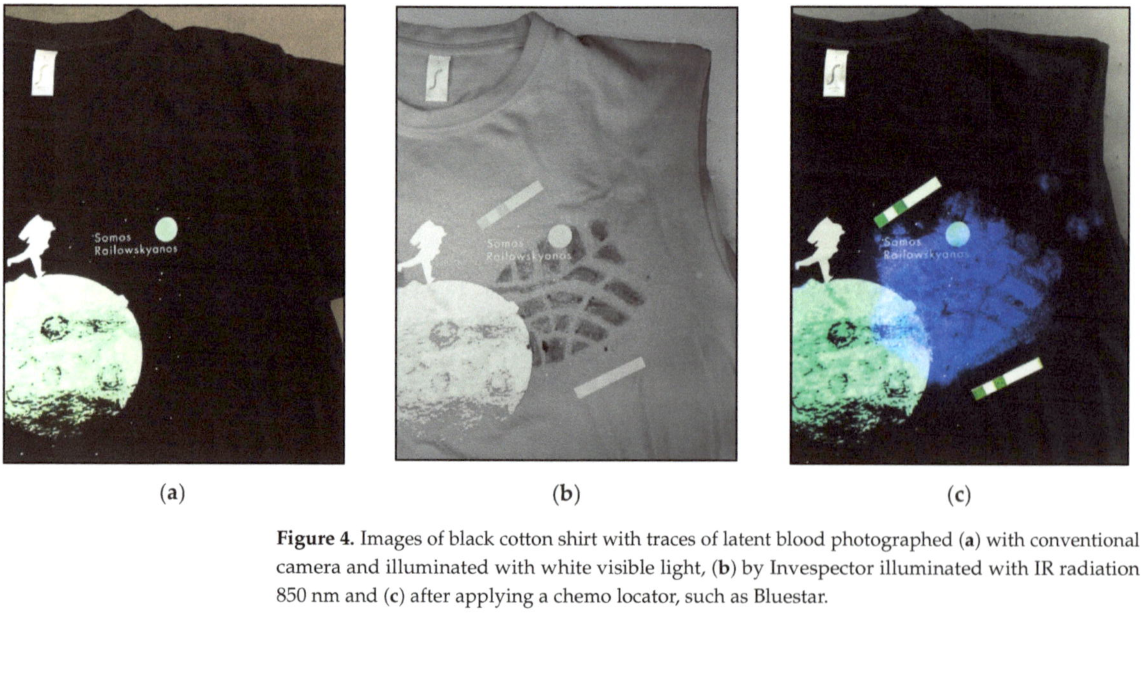

(**a**) (**b**) (**c**)

Figure 4. Images of black cotton shirt with traces of latent blood photographed (**a**) with conventional camera and illuminated with white visible light, (**b**) by Invespector illuminated with IR radiation 850 nm and (**c**) after applying a chemo locator, such as Bluestar.

Although optical procedures of localization in multispectral bands are less sensitive to the presence of blood than the classic chemical procedures, optical procedures offer several advantages:

- It is a non-invasive procedure that is with no contact, so it is possible to conduct subsequent comparative morphological studies.
- It preserves the morphology of the latent blood stain, for example, in this case, the footprint of the footwear. It can be shown that only Figure 4b can be correlated with the footprint provided in Figure 5.

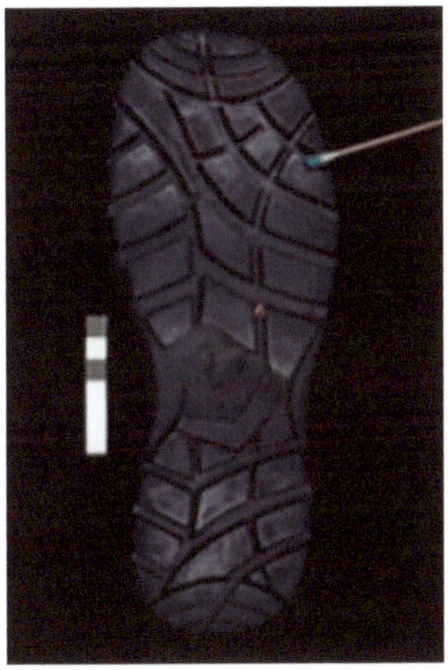

Figure 5. Footprint of the shoe used in the T-shirt under inspection.

- It does not interfere with the subsequent DNA study of the localized sample.
- It allows the subsequent search for lophoscopic traces if the surface is adequate.
- The high penetration capacity of IR radiation on certain surfaces. An example is the location of blood settled under layers of up to 5 mm of acrylic paint, wallpaper or cotton or nylon fabrics, as it can be seen in the images of the Figure 6.

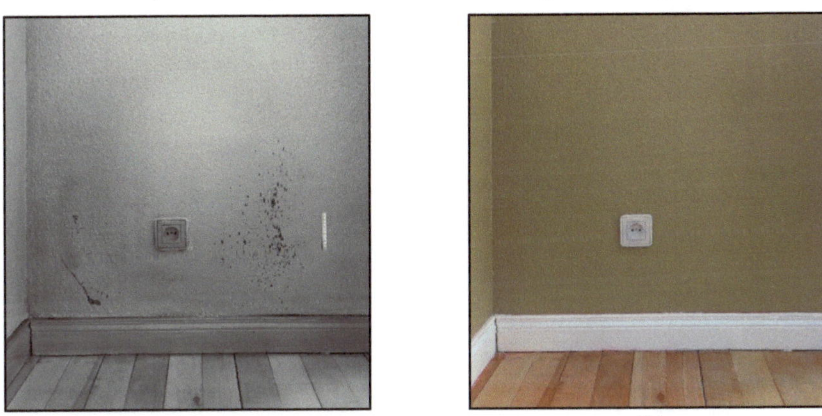

(**a**) (**b**)

Figure 6. Images of a wall with traces of latent blood under a layer of acrylic paint: (**a**) photographed with a conventional camera and illuminated with white visible light and (**b**) by Invespector illuminated with IR radiation 850 nm.

Additionally, in order to locate latent blood, UV radiation of 365 nm wavelength has been shown to be equally effective [29]. The two images that make up Figure 7 correspond to a fragment of cotton sheet from the scene of a violent crime. In this case, despite the fact that the sheet had been washed, using the multispectral device it was possible to locate remains of what a laboratory later certified to be human blood. To do this, the effect was illuminated with UV radiation of 365 nm with an angle of incidence of 90°, attaching to the camera a fluorite lens that provides permeability to the UV radiations and a UV shortpass filter Baader UV/IR-cut. Under these circumstances, and despite the fact that the amount of vestige was very small, after the sheet was subjected to washing, the blood vestige could be detected without the need to apply chemical locators for it.

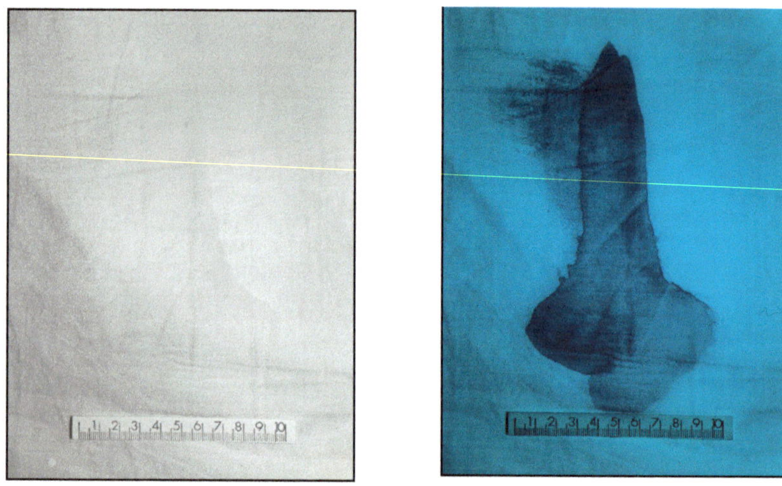

(a) (b)

Figure 7. Images of a cotton sheet with traces of latent blood photographed: (**a**) with a conventional camera illuminated with white visible light and (**b**) Invespector system using UV illumination 365 nm.

We have run multiple tests to show the capabilities of the multispectral device. Table 1 shows the right configurations for the detection of a drop of blood on different surfaces. For example, to search for blood stains in a denim-fabric, the best configuration is a Crown-glass lens with a Hoya R72 filter, illuminating the scene with a 720 nm UV source, capturing the sample in vertical (90°).

Table 1. Test results for the detection of a drop of human blood -0.05 ml- considering different surfaces.

Surface Configuratio	On Steel	On Washed Cotton	On Wood	On Tile	On Denim Fabric	On ABS Plastic	On Cotton	Under Acrilic Paint	On Nylon
Lens type	quartz	quartz	Crown glass	Crown glass	Crown glass	Crown glass	Crown glass	Crown glass	Crown glass
Filter type	Kolari UV	Baader UVIR	none	none	Hoya R72	Hoya R72	Schott RG9	Schott RG9	Wratten 87
Illumination	365 nm	365 nm	415 nm	455 nm	720 nm	750 nm	850 nm	850 nm	950 nm
Capturing angle	25°	90°	45°	45°	90°	45°	45°	90°	45°
Surface's reaction	reflectance	reflectance	reflectance	reflectance	transmittance	absorbance	reflectance	transmittance	transmittance

3.2. Detection of the Unlocking Pattern of Mobile Devices

Another application in which the Invespector multispectral device has been shown to be effective is in the detection of unlocking patterns of mobile phones not visible to the human-eye. As it can be seen in Figure 8, in this case the unlocking pattern could be detected, illuminating the phone with IR radiation of 950 nm wavelength at an angle of incidence of only 10°, attaching to the target of the multispectral device an IR longpass filter Wratten 87.

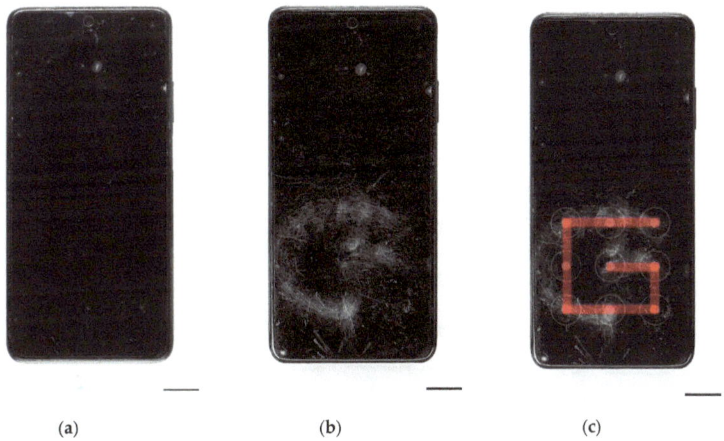

Figure 8. Images from a mobile phone photographed (a) with a conventional camera and illuminated with white visible light; (b) captured by Invespector camera and illuminated with IR radiation 950 nm; and (c) the pattern movement recovered to unlock the smartphone successfully.

The sebaceous secretions deposited by repetition of the movement on the screen of the mobile phone provoke the absorbance reaction to the IR radiation of 950 nm wavelength. Additionally, the outer layer of Indium Tin Oxide (ITO) [29] that coats the glass of the phone screen, as a consequence from the repeated friction of the user in the area of the unlocking pattern, modifies the reflectance reaction of the glass that is below the ITO coating making it more visible in the captured image. The Invespector device captures the contrast between the absorbance reaction of the sebaceous secretions and the ITO layer; and the reflectance of the glass below the ITO. Therefore, it is possible to recover the repeated finger movement to input a similar pattern and unlock the smartphone.

3.3. Locating Covert or Partially Erased Tattoos

Tattoos, like any other birth mark or supervening mark (wounds or scars), have been used as an authentication particular mark complementary to other identifying elements of the individual/person. In 1853, a court in Paris commissioned the pathologist Auguste Tardieu a study on the indelible condition of tattoo inks, the medico-legal interest in tattooing as a sign of identity capable of providing relevant information about its wearer gradually grew until it was incorporated, like any other distinctive physical characteristic, to the written and photographic police portrait of Alphonse Bertillon. However, as Tardieu claimed in his forensic study, *"despite the permanent character of the tattoo, they can always be made less noticeable, altered or increased by time"* [30].

Most of the pigments that make up the inks used in tattoos are derived from charcoal (black ink) or metals such as iron (black ink), mercury (red ink), cobalt (blue ink), chromium (green ink), titanium/zinc oxide (white ink) or cadmium (yellow ink) [31]. These substances have the property of reacting by transmitting or absorbing, to a greater or lesser extent, the wavelengths of the IR spectrum up to 1050 nm [32], which makes the tattoo either disappear or darken and become more visible, by increasing its contrast with human skin, which in turn reacts by reflecting NIR (Near Infrared) radiation.

Figure 9 shows how the multispectral camera coupled with an IR longpass filter Wratten 87 is capable to capture the transmittance and absorbance reactions of the different components of the inks used in the tattoo. In this particular case, this individual has hidden a small latent tattoo, with the now readable text "caliente", under another tattoo of the shield of the "Valencia CF". Although the ink is injected between 1.5 and 2 mm under the epidermis, the high penetration capacity of IR radiation −850 nm wavelength at an angle of incidence of 90°—cause certain pigments that make up the inks of this second tattoo—cobalt in the blue ink, cadmium in the yellow, chromium in the red—react by transmitting/reflecting the IR radiation, so they partially disappear in the final image. This make the underlying tattoo perceptible as its ink composition—iron of the black ink—reacts by absorbing the IR radiation, that is, darkening the text.

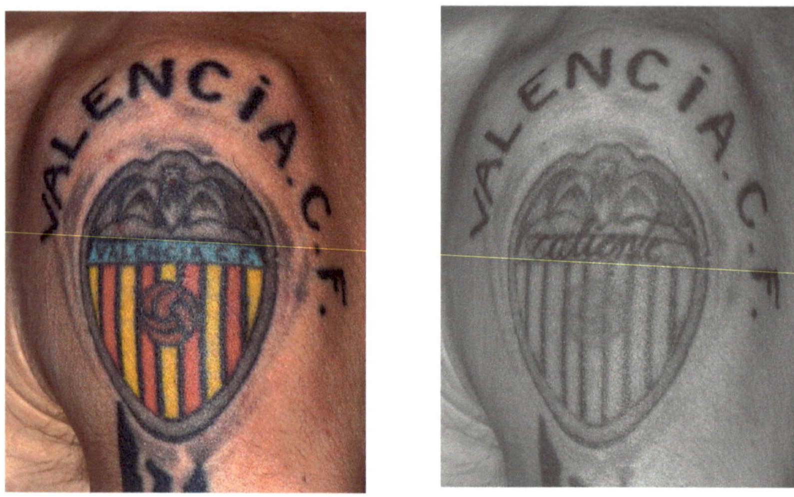

(a) (b)

Figure 9. Images of tattooed shoulder: (**a**) captured with a conventional camera and illuminated with white visible light, and (**b**) captured by Invespector and illuminated with IR radiation 850 nm.

As a conclusion, it can be said that this technique is highly effective to visualize hidden elements, partially erased or covered under external layers, when the elements located in these upper layers react according to their own composition modifying the transmittance/absorption of current spectrum illumination.

Multiple tests and combinations of the filter and illumination modules have been done to show the capabilities of the multispectral device to detect skin tattoos. We have concluded that, in the case of eternal ink, the best option is to use a crown glass lens, a Wratten 87 filter, illuminate the scene with an IR source (850 nm) and capture the image with a 90° angle. Table 2 shows the ink's reaction for the infrared radiation based on the components for each ink colour.

Table 2. Reactions of different tattoo inks to infrared radiation.

Tattoo Inks	Black	Red	Blue	Green	White	Yellow	Orange
components	Iron/charcoal	Mercury	cobalt	chromium	zinc oxide/titanium	cadmium	Iron/cadmium
Ink's reaction	absorbance	reflectance	transmittance	transmittance	reflectance	reflectance	absorbance

4. Conclusions

In this paper, an effective procedure has been shown to detect latent vestiges by modifying a conventional camera, from those assigned to the scientific police units in Spain to its multispectral version, thus increasing its curve of luminous efficiency and nominal sensitivity.

Several processing results are shown to demonstrate the efficiency of such a system considering not only the multispectral camera but different ways of filtering the light captured from IR/UV illumination considering both the radiation reflected/absorbed by the scene or vestige.

The effectiveness of this multispectral procedure has been shown in the location and graphic documentation of latent vestiges present in the crime scene itself, such as traces of stains or blood patterns, and in the pieces of evidence to be analysed in the laboratory, such as the detection of hidden or partially erased tattoos, or the visualization of unlocking patterns on mobile devices. Similar procedures can be carried out either during the technical inspection of the crime scene itself or later in the forensic laboratory.

Currently in forensic sciences, the usefulness of multispectral devices on the market lies in the fact that they allow to visualize the reactions that, in the face of electromagnetic radiation, present certain intrinsic and extrinsic components of these vestiges to be located. However, its high cost and an often poorly adaptation to field work reduces its use. The modified digital SLR camera, with a nominal sensitivity beyond the visible spectrum, open the possibility of incorporating this technology to police departments that currently lack it, complementing these new non-invasive procedures for locating latent vestiges.

Author Contributions: Conceptualization, S.M.-M.; Methodology, S.M.-M., B.A. and A.G. Research and experiment conduction, S.M.-M.; Validation, S.M.-M., B.A. and A.G. Writing—original draft preparation, S.M.-M.; Writing—review and editing, B.M and A.G. All authors have read and agreed to the published version of the manuscript.

Funding: This research was funded by Instituto Universitario de Investigación en Ciencias Policiales (IUICP) from University of Alcala (SPAIN), grant number Ref: IUICP/PI2019/009-2019/00377/001, project INVESPECTOR.

Institutional Review Board Statement: Not applicable.

Informed Consent Statement: Not applicable.

Data Availability Statement: Not applicable.

Conflicts of Interest: The authors declare no conflict of interest.

References

1. Dufour, D. *Images of Conviction, the Construction of Visual Evidence*; Editions Xavier Barral: Paris, France, 2016; ISBN 978-2365110839.
2. Kawase, M.; Shinoda, K.; Hasegawa, M. Demosaicking Using a Spatial Reference Image for an Anti-Aliasing Multispectral Filter Array. *IEEE Trans. Image Process.* **2019**, *28*, 4984–4996. [CrossRef] [PubMed]
3. Feng, K.; Zhao, Y.; Chan, J.C.-W.; Kong, S.G.; Zhang, X.; Wang, B. Mosaic Convolution-Attention Network for Demosaicing Multispectral Filter Array Images. *IEEE Trans. Comput. Imaging* **2021**, *7*, 864–878. [CrossRef]
4. Wang, P.; Wang, L.; Leung, H.; Zhang, G. Super-Resolution Mapping Based on Spatial-Spectral Correlation for Spectral Imagery. *IEEE Trans. Geosci. Remote Sens.* **2021**, *59*, 2256–2268. [CrossRef]
5. Bian, L.; Wang, Y.; Zhang, J. Generalized MSFA Engineering with Structural and Adaptive Nonlocal Demosaicing. *IEEE Trans. Image Process.* **2021**, *30*, 7867–7877. [CrossRef] [PubMed]
6. Ron, A.M. *El Ojo Desnudo*; Editorial Critica: Barcelona, Spain, 2016; p. 10. ISBN 978-84-9199-379-7.
7. Bartolić, D.; Mutavdžić, D.; Carstensen, J.M.; Stanković, S.; Nikolić, M.; Krstović, S.; Radotić, K. Fluorescence spectroscopy and multispectral imaging for fingerprinting of aflatoxin-B1 contaminated (*Zea mays* L.) seeds: A preliminary study. *Sci. Rep.* **2022**, *12*, 1–8. [CrossRef] [PubMed]
8. Althouse, M.L.; Chang, C.-I. Chemical vapor detection and mapping with a multispectral forward-looking infrared (FLIR). In *Optical Instrumentation for Gas Emissions Monitoring and Atmospheric Measurements*; SPIE: Bellingham, WA, USA, 1995; Volume 2366. [CrossRef]
9. Zapata, F.; López-López, M.; Amigo, J.M.; García-Ruiz, C. Multi-spectral imaging for the estimation of shooting distances. *Forensic Sci. Int.* **2018**, *282*, 80–85. [CrossRef] [PubMed]
10. Shang, X.; Song, M.; Wang, Y.; Yu, C.; Yu, H.; Li, F.; Chang, C.-I. Target-Constrained Interference-Minimized Band Selection for Hyperspectral Target Detection. *IEEE Trans. Geosci. Remote. Sens.* **2020**, *59*, 6044–6064. [CrossRef]
11. Suárez, E.R.; Fotografía General y Policial. *Estudios de Policía Científica*; División de Formación y Perfeccionamiento: Madrid, Spain; ISBN 84-86380-38-3. Available online: https://www.policia.es/_es/tupolicia_conocenos_estructura_rrhh_divisionformacion.php (accessed on 24 November 2022).

12. Khan, M.J.; Yousaf, A.; Khurshid, K.; Abbas, A.; Shafait, F. Automated forgery detection in multiespectral document images using fuzzy clustering. In Proceedings of the 2018 13th IAPR International Workshop on Document Analysis Systems (DAS), Vienna, Austria, 24–27 April 2018.
13. Silván-Cárdenas, J.L.; Caccavari-Garza, A.; Quinto-Sánchez, M.E.; Madrigal-Gómez, J.M.; Coronado-Juárez, E.; Quiroz-Suarez, D. Assessing optical remote sensing for grave detection. *Forensic Sci. Int.* **2021**, *329*, 111064. [CrossRef] [PubMed]
14. Morris, N.; Buszka, J. *Alternate Light Source Imaging: Forensic Photography Techniques*; Routledge: London, UK, 2013.
15. Coffey, V.C. Multispectral Imaging Moves into the Mainstream. *Opt. Photon. News* **2012**, *23*, 18–24. [CrossRef]
16. Multiespectral Imaging for Medical and Industrial Machine Vision Systems, Test Guide Jai See the Possibilities. Available online: https://multipix.com/wp-content/uploads/2021/01/JAI_PDF_Multispectral-imaging_EN.pdf (accessed on 10 October 2022).
17. Lira, J. *La Percepción Remota: Nuestros Ojos Desde el Espacio*; Fondo de Cultura Económica: Ciudad de México, México, 2021.
18. Geert, J.; Verhoven, J. Imaging the Invisible—Using Modified Digital Still Cameras for Straightforward and Low-Cost Archaeological Near-Infrared Photography. *J. Archaeol. Sci.* **2008**, *35*, 3087–3100.
19. McLeod-Henning, D. *Brass Board Forensic Crime Scene Survey Camera*; US Department of Justice: Washington, DC, USA, 2015; p. 10.
20. Edelman, G.; Gaston, E.; van Leeuwen, T.; Cullen, P.; Aalders, M. Hyperspectral imaging for non-contact analysis of forensic traces. *Forensic Sci. Int.* **2012**, *223*, 28–39. [CrossRef] [PubMed]
21. Home Office; Centre for Applied Science and Technology. *Fingermark Visualisation Manual*; Centre for Applied Science and Technology: London, UK, 2014; ISBN 978-17-82-46234-7.
22. Uzal, J.M.P. *Diagnóstico por Imagen en Bandas no Visibles Sobre Patrimonio Cultural*; Books on Demand GmbH: Leipzig, Gerrmany, 2019; pp. 5–11. ISBN 978-84-13-26619-0.
23. Goicoechea-Telleria, I.; Kiyokawa, K.; Liu-Jimenez, J.; Sanchez-Reillo, R. Low-Cost and Efficient Hardware Solution for Presentation Attack Detection in Fingerprint Biometrics Using Special Lighting Microscopes. *IEEE Access* **2019**, *7*, 7184–7193. [CrossRef]
24. Crowther, J. Monochrome Camera Conversion: Effect on Sensitivity for Multispectral Imaging (Ultraviolet, Visible, and Infrared). *J. Imaging* **2022**, *8*, 54. [CrossRef] [PubMed]
25. Best Practice Manual for Forensic Image and Video Enhancement. ENFSI-BPM-DI-02. 2018. Available online: https://enfsi.eu/wp-content/uploads/2017/06/Best-Practice-Manual-for-Forensic-Image-and-Video-Enhancement.pdf (accessed on 10 October 2022).
26. Hart, A.J.; Barnes, G.C.; Fuller, F.; Cornwell, A.M.; Gyula, J.; Marsh, N.P. Finding blood in the dark: A comparison of infrared imaging devices for the detection of bloodstains on dark fabrics based on their resolution. *Forensic Sci. Int.* **2021**, *330*, 111124. [CrossRef] [PubMed]
27. Li, B.; Beveridge, P.; O'Hare, W.T.; Islam, M. The application of visible wavelength reflectance hyperspectral imaging for the detection and identification of blood stains. *Sci. Justice* **2014**, *54*, 432–438. [CrossRef] [PubMed]
28. Yang, J.; Mathew, J.J.; Dube, R.R.; Messinger, D.W. Spectral feature characterization methods for blood stain detection in crime scene backgrounds. In *Algorithms and Technologies for Multispectral, Hyperspectral, and Ultraspectral Imagery XXII*; SPIE: Bellingham, WA, USA, 2016; Volume 9840, pp. 73–86.
29. Finnis, J.; Lewis, J.; Davidson, A. Comparison of methods for visualizing blood on dark surfaces. *Sci. Justice* **2013**, *53*, 178–186. [CrossRef] [PubMed]
30. Lamprey, J. *On a Method of Measuring the Human Form*; Journal of the Ethnological Society of London: London, UK, 1869.
31. Miranda, M.D. *Forensic Analysis of Tattoos and Tattoo Inks*; CRC Press: Boca Raton, FL, USA, 2021.
32. Pronti, L.; Ferrara, P.; Uccheddu, F.; Pelagotti, A.; Piva, A. Identification of pictorial materials by means of optimized multispectral reflectance image processing. In Proceedings of the 2015 IEEE International Workshop on Information Forensics and Security (WIFS), Rome, Italy, 16–19 November 2015. [CrossRef]

sensors

MDPI

Article

Development and Validation of a Tunable Diode Laser Absorption Spectroscopy System for Hot Gas Flow and Small-Scale Flame Measurement

Ran Tu [1], Junqing Gu [2], Yi Zeng [1,*], Xuejin Zhou [1], Kai Yang [1], Jiaojiao Jing [1], Zhihong Miao [1] and Jianhong Yang [1]

1 College of Mechanical Engineering and Automation, Huaqiao University, Xiamen 361021, China
2 Department of Modern Mechanics, University of Science and Technology of China, Hefei 230026, China
* Correspondence: antigy@hqu.edu.cn

Abstract: TDLAS (tunable diode laser absorption spectroscopy) is an important gas analysis method that can be employed to obtain characteristic parameters non-invasively by the infrared absorption spectra of tracer molecules such as CH_4, H_2O and O_2. In this study, a portable H_2O-based TDLAS system with a dual optical path was developed with the aim of assessing the combustion characteristics of flammable gases. Firstly, a calculation method of gas characteristics including temperature and velocity combining absorption spectra and a HITRAN database was provided. Secondly, to calibrate and validate this TDLAS system precisely, a pressure vessel and a shock tube were introduced innovatively to generate static or steady flow fields with preset constant temperatures, pressures, or velocities. Static tests within environment pressures up to 2 MPa and steady flow field tests with temperatures up to 1600 K and flow velocities up to 950 m/s were performed for verification. It was proved that this system can provide an accurate values for high temperature and velocity gas flows. Finally, an experimental investigation of CH_4/air flames was conducted to test the effectiveness of the system when applied to small diffusion flames. This TDLAS system gave satisfactory flame temperature and velocity data owing to the dual optical path design and high frequency scanning, which compensated for scale effects and pulsation of the flame. This work demonstrates a valuable new approach to thermal hazard analysis in specific environments.

Keywords: tunable diode laser absorption spectroscopy (TDLAS); gas flow field; H_2O-based TDLAS; doppler-shift; small-scale flame

Citation: Tu, R.; Gu, J.; Zeng, Y.; Zhou, X.; Yang, K.; Jing, J.; Miao, Z.; Yang, J. Development and Validation of a Tunable Diode Laser Absorption Spectroscopy System for Hot Gas Flow and Small-Scale Flame Measurement. *Sensors* **2022**, *22*, 6707. https://doi.org/10.3390/s22176707

Academic Editor: Vittorio M.N. Passaro

Received: 4 August 2022
Accepted: 3 September 2022
Published: 5 September 2022

Publisher's Note: MDPI stays neutral with regard to jurisdictional claims in published maps and institutional affiliations.

Copyright: © 2022 by the authors. Licensee MDPI, Basel, Switzerland. This article is an open access article distributed under the terms and conditions of the Creative Commons Attribution (CC BY) license (https://creativecommons.org/licenses/by/4.0/).

1. Introduction

The measurement of hot gas characteristic parameters in complex environments—e.g., those involving high flow rates, vibration, or combustion—has posed a long-standing challenge to the study of aero-engines and industrial furnaces [1]. In recent decades, there has been positive exploration concerning the non-contact assessment of combustion fields using spectral imaging technologies, including emission, absorption, Raman spectrum, and LIF/PLIF (laser induced fluorescence/planar laser induced fluorescence). Of all these methods, TDLAS (tunable diode laser absorption spectroscopy) has emerged an effective yet inexpensive technique providing a rapid response [2].

TDLAS technology has undergone significant development attributable to the progress in semiconductor lasers. As a consequence, this technique is now widely used for the non-contact measurement of temperature, velocity, or species concentrations for environmental engineering, petroleum engineering, and hazardous chemicals detection. Additionally, when incorporating DFBs (distributed feedback lasers) and VCSELs (vertical cavity surface emitting lasers), a TDLAS system becomes more portable and easier to establish [3,4].

Early research regarding the application of TDLAS to the analysis of combustion was reported by Hanson and Allen et al. [1,5–7]. Characteristic parameters of hot gases in combustion flow fields involving high temperatures and pressures were assessed by

TDLAS based on the absorption spectra of O_2 and H_2O. This method was then extended to the monitoring of waste gases such as NO_2 and NH_3 for the process control of industrial combustion systems. With the development of tunable diode laser sources, the useable wavelength range associated with this method was broadened to approximately 700–3300 nm, allowing the analysis of H_2O, CO, CO_2, CH_4, and other gases [8,9]. The data obtained from TDLAS have provided new perspectives concerning the monitoring of combustion processes [10].

An effective component for flame measurement by TDLAS now well recognized is H_2O [11–13]. The most important reason is that H_2O is commonly one of the primary combustion products and can reflect the progress of combustion based on the local thermodynamic equilibrium hypothesis. Although the measurement by gas molecular absorption spectra with TDLAS seems to be a proven technology, there are still significant deficiencies related to this technique. Specifically:

(1) Generally, a typical TDLAS analysis provides only the average value of a selected parameter along the optical path.

(2) Since it is challenging to generate a constant flow field with a given temperature, pressure, and velocity, the calibration of a TDLAS system is relatively difficult.

(3) H_2O-based TDLAS measurement can also be disturbed by the presence of ambient water vapor, leading to underestimations of temperature and overestimations of component concentrations.

(4) The Doppler effect can interfere with TDLAS velocity measurements in high-speed flow fields.

The aim of the study is to develop a portable TDLAS system for combustion analysis and to provide an associated means of calibrating this system. In this paper, a TDLAS system was designed and applied to the assessments of gas flow and CH_4/air co-flow flame including temperature and velocity measurements. A static flow field vessel and a shock tube were employed to generate a steady flow field along with specific constant temperature or velocity for precise calibration. The measurement results obtained from this system were verified. Finally, experimental investigation of the CH_4/air flame was conducted to test the effectiveness of the new TDLAS system in combustion diagnosis.

2. Experimental Methodology

2.1. Absorption Fundamentals

The principle of TDLAS measurement is on the basis of the Beer–Lambert law, depicting the relationship of incident light, transmitted light, molecule temperature, and concentration [14] as

$$I(\nu) = I_0(\nu)e^{-\sum_i \alpha_i(\nu)L}, \text{ and } \alpha_i(\nu) = p_i\varphi_i(\nu - \nu_0)S_i(T) \quad (1)$$

where I and I_0 are the transmitted light intensity and incident light intensity, respectively, which are corresponding to the output frequency of laser, ν. $\alpha_i(\nu)$ is the absorptivity of absorbing species i# and L is the path length. Moreover for $\alpha_i(\nu)$, p_i, and $\varphi_i(\nu - \nu_0)$ are the partial pressure and absorption line shape function of species i#, such as Gaussian, Lorentzian, or Voigt line profile (ν_0 is the center frequency of the absorption spectrum) [15]. In addition, $S_i(T)$, as a function of temperature, is the molecular absorption line strength of species i#. Consequently, integrated absorbance of independent transition line can be obtained as

$$A_i = \int_{-\infty}^{+\infty} -\ln(I/I_0)dv = \int S_i(T)\varphi_i(\nu - \nu_{0i})d\nu \quad (2)$$

Based on the theory of molecular spectroscopy [16], the absorption line strength at temperature T can be expressed as

$$S(T) = S(T_0)\frac{Q(T_0)}{Q(T)} \cdot \exp[-\frac{hcE}{k}(\frac{1}{T} - \frac{1}{T_0})] \cdot \frac{1-\exp(-hcE/kT)}{1-\exp(-hcE/kT_0)} \quad (3)$$

where T_0 is a reference temperature. ε_0, h, c and k are four constant numbers, including the dielectric constant, Planck's constant, speed of light and Boltzmann's constant, respectively. E is the energy of the lower transition state, and $Q(T)$ is the molecular internal partition function at temperature T, which could be obtained from the HITRAN database [17,18]. These relationships enable the non-invasive experimental analysis of hot gases.

2.2. Measuring Principle of Gas Temperature and Velocity

The absorption line shape, bandwidth and intensity of absorption spectral features can be obtained using a TDLAS system together with wavelength-scanning. Furthermore, the gas temperature can be calculated from the ratio, R (the integrated absorbances of two transition [7]), as

$$R = \frac{\int S_1(T)\varphi_1(\nu - \nu_{01})d\nu}{\int S_2(T)\varphi_2(\nu - \nu_{02})d\nu} = \frac{S_1(T_0)}{S_2(T_0)}\exp[-\frac{hc(E_1 - E_2)}{k}(\frac{1}{T} - \frac{1}{T_0})] \tag{4}$$

where subscripts 1 and 2 denote the two absorption transitions selected as shown in Figure 1a. $\Delta E = E_1 - E_2$ is the energy separation of the absorbing states. Based on the use of a near-IR laser (Figure 1b), the gas temperature can then be obtained from the relationship [19]

$$T = \frac{-\frac{hc}{k}\Delta E}{\ln R + \ln\frac{S_2(T_0)}{S_1(T_0)} - \frac{hc}{kT_0}\Delta E} \tag{5}$$

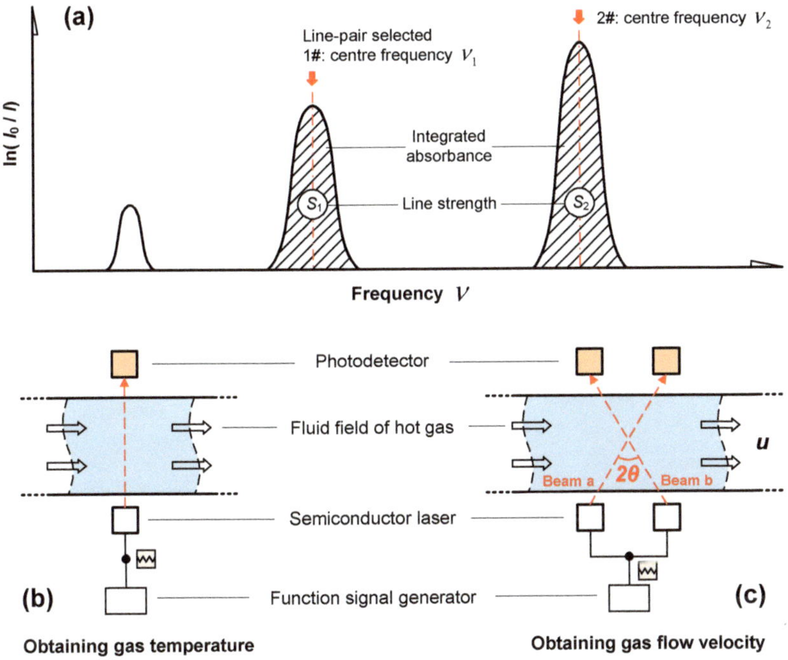

Figure 1. (**a**) Integrated absorbances of two selected transitions and diagrams summarizing the processes used to measure (**b**) gas temperature and (**c**) flow velocity.

On the other hand, the gas flow velocity can be calculated according to Doppler-shift as shown in Figure 1c [20]. Assuming that a monochromatic laser beam with a frequency

ν_0 (the same as the center frequency of the absorption spectrum) is used here, the Doppler-shifted center frequencies for laser beams a and b would change into

$$\nu_1 = \frac{\nu_0 c}{c + u \sin\theta} \tag{6}$$

$$\nu_2 = \frac{\nu_0 c}{c - u \sin\theta} \tag{7}$$

Considering that $c >> u$, the flow velocity relationship can be simplified to

$$\Delta\nu = \nu_2 - \nu_1 = \frac{2u\nu_0 c \cdot \sin\theta}{c^2 - u^2 \sin^2\theta} \sim \frac{2u\nu_0 \cdot \sin\theta}{c} \tag{8}$$

$$\text{or } u = \frac{c \cdot \Delta\nu}{2\nu_0 \cdot \sin\theta} \tag{9}$$

2.3. The Experimental TDLAS System

The TDLAS system designed in this work mainly comprised an infrared laser, modulation unit, optical path, photodetectors, and high-speed data acquisition and processing module, as shown in Figure 2. An OEM VCSEL driver (VITC002 from Thorlabs, Newton, NJ, USA) with a temperature controller was applied for laser modulation. The functional parameters of the laser (VCSEL from Vertilas, München, Germany) and function signal generator (DG-1022 from Rigol, Beijing, China) are listed in Table 1.

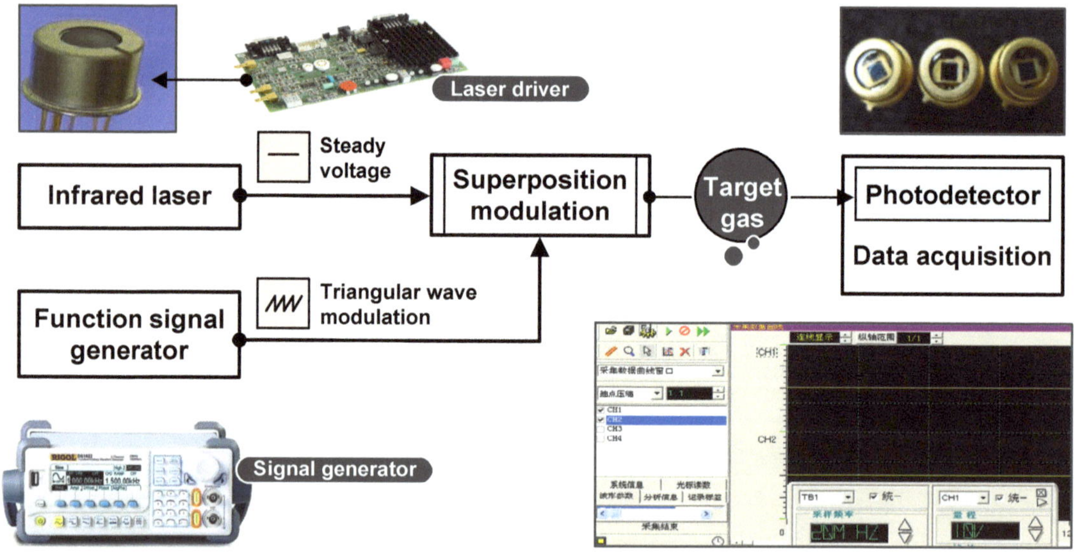

Figure 2. Schematic of the TDLAS system designed in the present work.

Table 1. Main parameters of laser and function signal generator selected.

Infrared Laser (VCSEL from Vertilas)	
Center wavelength (nm)	1392 *
Operating current range (mA)	0.5–8.5
Operating temperature range (°C)	15–35
Wavelength range at 20 °C (nm)	1389.16–1392.96
Wavelength range at 30 °C (nm)	1390.20–1393.70
Function signal generator (DG-1022 from Rigol)	
Maximum output frequency (MHz)	20
Frequency of sampling (MSa/s)	100
Frequency resolution (Hz)	1μ

* Center wavelength is chosen according to the absorption spectrum of H_2O, which will be discussed later.

Because of the advantages provided by analyzing water vapor, H2O-based TDLAS measurement assessments were used in this study but with more accurate experimental validation and a specially designed optical path intended for the monitoring of combustion processes. Water will produce intense absorbance bands in the near-IR region 1400, 1800, and 2700 [16,21]; and to avoid any interference by other species (such as C-H radical), 1392 nm was selected as the center wavelength for water vapor detection. Photodetectors (PN-2000 from Lightsensing Technologies, Beijing, China) with a response range of 900–1650 nm were used to determine the transmitted light intensity. Data were obtained using a data acquisition card (PCI-20612 from TDEC, Sichuan, China) with four channels, operating at 32 bits and a maximum rate of 50 MSa/s.

2.4. Experimental Design for Validation

The functioning of the TDLAS system was calibrated or examined in three ways, as shown in Figure 3, using a pressure vessel, a shock tube, and a co-flow combustion platform.

(1) Firstly, normal pressure and temperature were applied to the pressure vessel (Figure 3a) with standing air to calibrate the basic performance of the system. This vessel was made of stainless steel with optical glasses on both sides. The optical path in this device had a maximum length of 0.4 m and a 532 nm green laser was employed to adjust the path.

(2) Secondly, the shock tube was intended to provide determinable high-temperature and high-speed water vapor flow to permit the precision and response rate of the measurement system to be ascertained. As shown in Figure 3b, the shock tube was comprised a high-pressure section, a low-pressure section, a gas circuit, and an electronically controlled diaphragm. Prior to each test, the low-pressure section was charged with air to a preset pressure. Following this, the high-pressure section was also slowly filled with air until the diaphragm instantaneously ruptured to create a shock wave, thus producing a high-temperature/pressure, high-speed flow field.

(3) Finally, the calibrated TDLAS system was used for the CH_4/Air flame temperature and hot gas velocity measurements, as shown in Figure 3c. A co-flow CH_4/air burner was made to generate a stable diffusion flame with preset initial conditions [22,23]. High precision mass flowmeters (KM7100 from Alicat, Tucson, AZ, USA) were used to dispense the combustible gases. To avoid the disturbance by H_2O absorption in the non-flame zone (i.e., a background signal resulting from atmospheric H_2O), a beam splitter (50%:50%) was used to subtract the background interference. As noted, the flame width (absorption length about 2–3 cm) was relatively short, thus a reflector was added to obtain a stronger absorption signal.

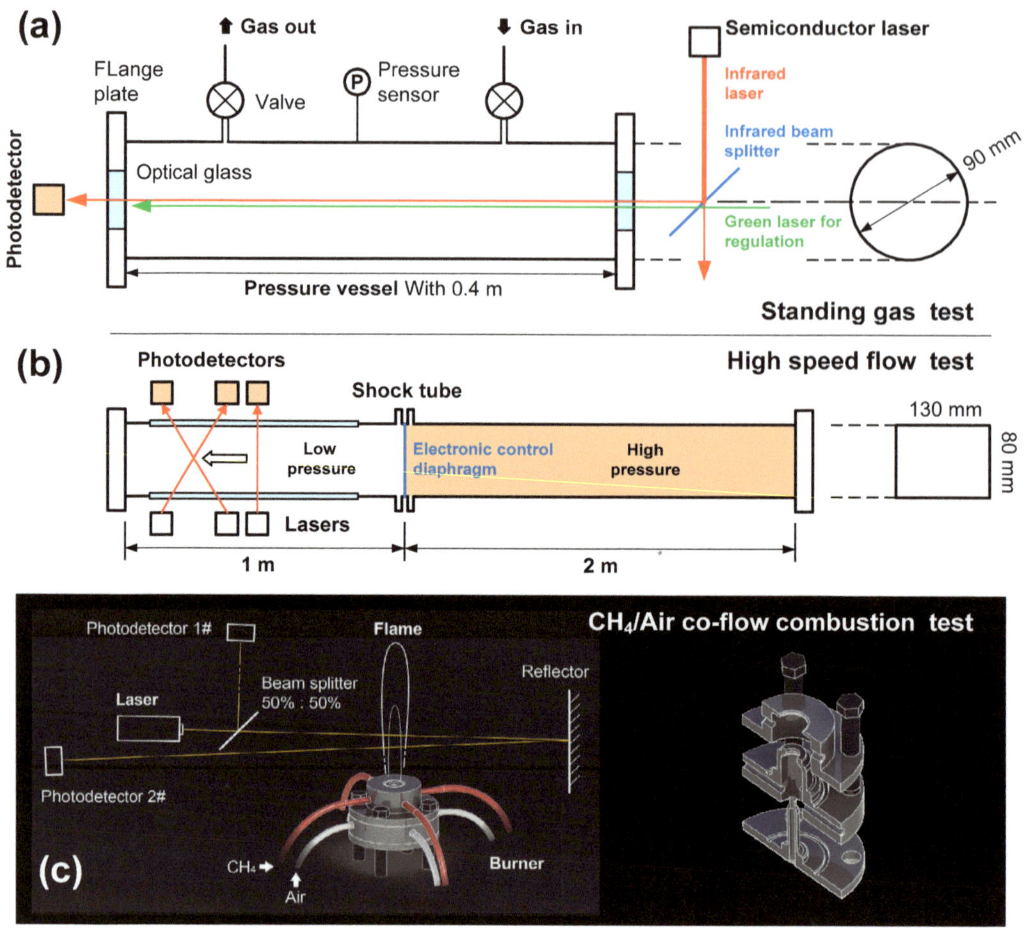

Figure 3. Experimental facilities for TDLAS tests, including (**a**) a pressure vessel, (**b**) a shock tube, and (**c**) a co-flow flame burner.

The experimental conditions are summarized in Table 2. All the tests were repeated 20 times to ensure reproducible results.

Table 2. Experimental conditions for tests by three facilities.

No.	Test Facility	Object	Initial Conditions
1	Pressure vessel	Air temperature	Room temperature Pressure: 90 kPa–2 MPa
2	Shock tube	Flow velocity	Velocity: 500–950 m/s
3	Combustion platform	Flame temperature and velocity	Room temperature at 1 atm CH_4 flow rate: 0.2–2.0 sl/min

3. Results and Analysis

3.1. Room Temperature Measurement by TDLAS

Absorption spectra of the contents of the pressure vessel (see Figure 3a) could be obtained on the basis of comparisons between the laser output and absorption line strength using the HITRAN [17] data, as shown in Figure 4 with an example at initial pressure of

1 atm. Figure 4a shows the voltage U variation of function signal generator output used for driving laser during a half cycle, and Figure 4b presents the transmitted light intensity after absorption. Furthermore, clear positions and strengths of absorption peaks could be found in Figure 4c. The line strength of water vapor vs. wavelength is plotted in Figure 4d with independent absorption line.

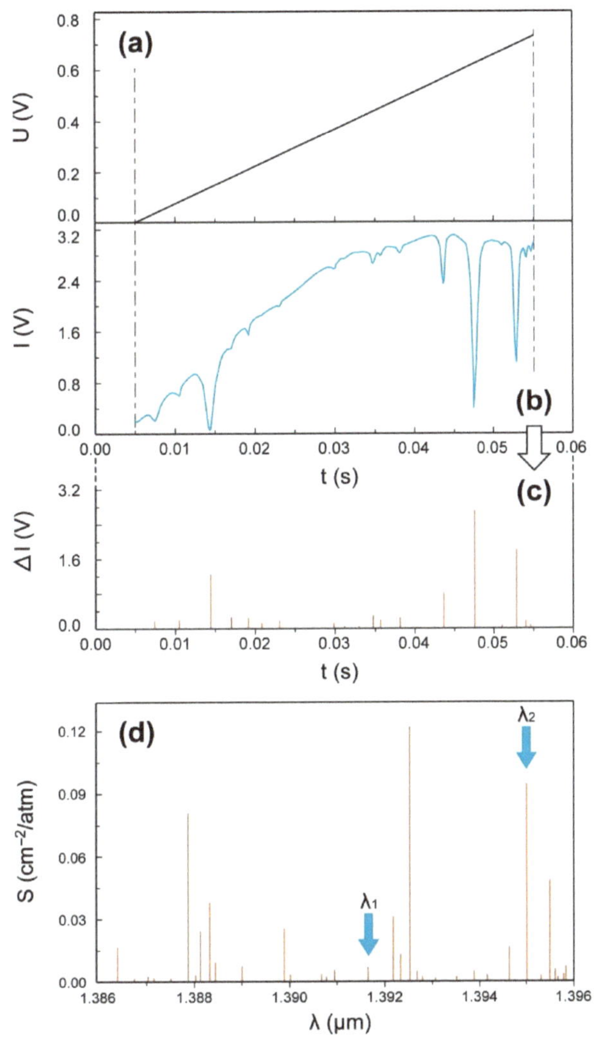

Figure 4. Variation in (**a**) laser driving voltage and (**b**) transmitted light intensity in a half circle with (**c**) peak positions and strengths, and (**d**) the line strength distribution calculated from the HITRAN [17] database.

To calculate the vapor temperature, the time-domain of transmitted light intensity (I vs. t) should be transformed to frequency-domain (that is, I vs. ν or I vs. λ) at first. Based on the approximately linear relationship between λ and U, two reference wavelength-time points were selected: (λ_1, t_1) and (λ_2, t_2). Then we obtained

$$\lambda = \frac{t - t_2}{t_1 - t_2}(\lambda_1 - \lambda_2) + \lambda_2 \tag{10}$$

This simplified linear fitting was considered a reasonable approximation over short time spans. Note that λ vs. t would not be a continuous function, due to the discreteness of λ. Furthermore, the wavelengths that were selected for these calculations (λ_1 and λ_2) should be a certain distance apart to reduce the error caused by uncertainties in determining the positions of the absorption peaks. Hence, λ_1 = 1391.67275 nm and λ_2 = 1395.00424 nm were selected in the present work for the purpose of wavelength calibration.

Consequently, as $\nu \sim 1/\lambda$, the correlation between I and ν could be obtained by combining Equation (10) with the data in Figure 4, as shown in Figure 5a. The baseline in Figure 5a was fitted by using the polynomial $I_0 = a_0 + a_1\nu + a_2\nu^2 + a_3\nu^3$ and employing data within the non-absorption region. The curve of $\ln(I/I_0)$ vs. ν, as the key relation for temperature calculation deduced in Equations (4) and (5), could be further illustrated in Figure 5b,c.

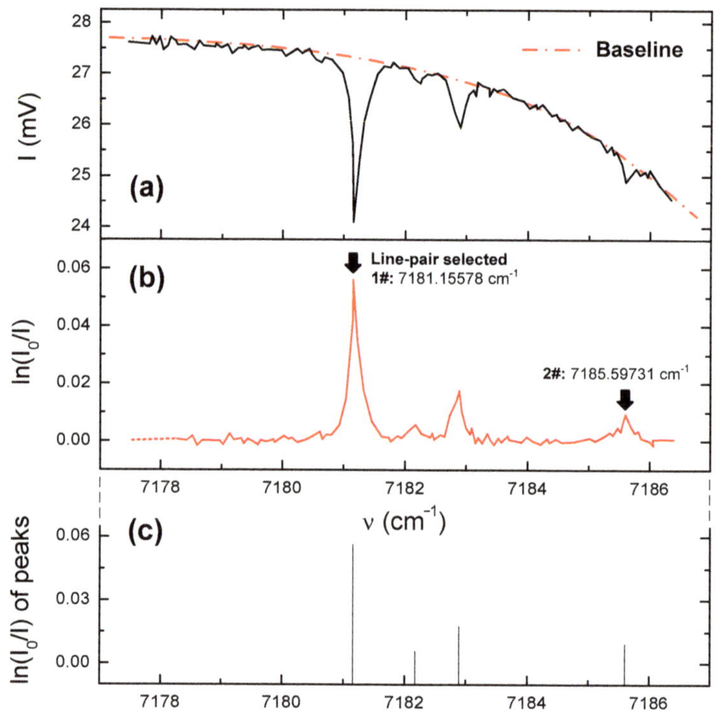

Figure 5. (**a**) Absorption frequency-domain diagrams of the transmitted light intensity and (**b**) the absorption ratio with (**c**) main peak positions and strengths.

It is necessary to take into account that the line-pair selection had to meet certain conditions, meaning that there was no interference by other spectral lines and these lines were positioned near the central wavelength of the laser. Furthermore, the lines had to be separated by a suitable distance to avoid overlap. Therefore, we chose ν = 7181.15578 cm^{-1} and ν = 7185.59731 cm^{-1} under overall consideration. Other important parameters related to the HITRAN database [17] are provided in Table 3. Combining Equations (4) and (5) provide T = 302 ± 1.4 K, and this value—compared with the average experimental value of 301.14 ± 0.8 K by the thermocouples—provides a measurement error of less than 0.3%.

Table 3. Parameters queried and calculated from HITRAN.

Frequency (cm^{-1})	Absorption Line Strength under Room Temperature (cm^{-2}/atm)	Energy of the Lower Transition State (cm^{-1})
7181.15578	0.12280	136.7617
7185.59731	0.00648	1045.0579

The pressure effects on measurement accuracy were investigated subsequently. The results showed that, with the enlarged initial test pressure, pressure broadening occurred generally and became dominant due to the increasing frequency of molecular collisions. In addition, spectral interference resulting from line overlap became evident at pressures exceeding 0.8 MPa. It should be noted that the measurement errors related to temperature and concentration could be larger than 10% at pressures above 1 MPa.

3.2. Velocity of High Speed Air Flow in Shock Tube

As shown in Figure 3b, a uniform and controllable air flow field could be generated in the shock tube. The velocity values measured by the TDLAS system were compared with those by both ICP (integrated circuits piezoelectric) shock wave pressure sensors (102B15 from Dibeiqi Electronic Technology, China) and theoretical calculations. Two ICP sensors with a distance of 120 mm were mounted along the tube as shown in Figure 6a, and typical results were plotted in Figure 6b. The shock wave velocity was determined from these data as

$$u_{shock} = \frac{120\ \text{mm}}{\Delta t_{ICP}} \tag{11}$$

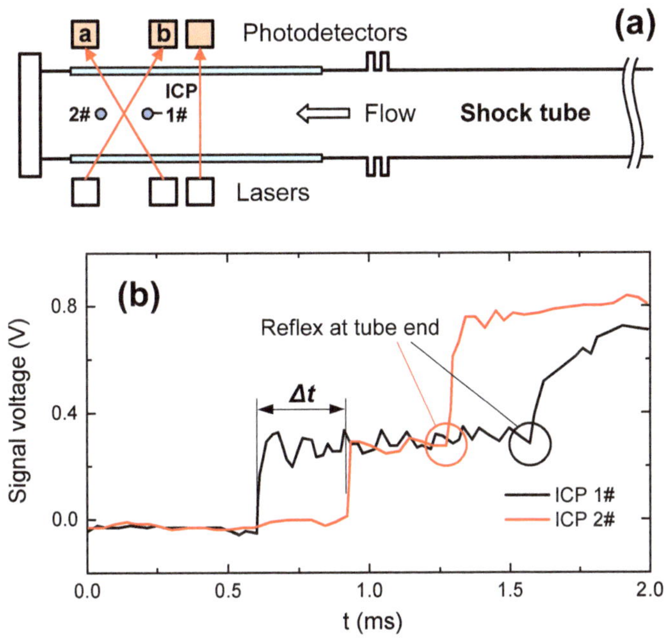

Figure 6. (**a**) ICP sensor locations and (**b**) responses for a pressure wave in the shock tube.

Based on assuming isentropic flow, the flow velocity could be simplified as

$$u = M\sqrt{\gamma R_{gas} T} \tag{12}$$

where M is the Mach number of the shock wave determined by u_{shock} in Equation (11) and sound velocity, γ is the adiabatic exponent, and R_{gas} is the universal gas constant.

The signals obtained from laser beams a and b during the TDLAS analyses are shown in Figure 7. To avoid miscalibaration, the synchronization of the initial absorption peaks of the two lasers under static air condition were performed. According to the correlation of t with λ (or Δt vs. $\Delta\lambda$) in Equation (10), the frequency shift resulting from the Doppler effect could be

$$|\Delta\nu| = \left|\Delta\frac{1}{\lambda}\right| \sim \frac{|\Delta\lambda|}{\lambda_0^2} \tag{13}$$

and combining this relationship with Equation (9) allowed the flow velocity to be determined from the TDLAS results.

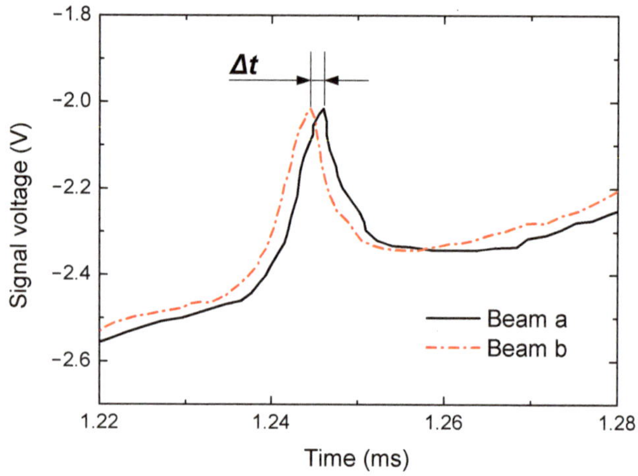

Figure 7. Time difference in the TDLAS signal due to the frequency shift induced by the Doppler effect.

In theory, the flow velocity and temperature could also be predicted by shock wave propagation equations numerically [24]. It is helpful to summarize the characteristics of the three measurement methods:

(1) The ICP sensors can capture the shock wave movement and time interval, which are the key parameters to calculate flow velocity with shock wave theory.
(2) As noted above, the velocity by TDLAS measurement is actually that of H_2O molecule.
(3) Numerical prediction is based on both the Mach number (obtained according to the pressure ratio between the high-pressure and low-pressure sections) and the shock wave propagation equations.

To validate these methods, tests were conducted applying an initial pressure in the range of 100–650 kPa for the low-pressure section, and a comparison of the results is provided in Figure 8.

It is found that the results obtained from the ICP sensors and the numerical predictions were similar, presumably because both methods are on the basis of Mach number calculations and shock wave theory. However, the manner in which the Mach number is obtained is very different between the two. In the case of the ICP sensors method, the Mach number was deduced by shock wave velocity and sound velocity of the wave front, whereas, for numerical prediction, the Mach number was calculated from the aforementioned pressure ratio and iterative solution of gas dynamics relationships. Moreover, we should note that some assumptions had been made in shock wave theoretical analysis that may have led to a slight overestimation of the flow velocity. Specifically,

(1) Air viscosity effects and the propagation of waves in more than one dimension were ignored.

(2) The attenuation of the shock wave along shock tube was not considered.

(3) Considering the velocity range of gas flow, the re-absorption effect was also ignored [25].

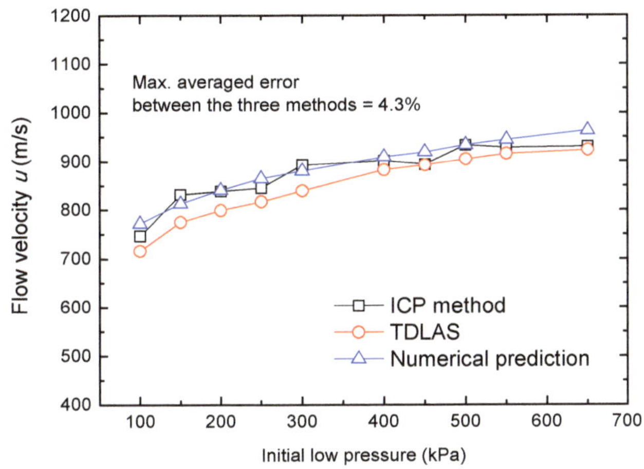

Figure 8. Gas flow velocities in the low-pressure section of the shock tube as determined using three methods.

In general, for high-speed gas flow velocity, the TDLAS measurements show good agreement the results obtained using the other two methods, with a maximum averaged error of 4.3%. The temperature of the reflected shock wave was also monitored, and a comparison between the TDLAS results and numerical predictions is shown in Figure 9. It is proved that the TDLAS system we developed in the present work was able to accurately capture the variation of temperature and velocity for high-speed gas flow.

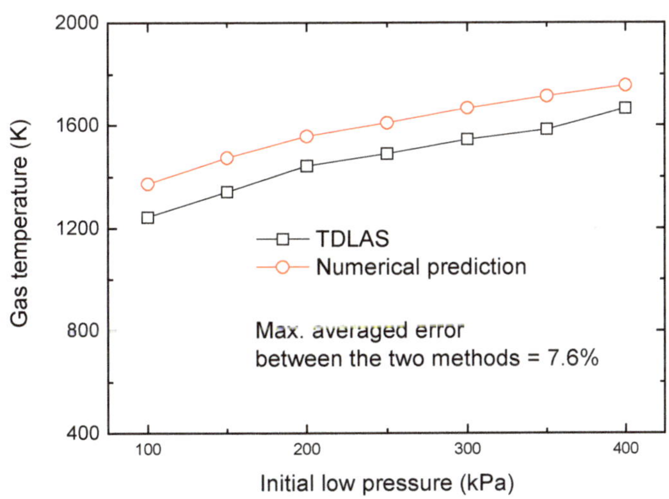

Figure 9. Comparison of gas temperature after reflected shock wave by two measurement methods.

3.3. Local Temperature and Velocity of CH_4/Air Flame

As one of the most important combustion products, the H_2O is an ideal tracer for flame measurement by the TDLAS system. Here, variations in $S(T)$ were evaluated using a thin Pt–Rh thermocouple wire with a diameter of 0.1 mm to estimate the approximate flame temperature range (in the vicinity of 1500 K). These data were subsequently used to validate the results by TDLAS as shown in Figure 10a,b.

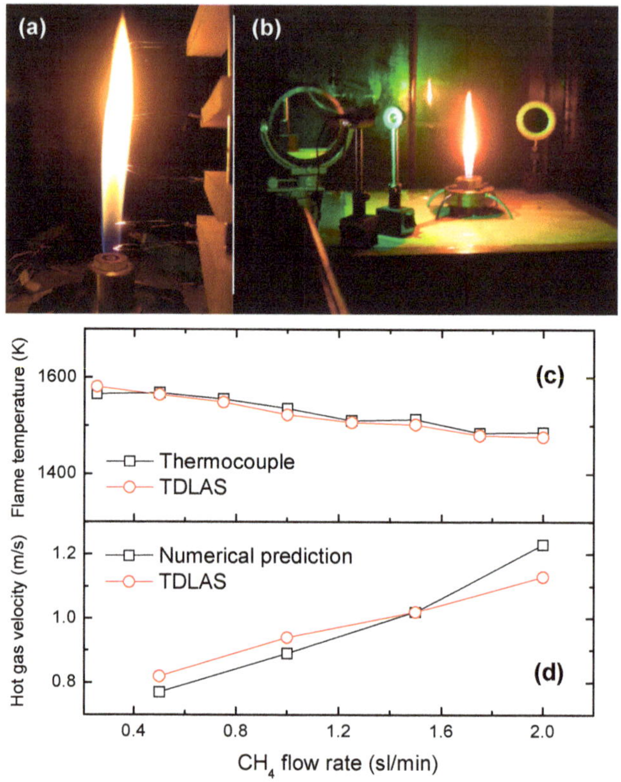

Figure 10. CH_4/air flame data acquired using (**a**) thin Pt-Rh thermocouples, (**b**) the TDLAS system, and validations of the (**c**) flame temperature and (**d**) velocity results.

The line-pair of ν = 7181.15578 cm^{-1} and ν = 7185.59731 cm^{-1} was again used in this assessment. During the trials, the fuel mass (or volume) flow rate of CH_4/air diffusion flame was controlled by mass flowmeters to obtain an air flow rate of 2 standard liters per minute (slpm) and CH_4 flow rates in the range of 0.25–2 slpm under an ambient pressure of 1 atm.

Figure 10c compares the temperature results between the thermocouples and the TDLAS system. The tendencies and actual values of the two curves are very similar, meanwhile, velocities by numerical simulations and the TDLAS process are also compared in Figure 10d with good agreement. It is interesting that although there were pulsations of the flame during testing, it can be ignored when using a driver scanning frequency above 1 kHz, meaning that approximately steady state flame information could be acquired at high scanning and sampling frequencies.

4. Conclusions

With the goal of performing combustion diagnostics and assessing the thermal characteristics of flammable gases, a portable H2O-based TDLAS system was designed. The measuring principle was revealed theoretically, and experimental calibration and validation were performed using a pressure vessel, shock tube, and CH_4/air diffusion flame, respectively. The main conclusions from this work are:

(1) A method of calculating gas characteristics including temperature, velocity, and species concentration was provided using H_2O absorption spectra and the HITRAN database in conjunction with TDLAS system. In addition, the line-pair selection criteria was verified. This method was demonstrated to be a viable approach to hot gas analyses.

(2) The dual optical path TDLAS system established can eliminate the background interference effectively when applied to a relatively small test target. The combined selection of an appropriate center wavelength, wavelength calibration between the time and frequency domains, and use of a specific scanning or sampling frequency allowed this system to be used for the assessment of small flames.

(3) The temperature calibration and pressure broadening effect of absorption spectra of this TDLAS system were studied using a pressure vessel. It is found that this system can provide an accurate measurement within environment pressure of 0.8 MPa.

(4) A shock tube was built to provide controllable and steady gas flows with high temperature or high speed, which turned out to be an ideal experimental setup for the parameter calibration of TDLAS at extreme conditions. The TDLAS system was confirmed to accurately monitor variations in high temperature and velocity gas flows.

(5) A small scale CH_4/air diffusion flame burner was developed to validate the ability of the TDLAS system to monitor combustion characteristics. A comparison of the TDLAS results with thermocouple measurements and numerical simulations indicated that the TDLAS method provided satisfactory flame temperature and velocity values.

Future work will involve the application of TDLAS to the assessment of hydrocarbon combustion products such as CO_2 and C-H radicals to explore more precise measurements.

Author Contributions: Initial conceptualization, R.T. and J.G.; Data curation, J.G., Y.Z. and Z.M.; Formal analysis, R.T., X.Z. and J.J.; Funding acquisition, R.T., X.Z., K.Y. and J.Y.; Methodology, R.T. and J.G.; Project administration, R.T.; Writing—original draft, R.T. and Y.Z.; Writing—review and editing, J.G. and X.Z. All authors have read and agreed to the published version of the manuscript.

Funding: This research was supported by the National Natural Science Foundation of China (nos. 52076084 and 52175508), Collaborative Innovation Platform of Fuzhou-Xiamen-Quanzhou Independent Innovation Demonstration Area (no. 3502ZCQXT202002), and the Fundamental Research Funds for the Central Universities (no. ZQN-1001). The authors thankfully acknowledge all these supports.

Institutional Review Board Statement: Not applicable.

Informed Consent Statement: Not applicable.

Data Availability Statement: Data is contained within the article.

Acknowledgments: We thank S.L. Xu for assistance with the experimental design.

Conflicts of Interest: The authors declare no conflict of interest.

Nomenclature

A	Integrated absorbance of independent transition line
c	Speed of light (m s^{-1})
E	Energy of the lower transition state (J)
I	Signal of light intensity (V)
L	Path length (m)
p	Partial pressure (Pa)
S	Molecular absorption line strength (cm^{-2} atm^{-1})
T	Temperature (K)
t	Time (s)
u	Velocity (m s^{-1})
α	Absorptivity of absorbing species
ν	Frequency of laser
φ	Absorption lineshape function of species
θ	Angle between two lasers
λ	Wavelength

References

1. Philippe, L.C.; Hanson, R.K. Laser diode wavelength-modulation spectroscopy for simultaneous measurement of temperature, pressure, and velocity in shock-heated oxygen flows. *Appl. Opt.* **1993**, *32*, 6090–6103. [PubMed]
2. Alexander, K.; Oliver, W.; Volker, E. Rapid, Time-division multiplexed, direct absorption- and wavelength modulation-spectroscopy. *Sensors* **2014**, *14*, 21497–21513.
3. Lo, Y.H.; Grabbe, P.; Iqbal, M.Z.; Bhat, R.; Gimlett, J.L.; Young, J.C.; Lin, P.S.D.; Gozdz, A.S.; Koza, M.A.; Lee, T.P. Multigigabit/s 1.5 mu m lambda/4-shifted DFB OEIC transmitter and its use in transmission experiments. *IEEE Photonics Technol. Lett.* **1990**, *2*, 673–674.
4. Zah, C.E.; Amersfoort, M.R.; Pathak, B.; Favire, F.; Lin, P.S.D.; Rajhel, A.; Andreadakis, N.C.; Bhat, R.; Caneau, C.; Koza, M.A. Wavelength accuracy and output power of multiwavelength DFB laser arrays with integrated star couplers and optical amplifiers. *IEEE Photonics Technol. Lett.* **1996**, *8*, 864–866.
5. Nagali, V.; Hanson, R.K. Design of a diode-laser sensor to monitor water vapor in high-pressure combustion gases. *Appl. Opt.* **1997**, *36*, 9518–9527.
6. Mihalcea, R.M.; Baer, D.S.; Hanson, R.K. Advanced diode laser absorption sensor for in situ combustion measurements of CO_2, H_2O, and gas temperature. *Proc. Combust. Inst.* **1998**, *27*, 95–101.
7. Allen, M.G. Diode Laser Absorption Sensors for Gas-Dynamic and Combustion Flows. *Meas. Sci. Technol.* **1998**, *9*, 545–562.
8. Chang, A.Y.; Dirosa, M.D.; Davidson, D.F.; Hanson, R.K. Rapid tuning cw laser technique for measurements of gas velocity, temperature, pressure, density, and mass flux using NO. *Appl. Opt.* **1991**, *30*, 3011–3022.
9. Kranendonk, L.A.; Walewski, J.W.; Kim, T.; Sanders, S.T. Wavelength-agile sensor applied for HCCI engine measurements. *Proc. Combust. Inst.* **2005**, *30*, 1619–1627.
10. Webber, M.E.; Wang, J.; Sanders, S.T.; Baer, D.S.; Hanson, R.K. In situ combustion measurements of CO, CO_2, H_2O and temperature using diode laser absorption sensors. *Proc. Combust. Inst.* **2000**, *28*, 407–413.
11. Torek, P.V.; Hall, D.L.; Miller, T.A.; Tiffany, A.; Wooldridge, M.S. H_2O absorption spectroscopy for determination of temperature and H_2O mole fraction in high-temperature particle synthesis systems. *Appl. Opt.* **2002**, *41*, 2274–2284.
12. Zhou, X.; Sanders, S.T.; Jeffries, J.B.; Hanson, R.K.; Jenkins, T.P. Combustion temperature and H_2O concentration sensor using a single diode laser. In Proceedings of the 40th AIAA Aerospace Sciences Meeting & Exhibit, Reno, NV, USA, 14–17 January 2002; Volume 395.
13. Zhou, X.; Liu, X.; Jeffries, J.B.; Hanson, R.K. Development of a sensor for temperature and water concentration in combustion gases using a single tunable diode laser. *Meas. Sci. Technol.* **2003**, *14*, 1459.
14. Fuwa, K.; Valle, B.L. The physical basis of analytical atomic absorption spectrometry. The pertinence of the Beer-Lambert law. *Anal. Chem.* **1963**, *35*, 942–946.
15. Gao, N.; Du, Z.H.; Qi, R.B.; Ma, Y.W.; Wang, Y. Lineshape deviation and correction in tunable diode laser absorption spectroscopy technology. *J. Optoelectron. Laser* **2011**, *22*, 893–896.
16. McHale, J.L. *Molecular Spectroscopy*; Prentice Hall: Hoboken, NJ, USA, 1998.
17. High-Resolution Transmission Molecular Absorption Database. Atomic and Molecular Physics Division, Harvard-Smithsonian Center for Astrophysics, 2012. Available online: http://cfa-www.harvard.edu/HITRAN/ (accessed on 2 May 2013).
18. Rothman, L.S.; Barbe, A.; Chris, B.D.; Brown, L.R.; Camy-Peyret, C.; Carleer, M.R.; Chance, K.; Clerbaux, C.; Dana, V.; Devi, V.M.; et al. The HITRAN molecular spectroscopic database: Edition of 2000 including updates through 2001. *J. Quant. Spectrosc. Radiat. Transf.* **2003**, *82*, 5–44.
19. Allen, M.G.; Kessler, W.J. Simultaneous water vapor concentration and temperature measurements using 1.31 μm diode lasers. *AIAA J.* **1996**, *34*, 483–488.
20. Miller, M.F.; Kessler, W.; Allen, M.G. Diode laser-based air mass flux sensor for subsonic aeropropulsion inlets. *Appl. Opt.* **1996**, *35*, 4905–4912.
21. Rieker, G.B.; Liu, X.; Li, H.; Jeffries, J.B.; Hanson, R.K. Measurements of near-IR water vapor absorption at high pressure and temperature. *Appl. Phys. B* **2006**, *87*, 169–178.
22. Zeng, Y.; Fang, J.; Wang, J.W.; Li, J.; Tu, R.; Zhang, Y. Momentum-dominated methane jet flame at sub-atmospheric pressure. *Procedia Eng.* **2013**, *63*, 924–931.
23. Wang, J.W.; Fang, J.; Guan, J.F.; Zeng, Y.; Zhang, Y.M. Flame volume and radiant fraction of jet diffusion methane flame at sub-atmospheric pressures. *Fuel* **2016**, *167*, 82–88.
24. Anderson, J.D. *Fundamentals of Aerodynamics*; McGraw Hill: New York, NY, USA, 2006.
25. Lundgaard, S.; Ng, S.H.; Cahill, D.; Dahlberg, J.; Allender, J.; Barber, M.; Stephens, J.; Juodkazis, S. Electrical breakdown spectroscopy of nano-/micro-thermites. *Technologies* **2021**, *9*, 34. [CrossRef]

sensors

MDPI

Article

Detection of Water pH Using Visible Near-Infrared Spectroscopy and One-Dimensional Convolutional Neural Network

Dengshan Li and Lina Li *

College of Mechanical Engineering and Automation, Huaqiao University, Xiamen 361021, China; 20013080028@stu.hqu.edu.cn
* Correspondence: lilina@hqu.edu.cn; Tel.: +86-13-395023485

Abstract: pH is an important parameter for water quality detection. This study proposed a novel calibration regression strategy based on a one-dimensional convolutional neural network (1D-CNN) for water pH detection using visible near-infrared (Vis-NIR) spectroscopy. Two groups of Vis-NIR spectral analysis experiments of water pH detection were employed to evaluate the performance of 1D-CNN. Two conventional multivariate regression calibration methods, including partial least squares (PLS) and least squares support vector machine (LS-SVM), were introduced for comparative analysis with 1D-CNN. The successive projections algorithm (SPA) was adopted to select the feature variables. In addition, the learning mechanism of 1D-CNN was interpreted through visual feature maps by convolutional layers. The results showed that the 1D-CNN models obtained the highest prediction accuracy based on full spectra for the two experiments. For the spectrophotometer experiment, the root mean square error of prediction (RMSEP) was 0.7925, and the determination coefficient of prediction (R_p^2) was 0.8515. For the grating spectrograph experiment, the RMSEP was 0.5128 and the R_p^2 was 0.9273. The convolutional layers could automatically preprocess the spectra and effectively extract the spectra features. Compared with the traditional regression methods, 1D-CNN does not need complex spectra pretreatment and variable selection. Therefore, 1D-CNN is a promising regression approach, with higher prediction accuracy and better modeling convenience for rapid water pH detection using Vis-NIR spectroscopy.

Keywords: visible; near-infrared; pH detection; one-dimensional convolutional neural network; multivariate regression calibration

Citation: Li, D.; Li, L. Detection of Water pH Using Visible Near-Infrared Spectroscopy and One-Dimensional Convolutional Neural Network. *Sensors* **2022**, *22*, 5809. https://doi.org/10.3390/s22155809

Academic Editor: Simone Borri

Received: 28 June 2022
Accepted: 2 August 2022
Published: 3 August 2022

Publisher's Note: MDPI stays neutral with regard to jurisdictional claims in published maps and institutional affiliations.

Copyright: © 2022 by the authors. Licensee MDPI, Basel, Switzerland. This article is an open access article distributed under the terms and conditions of the Creative Commons Attribution (CC BY) license (https://creativecommons.org/licenses/by/4.0/).

1. Introduction

Water is the basic resource for human survival and ecosystem evolution. Water safety is an issue that cannot be overlooked and that directly affects the earth's ecological environment and human survival. However, with the development of society and economics, industrial waste is emerging as a principal source of water contamination. Water quality parameters, such as pH, dissolved oxygen, turbidity, sediments, chloride ions, potassium ions, and so on, characterize the quality of the water environment. pH is one of the important water quality parameters, as it involves the water's natural phenomena, chemical changes, and production process [1]. Therefore, research on the rapid detection of water pH is significant.

Traditionally, the determination of water pH mainly adopts the glass electrode method [2]. The measurements of this method are accurate, but the process is cumbersome and needs the pH buffers to calibrate. Visible near-infrared (Vis-NIR) spectroscopy is a fast, non-destructive and qualitative analysis technique which has been widely used in the rapid detection of chemical composition [3–5]. In the Vis-NIR region, the absorption of the spectrum mainly comes from the frequency doubling and combination band of hydrogen-containing groups (such as O-H, C-H, and N-H) [6]. Vis-NIR spectroscopy combined with chemometrics is one of the promising techniques for rapid water pH detection. However, since the prediction accuracy is not satisfactory, there are few reports about the use of this technology to detect the water pH value [7].

High-accuracy and effective multivariate regression calibration models play a key role in Vis-NIR spectral analysis. A lot of multivariate calibration algorithms have been developed to build the relationship between the Vis-NIR spectra and the target attributes, including partial least squares (PLS), least squares support vector machine (LS-SVM), artificial neural network (ANN), and extreme learning machine (ELM) [8–10], etc. Among these methods, the PLS (linear) and LS-SVM (nonlinear) are the most commonly used multivariate calibration methods [11]. The high dimension and high complexity are the main characteristics of modern spectroscopy. The spectra data contains a large number of uninformative variables and noise, which is the challenge of spectra analysis. In order to reduce the dimension of spectra and improve the prediction accuracy, a large amount of characteristic wavelength variable selection methods were proposed [12]. However, many variable selection algorithms are of low reproducibility and have the risk of overfitting, such as competitive adaptive reweighted sampling (CARS), genetic algorithm (GA), and variable combination population analysis (VCPA) [13].

In recent years, convolutional neural networks (CNN) have provided new insights for spectra analysis. CNN is a representative deep learning algorithm which was initially used to solve the spectra classification problems. Recent studies have found that CNN is feasible in spectra multivariate regression calibration, and the prediction performance of CNN is better than PLS. Cui et al. (2018) proposed a unified CNN structure used for multivariate regression, and the results indicated that the CNN models achieved the highest prediction accuracy [14]. Li et al. (2022) explored the potential of using the short-wave infrared (SWIR) hyperspectral imaging (HSI) technique combined with a one-dimensional convolutional neural network (1D-CNN) for predicting the SSC of dried Hami jujube, and the results indicated that the prediction accuracy of 1D-CNN was satisfactory [11]. Mishra et al. (2022) presented a strategy of 1D-CNN modeling for the multi-response prediction for spectral data of fresh fruit, they found that the prediction performance of 1D-CNN is better than PLS [15]. Moreover, CNN modeling has no need for complex spectra pretreatment and variable selection, since the CNN with convolutional layers can automatically extract the features of spectra [13]. Thus, CNN is expected to be a widely used multivariate calibration technology in Vis-NIR spectroscopy analysis.

Although the CNN model has outperformed the traditional calibration methods, it is hard to explain how CNN extracts efficient information from the input spectra and how CNN establishes the relationship between the spectral and the objective attribute. For linear regression strategies such as principal component regression (PCR) and PLS, the absolute value of regression coefficients indicates the importance of the corresponding wavelength variables. The nonlinear regression methods, such as LS-SVM, ELM, and CNN, have more complex calculation processes and structures, and it is difficult to explain the relevance between the spectral data and the attribute of interest. The CNN model is a black box due to the complex network structure and multilayer operations. Interpreting the learning mechanism of CNN is critical for further application. In the field of spectra classification, CNN's visual explanation method has already been reported. Fukuhara et al. (2019) proposed a feature visualization method that calculates important regions in the spectra from weights in pooling and fully-connected layers [16]. This approach focuses on explaining the characteristic variables extraction but ignores the explanation of how the convolutional layers process the spectra and learn the spectra features. Zhang et al. (2020) and Ng et al. (2020) visualized the feature maps of convolutional layers to explain how the CNN model processes the spectra and extracts the feature variables [13,17]. This method explains the contribution of each variable to the classification results and interprets how the convolutional layers process the spectra and learn the spectra features. However, in the field of spectra regression, there are few reports about interpreting the CNN regression model.

Therefore, a novel calibration regression strategy based on 1D-CNN for water pH detection is proposed in this study. Two experiments of water pH value detection using Vis-NIR spectral analysis are adopted to evaluate the performance of 1D-CNN. PLS and LS-SVM are introduced for comparative analysis with 1D-CNN. The main goal of this study are the following: (1) explore the potential of using Vis-NIR spectroscopy for water pH detection; (2) establish and optimize traditional models (PLS and LS-SVM) through spectra pretreatment methods and characteristic wavelength variables algorithms; (3) construct the 1D-CNN architectures and optimize the parameters of 1D-CNN for modeling to predict water pH; (4) visualize the feature maps by convolutional layers and interpret the internal feature representations of 1D-CNN; (5) compare the prediction performance of 1D-CNN and traditional multiple regression algorithms, the model prediction accuracy, and convenience of modeling are discussed.

2. Materials and Methods

2.1. Experiment

In this study, two independent experiments were conducted to evaluate the effectiveness and applicability of the 1D-CNN regression method for water pH detection using Vis-NIR spectroscopy. In these two experiments, we adopted two Vis-NIR spectrometers with different spectroscopic principles to measure the absorbance of samples.

Sample preparation: Randomly added the 0.1 mol/L HCl solution or 0.1 mol/L NaOH solution into distilled water to prepare samples with different pH values.

Reference pH measurement: Use the pH meter (Shanghai Lichen-BX Instrument Co., Ltd. Shanghai, China) to measure the reference pH of each sample. The measurement accuracy of the pH meter is ± 0.01 pH unit.

2.1.1. Spectrophotometer Experiment: Detecting the pH Value of Distilled Water with a Spectrophotometer

Sample description: A total of 34 samples were used in this experiment. The reference pH range was 3.19~10.63, and the average value and standard deviation were 6.15 and 2.08, respectively.

Spectra acquisition: A spectrophotometer (V-1800PC, Shanghai Mapada Instrument Co., Ltd. Shanghai, China) with a resolution of 2 nm was employed to measure the Vis-NIR spectra in the range of 400~1100 nm. Each sample was placed in a 10 mm quartz cell. Adopted transmission measurements to obtain the spectra and record the absorbance as the spectra data. The original Vis-NIR spectrum of the 34 distilled water samples are shown in Figure 1a.

Sample grouping: In order to ensure that the multivariate calibration model has a good generalization ability, the distilled water samples were divided into a calibration set (24 samples) and a prediction set (eight samples) by the descending sorting method [18]. The calibration set is used for modeling, and the prediction set is adapted to test the performance of the multivariate calibration model.

2.1.2. Grating Spectrograph Experiment: Detecting the pH Value of Distilled Water with a Grating Spectrograph

Sample description: A total of 32 samples were used in this experiment. The reference pH range was 1.64~9.31, and the average value and standard deviation were 6.27 and 2.03, respectively.

Spectra acquisition: The spectra acquisition system consists of a light source (HL-2000-LL), two transmission fibers, and a grating spectrograph (FLAME-T-XR1-RS, Ocean Optics Inc., Orlando, FL, USA). The wavelength range was 400~1049 nm and the spectra resolution were 2860 pixels. Each sample was placed in a 10 mm quartz cell. Spectrometer parameters include an integration time of 34 ms, smoothing 3, and average times 200. Adopted transmission measurements to obtain the spectral and record the absorbance value as the spectra data. The original Vis-NIR spectrum of the 32 distilled water samples is shown in Figure 1b.

Sample grouping: The 32 distilled water samples were partitioned into two parts by the descending sorting method, including a calibration set (24 samples) and a prediction set (8 samples).

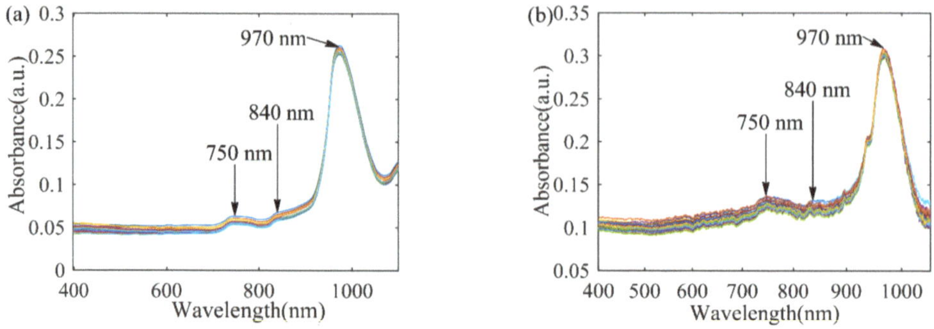

Figure 1. Original Vis-NIR spectra of distilled water samples: (**a**) spectrophotometer experiment; (**b**) grating spectrograph experiment.

2.1.3. Spectral Reference Characteristics

As shown in Figure 1, the key chemical features appear near the ~970 nm, which is related to the third overtones of O-H bonds in water [19]. The small mounds near ~840 nm and ~750 nm, which can be correlated to $2v_1 + v_2 + v_3$ and $3v_1 + v_3$ combination transition, respectively, where v_1 is the symmetric O-H stretch, v_2 is the O-H bending mode, and v_3 is the antisymmetric O-H stretch [20].

2.2. *Traditional Modeling Methods*

2.2.1. Spectral Preprocessing Methods

The spectra data is easily disturbed from instrumental drifts, measurement modality, sample state, and other environmental factors in the acquisition process [21]. Three commonly used preprocessing strategies are applied in this study, including Savitzky-Golay (S-G) smoothing, standard normal variate (SNV), and Z-score normalization. The S-G smoothing with a second-order polynomial and seven smoothing points (SG (2,7)) was used to improve the spectral smoothness and reduce noise interference. The purpose of SNV is to remove scatter noise [22]. The spectra data by Z-score normalization conforms to the standard normal distribution [23].

2.2.2. Modeling Algorithms

Two traditional calibration algorithms, PLS and LS-SVM, are adopted for comparative analysis with 1D-CNN. These two calibration algorithms are briefly introduced as follows.

(1) Partial least squares regression is one of the most commonly used calibration methods, which establishes a linear connection between the spectra data matrix (x) and the target attributes (y). PLS extract uncorrelated principal components (PCs) from the spectra to construct the calibration models. For more details about PLS, please refer to reference [24]. In this study, according to the root mean square error of cross-validation (RMSECV), we chose the optimum number of PCs (nPCs) [18].

(2) Least squares support vector machine is a commonly-used machine learning algorithm which exhibits high prediction accuracy in addressing linear and nonlinear problems [25]. LS-SVM employs a kernel function to transform the original spectra data into a high-dimensional space. Then support vectors are obtained by a set of

linear equations. For more details about LS-SVM, please refer to reference [8]. The prediction results of LS-SVM can be expressed as Equation (1):

$$\hat{y} = \sum_{i=1}^{n} \boldsymbol{\alpha}_i \cdot K\,(\boldsymbol{x},\, \boldsymbol{x}_i) + b \quad (1)$$

where $\boldsymbol{\alpha}_i$ correspond to the Lagrange multiplier called support vector, $K\,(\boldsymbol{x}, \boldsymbol{x}_i)$ represents the kernel function, $\boldsymbol{x}$ refers to the spectra data in the prediction set, and b is the bias.

Before building the LS-SVM model, three crucial factors were considered, including the optimal input features, kernel function, and optimal model parameters [26]. Firstly, the raw or the preprocessed spectra was adopted as the input dataset. Secondly, the radial basis function (RBF) was selected as the kernel function. The RBF kernel can reduce the computational complexity and has an excellent performance in dealing with nonlinear problems [25]. Finally, two model parameters need to be optimized, including the regularization parameter (γ) and the parameter (δ^2) of RBF.

This research employed the particle swarm optimization (PSO) algorithm to search the optimal γ and δ^2. The range of γ and δ^2 within 10^{-2} to 10^6 [25,26]. PSO algorithm to find the optimal solution by iteration, according to different problems, the corresponding fitness function is used to evaluate the quality of the solution [27]. Compared with other optimization methods, the advantage of PSO is easy to realize, and has high solution accuracy and calculation speed. For the PSO, parameters include the population size of particle swarm (20), learning factors $c_1 = c_2 = 1.5$, inertia weight $w_{max} = 1.2$ and $w_{min} = 0.8$, and the velocity $v_{max} = -v_{min} = 20$. The fitness function of the PSO algorithm is the root mean square error of calibration (RMSEC). After 200 iterations, the optimal parameters of LS-SVM were obtained.

2.2.3. Successive Projection Algorithm (SPA)

Characteristic wavelength variables selection can eliminate the noise and uninformative variables in the original spectra. Models established based on the small numbers of characteristic wavelength variables are less costly and easy to interpret.

The successive projection algorithm is a forward variable selection method [7]. SPA has a good performance in solving the collinearity problems of spectra data. The objective of SPA is to select a subset of variables whose information content is at least redundant [28]. Firstly, SPA projects the variable onto other variables, and the variable of the largest projection vector is selected as the characteristic variable. It then incorporates a new one at each iteration until the best m variables are selected. It is worth noting that the m value is not more than the number of samples in the calibration set. More details about SPA can be found at reference [29].

All spectra pretreatment and traditional models were implemented using MATLAB R2016b (The MathWorks Inc., Natick, MA, USA).

2.3. *One-Dimensional Convolutional Neural Network (1D-CNN)*

2.3.1. Data Augmentation and Spectral Preprocessing

In this study, the number of distilled water samples is small for 1D-CNN modeling. The raw spectra data were augmented using the data augmentation algorithm proposed by Bjerrum et al. to avoid the overfitting phenomenon in the training process and improve the robustness of the 1D-CNN model [30]. This method adds random offset, multiplication, and slope into the original dataset. The offset was varied −0.1~0.1 times the standard deviation of the calibration set. The multiplication was done with 0.9~1.1 times the standard deviation of the calibration set. In addition, the slope was uniformly randomly adjusted between 0.95~1.05.

For the spectrophotometer experiment, the raw spectra were first processed by Z-score normalization. Then, each sample in the calibration set was done 10 times data augmentation and appended to the dataset. After data augmentation, the calibration dataset has a total of 264 spectra.

For the grating spectrograph experiment, the raw spectra were first processed by Z-score normalization. Then, each sample in the calibration set was done 10 times data augmentation and appended to the dataset. After data augmentation, the calibration set has a total of 264 spectra. Finally, the calibration set and prediction set were subjected to SNV preprocess.

2.3.2. 1D-CNN Architecture

The CNN basic architecture generally includes input layer, convolutional layer, pooling layer, activation function layer, flatten layer, fully connected layer, and output layer, etc. The main function of convolutional layers is to extract the characteristics of input spectra [13,31]. Take the first convolutional layer as an example, the convolutional layer with N same size filters. After the convolutional layer, the input spectra are transformed to N feature maps. However, the convolution is a linear operation. In order to implement nonlinearity transformation in the network, the feature maps are passed to an activation function layer [31]. The generally used activation function includes sigmoid function, rectified linear units (ReLU), and exponential linear units (ELU). The main purpose of the pooling layer is to reduce the dimensional of convolutional layer feature maps, which helps increase the calculation speed and prevent overfitting. The fully connected layer is a multi-layer perceptron, and each neuron in this layer is connected to all the elements in the previous layer [18].

Inspired by the classic CNN network structures of LeNet-5 [18], as shown in Figure 2, this study constructed a 1D-CNN framework to predict the water pH value. The 1D-CNN model consists of an input layer, three convolutional layers, three batch normalization layers, four activation function layers, three average pooling layers, a flatten layer, a dropout layer, a fully connected layer, and an output layer. The pooling size is 2×1 and the stride is 2. The purpose of the batch normalization layer is to standardize the data of each mini-batch and normalize the output into a standard normal distribution, which is an effective regularization strategy [18]. In order to prevent overfitting and improve the calculation speed, a batch normalization layer was added after each convolutional layer. We added a dropout layer after the flatten layer. The dropout layer randomly drops out nodes to further reduce overfitting. The output layer predicts the pH value. It is worth noting that the output layer is a fully connected dense layer with one node.

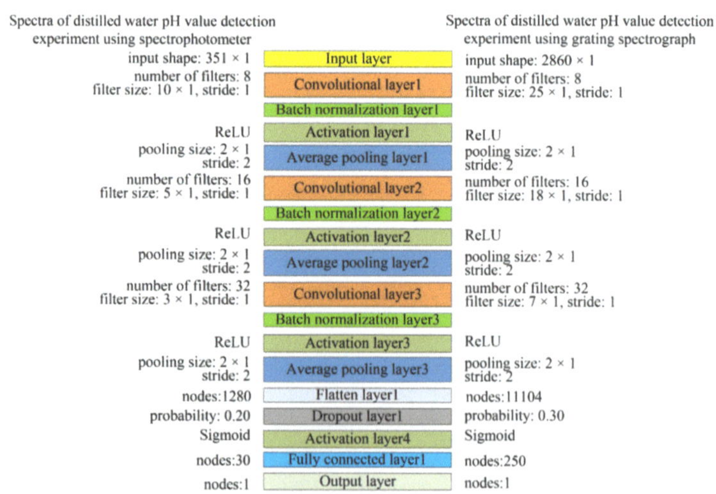

Figure 2. Architecture of 1D-CNN models.

For the spectrophotometer experiment, the input of the 1D-CNN model is one-dimensional spectra data which have a total of 351 wavelength variables, so the shape of the input layer is 351 × 1. The number of filters in the three convolutional layers is 8, 16, and 32, respectively. The ReLU is the activation function added between the batch normalization layer and the pooling layer. The sigmoid is the activation function added after the dropout layer.

For the grating spectrograph experiment, the shape of input layer is 2860 × 1. The number of filters in the three convolutional layers is 8, 16, and 32, respectively. The ReLU is the activation function added between the batch normalization layer and pooling layer, and the sigmoid is the activation function added after the dropout layer.

The hyperparameters of 1D-CNN models (filter size, node number of fully connected layer, and dropout probability) were optimized according to the number of samples, the number of convolutional layers, and the spectral dimension [13]. The optimal results of hyperparameters are shown in Figure 2.

2.3.3. Training of 1D-CNN

In the 1D-CNN model training process, mean squared error (MSE) was employed as the loss function and calculated in Equation (2) [11].

$$loss = MSE = \frac{\sum_{i=1}^{n}(\hat{y}_i - y_i)^2}{n} \quad (2)$$

where n is the number of water samples in the calibration set, $\hat{y}_i$ and y_i are the predicted pH and reference pH of the ith water samples, respectively. The initial learning rate was 0.001. We adopted the adaptive moment estimation (Adam) optimizer to minimize the MSE through the gradient descent algorithm.

The flowchart of water pH detection using Vis-NIR spectral analysis based on 1D-CNN is shown in Figure 3. The 1D-CNN model training process was as follows:

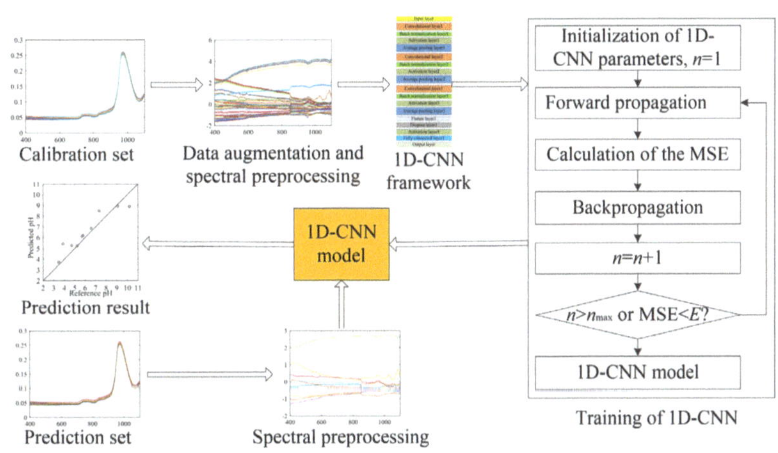

Figure 3. The flowchart of water pH detection using Vis-NIR spectral analysis based on 1D-CNN. In this figure, n is the training epochs and E is the expected value of MSE.

(1) Data augmentation and spectral preprocessing. Before 1D-CNN training, in order to improve the prediction accuracy and prevent overfitting. As previously described, after Z-score preprocess, the calibration set was augmented 10 times using the data augmentation method.

(2) The parameters of 1D-CNN, including all layer weight and biases, were initialized randomly.

(3) Forward propagation. The spectra in the calibration set as the input data of the 1D-CNN finally acquired the predicted pH values from the output layer.
(4) Calculate the MSE value between the predicted and the reference pH values by Equation (2).
(5) Backpropagation. Calculate the error gradient of the output layer, and use the back-propagation algorithm to calculate the error gradient of each weight. Then, use the gradient descent algorithm to update the weight value in each layer. The purpose of this step is to optimize the weight of 1D-CNN to minimize the MSE [32].
(6) Go to step (3) until the training epochs reach the maximum number of training epochs or the MSE value is less than the set value.

For the spectrophotometer experiment, the batch size was set to 24. The model has been trained for 180 epochs to make sure it was fully trained. For the grating spectrograph experiment, the batch size was set to 24, and the model has been trained for 80 epochs.

All 1D-CNN models were implemented using the Python (3.7.1) programming language and TensorFlow (1.14.0).

2.4. Criteria for Model Evaluation

The performance of the calibration models is evaluated according to the values of the root mean square error (RMSE) of calibration (RMSEC), prediction (RMSEP), and coefficient (R^2) of determination calibration (R_c^2) and prediction (R_p^2) [11,33]. Generally, the lower value of RMSE and the closer the R^2 is to 1 indicate that the prediction result from the calibration model is more reliable and the calibration model has a better-predicted performance [11].

2.5. Outlier Recognition

Outliers contained in the calibration set may have a significant effect on the calibration result [34]. Particularly for the 1D-CNN model, the MSE loss function is more sensitive to outliers [35,36]. In order to reduce the influence of outliers in the calibration model training process, leave-one-out cross-validation (LOOCV) with the 3σ criterion was employed for outlier recognition [37]. This algorithm firstly constructs the PLS model, then adopts the leave-one-out manner to calculate the standard deviation $\sigma(i)$ of the prediction error $e(i)$ for the ith sample. If the absolute value of $e(i)$ is larger than the absolute value of $3\sigma(i)$, the ith sample is an outlier sample and should be eliminated. For more details about LOOCV with the 3σ criterion, please refer to reference [37].

Figure 4a,b show the results of LOOCV with the 3σ criterion for two groups of water pH detection experiments. The results show that there are no outliers in the calibration set.

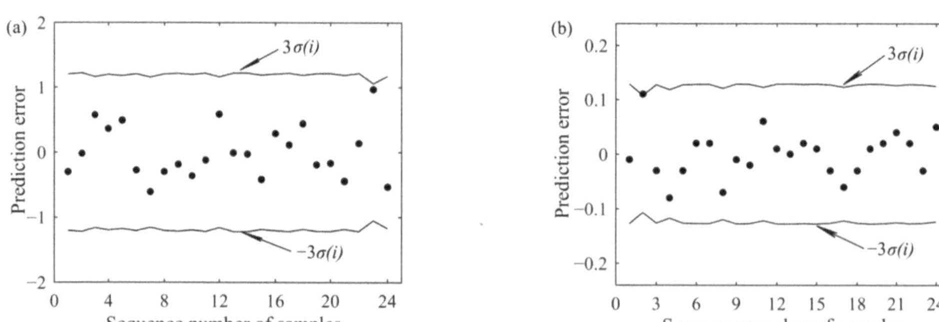

Figure 4. Results of LOOCV with the 3σ criterion: (**a**) spectrophotometer experiment; (**b**) grating spectrograph experiment.

3. Results and Discussion

3.1. Prediction Results Using Traditional Modeling Methods

To compare the effects of different preprocessing methods on the spectra analysis, PLS and LS-SVM calibration models based on different preprocesses were established.

For the spectrophotometer experiment, the prediction results are given in Table 1. Under the PLS, for the raw spectra, the RMSEP is 1.1381, and the R_p^2 is 0.6942. For other preprocessing methods, the prediction results show a decrease to some degree. Under the LS-SVM, for the raw spectra, the RMSEP is 1.0295, and the R_p^2 is 0.7495. With the comparison of the prediction results for different preprocessing methods, the SG (2,7) smoothing preprocess achieved the best prediction accuracy, the RMSEP is 1.0290, and the R_p^2 is 0.7498.

Table 1. Prediction parameters of PLS and LS-SVM with different preprocessing for the spectrophotometer experiment.

Model	Preprocessing	nPCs	γ	δ^2	Calibration Set		Prediction Set	
					RMSEC	R_c^2	RMSEP	R_p^2
PLS	Raw	9	-	-	0.3986	0.9621	1.1381	0.6942
	Smoothing	6	-	-	1.0119	0.7557	1.3038	0.5986
	SNV	3	-	-	1.6956	0.3141	1.8039	0.2318
	Z-score	5	-	-	1.1286	0.6961	1.1711	0.6762
LS-SVM	Raw	-	77,838.29	26,573.41	0.8957	0.8086	1.0295	0.7495
	Smoothing	-	85,781.59	29,349.56	0.9332	0.7923	1.0290	0.7498
	SNV	-	22,520.66	79,010.55	0.7688	0.8590	1.6613	0.3478
	Z-score	-	54,293.14	14,798.45	0.8296	0.8358	1.2398	0.6368

For the grating spectrograph experiment, the prediction results are given in Table 2. Under the PLS, for the raw spectra, the RMSEP is 1.1496 and the R_p^2 is 0.6569. Compared with the raw spectra, the prediction accuracy of the model based on SG (2,7) smoothing shows a slight increase. While the prediction results of the models based on other preprocessing methods show a decrease to some degree. Under the LS-SVM, for the raw spectra, the RMSEP is 1.1991, and the R_p^2 is 0.6025. Compared with the prediction results based on different preprocessing methods, the Z-score normalization acquired the best prediction accuracy, the RMSEP is 1.0228, and the R_p^2 is 0.7108.

Table 2. Prediction parameters of PLS and LS-SVM with different preprocessing for the grating spectrograph experiment.

Model	Preprocessing	nPCs	γ	δ^2	Calibration Set		Prediction Set	
					RMSEC	R_c^2	RMSEP	R_p^2
PLS	Raw	8	-	-	0.0424	0.9995	1.1496	0.6569
	Smoothing	8	-	-	0.0754	0.9985	1.1366	0.6646
	SNV	6	-	-	0.1354	0.9954	1.2530	0.5924
	Z-score	6	-	-	0.2187	0.9882	1.2879	0.5694
LS-SVM	Raw	-	29,195.21	3095.09	0.0022	0.9999	1.1991	0.6025
	Smoothing	-	92,829.47	99,301.32	0.0293	0.9998	1.2294	0.5821
	SNV	-	89,171.26	1500.91	0.0001	0.9999	1.3533	0.4936
	Z-score	-	35,097.78	3077.16	0.0018	0.9999	1.0228	0.7108

3.2. Characteristic Wavelength Selection and Validation

SPA was employed to select the characteristic wavelength variables and simplify the calibration models. For the spectrophotometer experiment, after extensive experimentation, using the raw spectra to select characteristic wavelength variables can acquire the best prediction accuracy. A total of 15 characteristic wavelength variables were selected. Figure 5a shows the distribution of the selected wavelength variables. The characteristic wavelength variables are 402 nm, 448 nm, 472 nm, 572 nm, 592 nm, 824 nm, 874 nm, 954 nm, 964 nm, 1010 nm, 1062 nm, 1068 nm, 1078 nm, 1088 nm, and 1100 nm.

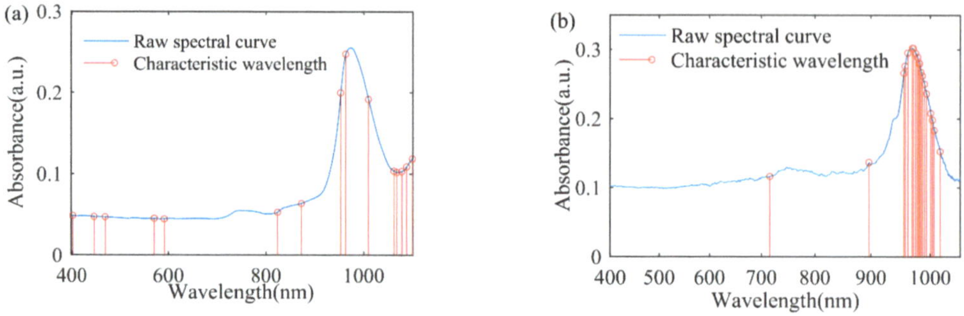

Figure 5. The distribution of wavelength variables selected by SPA: (**a**) spectrophotometer experiment; (**b**) grating spectrograph experiment.

For the grating spectrograph experiment, before characteristic wavelength selection, the spectra were preprocessed by the Z-score normalization and SNV. A total of 19 characteristic wavelength variables were selected. Figure 5b shows the distribution of the selected wavelength variables. The characteristic wavelength variables are 716 nm, 896 nm, and the band of 956~994 nm and 1004~1017 nm.

To evaluate the effectiveness of characteristic wavelength variables, SPA-PLS and SPA-LSSVM models were established by taking these characteristic wavelength variables as the input data matrix. For the spectrophotometer experiment, the prediction results are given in Table 3. Under the SPA-PLS, the RMSEP is 1.0209 and the R_p^2 is 0.7539. Under the SPA-LS-SVM, the RMSEP is 1.1286 and the R_p^2 is 0.6990.

Table 3. Prediction parameters of SPA-PLS and SPA-LS-SVM based on characteristic wavelength variables for the spectrophotometer experiment.

Model	nPCs	γ	δ^2	Calibration Set		Prediction Set	
				RMSEC	R_c^2	RMSEP	R_p^2
SPA-PLS	12	-	-	0.8760	0.9169	1.0209	0.7539
SPA-LS-SVM	-	98,472.83	2024.81	1.0019	0.7605	1.1286	0.6990

For the grating spectrograph experiment, the prediction results are given in Table 4. Under the SPA-PLS, the RMSEP is 0.5737 and the R_p^2 is 0.9145. Under the SPA-LS-SVM, the RMSEP is 0.5211 and the R_p^2 is 0.9249.

Table 4. Prediction parameters of SPA-PLS and SPA-LS-SVM based on characteristic wavelength variables for the grating spectrograph experiment.

Model	nPCs	γ	δ^2	Calibration Set		Prediction Set	
				RMSEC	R_c^2	RMSEP	R_p^2
SPA-PLS	8	-	-	0.1549	0.9941	0.5737	0.9145
SPA-LS-SVM	-	73,016.22	6037.97	0.0782	0.9985	0.5211	0.9249

3.3. *Prediction Results of 1D-CNN*

In order to estimate the effectiveness of 1D-CNN for water pH detection using Vis-NIR spectroscopy, 1D-CNN is introduced to build the calibration model. Table 5 lists the prediction results of 1D-CNN models. For the spectrophotometer experiment, the RMSEP is 0.7925 and the R_p^2 is 0.8515. For the grating spectrograph experiment, the RMSEP is 0.5128, and the R_p^2 is 0.9273.

Table 5. Prediction parameters of 1D-CNN models for the spectrophotometer experiment and grating spectrograph experiment.

Experiment	Calibration Set		Prediction Set	
	RMSEC	R_c^2	RMSEP	R_p^2
Spectrophotometer	0.7478	0.8715	0.7925	0.8515
Grating spectrograph	0.1337	0.9953	0.5128	0.9273

3.4. Interpreting the Feature Representations of Convolutional Layers

To further understand how the 1D-CNN extracts intricate features hierarchically, we took one sample from the prediction set to visualize the feature maps of each convolutional layer [13,17].

For the spectrophotometer experiment, Figure 6 shows the input spectral and feature maps of each convolutional layer. The first convolutional layer convolutes the input spectra to generate eight feature maps. Then, these eight feature maps are used as the input of different channels for the second convolutional layer to generate 16 feature maps. Similarly, the third convolutional layer generates 32 feature maps. Since the spatial invariance of the 1D-CNN, the feature map corresponds to the input spectral. Therefore, visualization of feature maps is helpful to understand the data transformation by convolutional layers [13].

The first convolutional layer (Conv1) mainly acts for the spectra preprocess and learns the shape characteristics of the spectra [17,30]. As shown in Figure 6, the effect of many filters (#1, #2, #3, #4, and #5) are similar to the commonly spectral preprocessing methods. The effect of the #3 filter is similar to the first derivative preprocessing. Three feature maps (filters #6, #7, and #8) show nearly zero activations in the whole spectrum, except for high responses in the band of near ~970 nm. The zero activations indicate that these filters are not sensitive to the input wavelength variables.

The second convolutional layer (Conv2) enhances the response of spectra peaks to extract the informative wavelength variables [30]. As shown in Figure 6, half the number of filters (#1, #2, #5, #6, #7, #14, #15, and #16) have high activations on variables between 960~980 nm, and the theoretical peak around 970 nm is related to the third overtones of O-H bonds in water [19]. Seven filters (#1, #2, #6, #7, #10, #15, #16) have high responses on variables between 840~860 nm, which around the O-H weak absorption bands [20]. The uninformative variables are reduced to zero.

Deeper in the network, the third convolutional layer (Conv3) becomes more complex. As shown in Figure 6, the feature maps show a stable increased activation on the spectra peaks. The higher activation value indicates more contribution to the final calibration regression result. On the contrary, the zero activations do not influence the final calibration regression result [13].

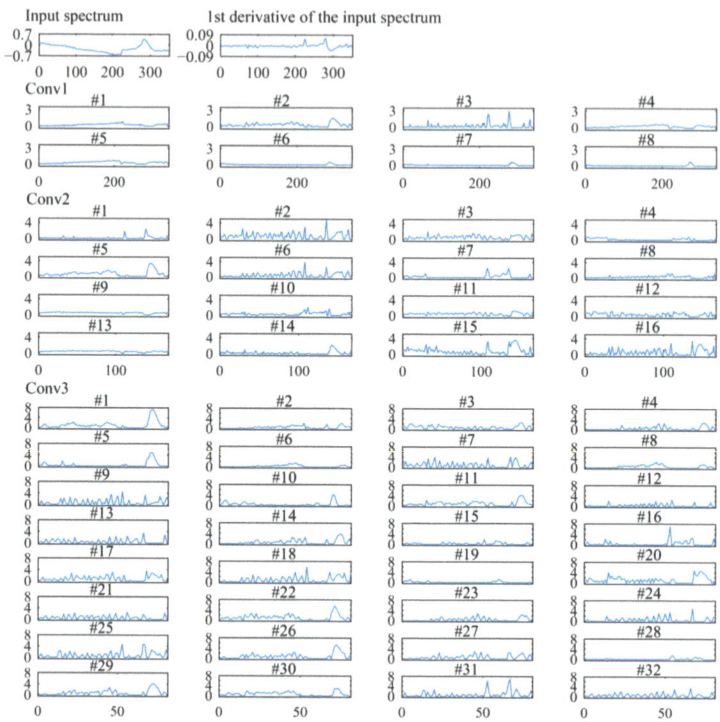

Figure 6. Visualization of the input spectral, the first derivative of the input spectral, and the feature maps of each convolutional layer for the spectrophotometer experiment.

For the grating spectrograph experiment, Figure 7 shows the input spectral and feature maps of each convolutional layer. In the first convolutional layer (Conv1), the effect of many filters is preprocessing the input spectra [13]. Six filters (#1, #3, #5, #6, #7, and #8) have high responses on the band of 400~500 nm and 960~1049 nm.

The second convolutional layer (Conv2) works something like feature variable selection. The Conv2 has enhanced the informative peaks on a few of the characteristic wavelength variables, while the uninformative wavelength variables were almost zero activations. Most filters have high activations on wavelength variables on 400~500 nm, 890~900 nm, and 960~1049 nm. In the third convolutional layer (Conv3), a stable increased activation on the wavelength band at 400~500 nm, 890~900 nm, and 960~1049 nm was noted. The results indicate that these wavelength variables contribute more greatly to the final calibration regression result.

3.5. Calibration Performance Comparisons Discussion of the Multivariate Calibration Models

3.5.1. Discussion of Model Prediction Accuracy

The best prediction results and scatter plots of predicted results and reference pH values obtained by different calibration models are shown in Figures 8 and 9, respectively.

As shown in Figure 8a, for the spectrophotometer experiment, the 1D-CNN model achieved the best prediction performance, the RMSEP is 0.7925 and the R_p^2 is 0.8515. Compared with the SPA-PLS model, the RMSEP is reduced by 22.37%, and the R_p^2 is increased by 12.95%. As shown in Figure 9a, the 1D-CNN model shows that more sample points are closed to the ideal regression line, which indicates that the prediction performance of 1D-CNN is better than the other four traditional regression models [11,38].

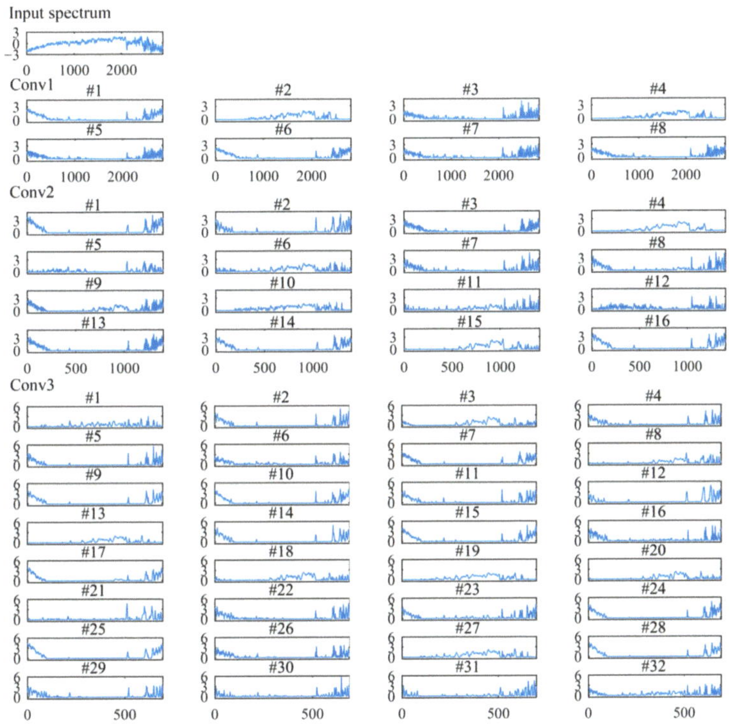

Figure 7. Visualization of the input spectral and feature maps of each convolutional layer for the grating spectrograph experiment.

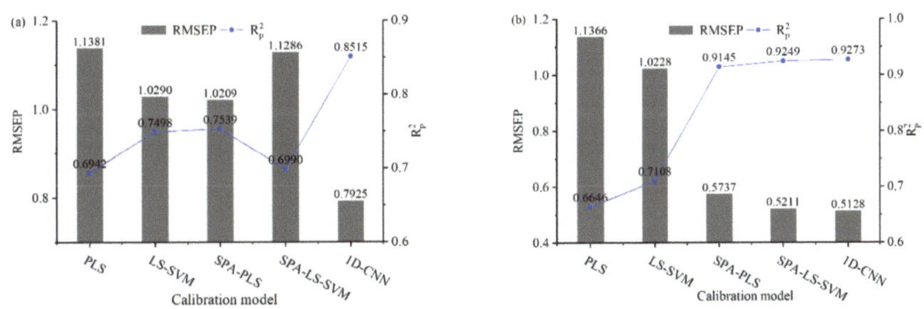

Figure 8. The best prediction accuracy of different calibration models: (**a**) spectrophotometer experiment; (**b**) grating spectrograph experiment.

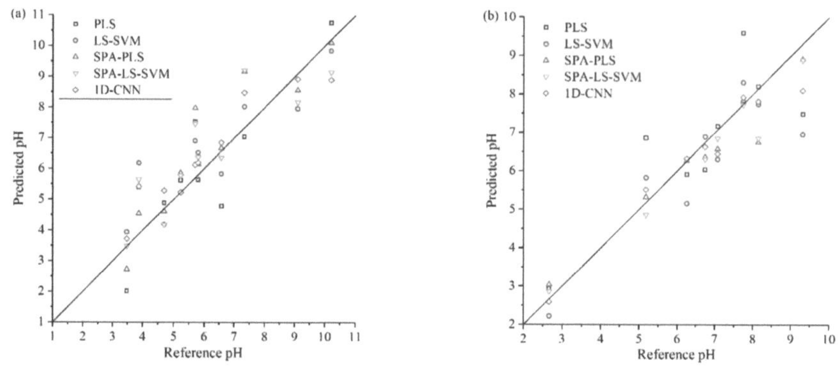

Figure 9. Scatter plots of reference pH and the predicted pH obtained by different regression models: (**a**) spectrophotometer experiment; (**b**) grating spectrograph experiment. The solid line in each figure represents the ideal regression line.

As shown in Figure 8b, for the grating spectrograph experiment, both the SPA-PLS, SPA-LS-SVM, and 1D-CNN models acquired good prediction accuracy. The 1D-CNN model achieved the best prediction performance, the RMSEP is 0.5128, and the R_p^2 is 0.9273. As shown in Figure 9b, both the sample points of SPA-PLS, SPA-LS-SVM, and 1D-CNN models are closed to the solid line, which indicates that these three models could predict the water pH accurately.

Compared with the prediction accuracy of different calibration models for the water pH detection experiments, the results show that the prediction performance of the 1D-CNN models is better than that of other traditional calibration models. The root cause of better prediction performance of 1D-CNN is the well-trained convolutional layers that can extract the spectra features more effectively and accurately [11,18]. As shown in Figures 6 and 7, compared with the characteristic variables selected by SPA, more variables are activated by the third convolutional layer in the 1D-CNN models. The activation function also plays a central role in the prediction performance of 1D-CNN models [11]. Convolution is a linear operation. The feature maps of the convolutional layer passed the activation function to implement the nonlinear transformation. Thus, 1D-CNN with the outperformance of extracting the linear and nonlinear features in the spectra [18,33].

In addition, for the SPA-PLS and SPA-LS-SVM models, although the variable selection can effectively improve the prediction performance, the SPA algorithm does not consider the effect of the variable combination. The variable combination also has a significant effect on the prediction performance [12]. However, few of the variable selection algorithms have solved this problem [39].

3.5.2. Impacts of Spectra Preprocessing on Calibration Models

After different preprocess for raw spectra, PLS and LS-SVM models were established. As shown in Table 1, for the spectrophotometer experiment, the LS-SVM model based on SG (2,7) smoothing preprocess achieved the highest prediction accuracy. Under some preprocessing methods, the prediction results decrease to some degree. As shown in Table 2, for the grating spectrograph experiment, the LS-SVM model based on Z-score normalization preprocess showed the best prediction performance. The results show that different preprocessing methods have quite different effects on PLS and LS-SVM models. Thus, spectra preprocessing plays an important role in traditional calibration modeling. Appropriate preprocessing methods can improve the prediction performance, but misuse of preprocessing methods will decrease the prediction performance.

For 1D-CNN models, the only required preprocessing is normalization [14]. As shown in Figures 6 and 7, the first convolutional layer can process the spectra automatically. For the water pH detection experiments, 1D-CNN acquired the highest prediction accuracy.

Therefore, the results indicate that the 1D-CNN regression strategy can achieve excellent prediction performance without the need for complex spectra pretreatment [11].

3.5.3. Impacts of Feature Selection on Calibration Models

As shown in Figure 8a, for the spectrophotometer experiment, the prediction performance of SPA-PLS model is better than that of the PLS model. Under the SPA-PLS model, the RMSEP is 1.0209 and the R_p^2 is 0.7539. Compared with the PLS model, the RMSEP is reduced by 10.30%, and the R_p^2 is increased by 8.59%. However, it is worth noting that, under the SPA-LS-SVM model, the prediction accuracy decreased to some degree. Compared with the LS-SVM model, the RMSEP is increased by 9.63%, and the R_p^2 is reduced by 6.74%. According to the principle of the SPA algorithm, the characteristic wavelength variables are selected based on linear regression. Therefore, using these characteristic wavelength variables to establish a nonlinear model may reduce the prediction accuracy.

As shown in Figure 8b, for the grating spectrograph experiment, both the SPA-PLS and SPA-LS-SVM models acquired good prediction accuracy. Under the SPA-PLS model, the RMSEP is 0.5737, and the R_p^2 is 0.9145. Compared with PLS model, the RMSEP is reduced by 49.52%, and the R_p^2 is increased by 37.60%. Under the SPA-LS-SVM model, the RMSEP is 0.5211, and the R_p^2 is 0.9249. Compared with LS-SVM model, the RMSEP is reduced by 49.05%, and the R_p^2 is increased by 30.12%. The results further indicated that variable selection could effectively extract the informative variables and eliminate the noise and uninformative variables from the original spectra [18]. There are five characteristic variables located at the wavelength near 970 nm, which is related to the third overtones of O-H bonds in water [19]. The result indicates that the O-H bond in water is helpful to water pH prediction.

For 1D-CNN models, as the results show in Figures 6 and 7, the convolutional layers can automatically extract the informative variables from full spectra. Uninformative variables are almost not activated, and information variables are highly activated [13]. Compared with the characteristic wavelength variables selected by SPA, the convolutional layers extract more hidden features of spectra. For both the spectrophotometer experiment and grating spectrograph experiment, 1D-CNN models based on full spectra acquired the best prediction accuracy. The results indicate that 1D-CNN can effectively extract the spectra features to build the relationship between spectra and water pH value.

3.5.4. Discussion of Calculation Rapidity

This subsection discussed the calculation rapidity in the prediction process of different calibration models. The original spectra of the prediction set were inputted into the trained calibration models to calculate the prediction results. All calculations were repeated five times, and the statistical results of the calculation time are shown in Table 6.

Table 6. Calculation time in the prediction process of different calibration models *.

Experiment	PLS (s)		LS-SVM (s)		SPA-PLS (s)		SPA-LS-SVM (s)		1D-CNN (s)	
	Mean	Std.	Mean	Std.	Mean	Std.	Mean	Std.	Mean	Std.
Spectrophotometer	0.0014	0.0005	0.0065	0.0004	0.0017	0.0002	0.0081	0.0009	0.0024	0.0005
Grating spectrograph	0.0068	0.0008	0.0065	0.0012	0.0186	0.0008	0.0232	0.0009	0.0082	0.0011

* All calculations were performed on a PC with an Intel® Core™ i5-9500 CPU (3.0 GHz) and 8.0 GB RAM, running a Windows 11 operating system.

As shown in Table 6, for the spectrophotometer experiment, 1D-CNN is faster than LS-SVM and SPA-LS-SVM in the prediction process. Although PLS and SPA-PLS are faster than 1D-CNN, PLS and SPA-PLS are not as accurate as 1D-CNN. For the grating spectrograph experiment, 1D-CNN is faster than SPA-PLS and SPA-LS-SVM. The PLS and LS-SVM are faster than 1D-CNN, but the prediction performance of 1D-CNN is better than PLS and LS-SVM. That is because 1D-CNN does not need the complex spectra preprocessing and variable selection, and 1D-CNN can effectively extract the linear and nonlinear features from the spectra. Therefore, 1D-CNN provides a better balance of prediction accuracy and calculation rapidity.

For the PLS and LS-SVM methods, in order to build a high-accuracy calibration model, a lot of experiments are needed to find the best preprocessing and variable selection methods. The spectra preprocessing and variable selection are repetitive and time-consuming. On the contrary, 1D-CNN only required preprocessing is normalization. Therefore, 1D-CNN could improve the convenience of modeling.

4. Conclusions

This study proposed a novel calibration regression strategy for distilled water pH detection using Vis-NIR spectroscopy, a 1D-CNN regression architecture with three convolutional layers constructed to optimize the prediction performance. Two groups of Vis-NIR spectral analysis experiments of water pH detection are employed to evaluate the performance of 1D-CNN. Two traditional multiple regression algorithms (PLS and LS-SVM) are introduced for comparison analysis with 1D-CNN. The conclusions are described as follows.

(1) The prediction performance of 1D-CNN based on full spectra is better than the traditional linear (PLS) and nonlinear (LS-SVM) approaches using full spectra and characteristic wavelength variables. For the spectrophotometer experiment, the RMSEP is 0.7925 and the R_p^2 is 0.8515. For the grating spectrograph experiment, the RMSEP is 0.5128 and the R_p^2 is 0.9273.

(2) (By visualizing the characteristic map through three convolution layers, we can understand how the convolution network converts one-dimensional spectral data into prediction results. The first convolutional layer acts for spectra pretreatment and learns the shape feature of input spectra. The second convolutional layer extracts the hidden features in the spectra. The third convolutional layer stably enhances the activations of the feature spectra peaks.

(3) 1D-CNN could effectively extract the spectra features. The number of activation variables of 1D-CNN is more than the feature variables selected by SPA, and the prediction accuracy of 1D-CNN is higher than that of SPA-PLS and SPA-LS-SVM for both experiments.

(4) 1D-CNN could improve the convenience of modeling. Compared with the traditional regression methods, 1D-CNN modeling only require preprocessing is normalization. 1D-CNN does not need complex spectra pretreatment and variable selection, which ensures the calculation rapidity of 1D-CNN.

This study indicates the 1D-CNN regression method is an alternative quantitative calibration technology for water pH detection using Vis-NIR spectroscopy. The future direction of our work is to explore the feasibility of the 1D-CNN regression method for other water quality parameters (such as chemical oxygen demand) detection and to develop a calibration transfer approach of water pH based on 1D-CNN.

Author Contributions: Conceptualization, D.L. and L.L.; methodology, D.L.; software, D.L.; validation, D.L. and L.L.; formal analysis, D.L.; investigation, D.L.; resources, L.L.; data curation, D.L.; writing—original draft preparation, D.L.; writing—review and editing, L.L.; visualization, D.L.; supervision, L.L.; project administration, L.L.; funding acquisition, L.L. All authors have read and agreed to the published version of the manuscript.

Funding: This research was funded by the National Natural Science Foundation of China (No. 51805177), and the Scientific Research Fund Project of Huaqiao University (No. 11BS413).

Institutional Review Board Statement: Not applicable.

Informed Consent Statement: Not applicable.

Data Availability Statement: The data presented in this study are available on request from the corresponding author. The data are not publicly available due to it is being used to apply for project.

Acknowledgments: The authors would like to thank the editors and the anonymous reviewers for their comments and suggestions.

Conflicts of Interest: The authors declare that they have no conflicts of interest.

References

1. Chen, Z.; Guo, Q.; Shi, Z. Design of WSN node for water pollution remote monitoring. *Telecommun. Syst.* **2013**, *53*, 155–162. [CrossRef]
2. Golan, R.; Gavrieli, I.; Lazar, B.; Ganor, J. The determination of pH in hypersaline lakes with a conventional combination glass electrode. *Limnol. Oceanogr. Methods* **2014**, *12*, 810–815. [CrossRef]
3. Song, J.; Li, G.L.; Yang, X.D.; Liu, X.W.; Xie, L. Rapid analysis of soluble solid content in navel orange based on visible-near infrared spectroscopy combined with a swarm intelligence optimization method. *Spectrochim. Acta Part A* **2020**, *228*, 117815. [CrossRef] [PubMed]
4. Wang, L.S.; Wang, R.J.; Lu, C.P.; Wang, J.; Huang, W. Rapid determination of moisture content in compound fertilizer using visible and near infrared spectroscopy combined with chemometrics. *Infrared Phys. Technol.* **2019**, *102*, 103045. [CrossRef]
5. Wang, K.; Bian, X.; Zheng, M.; Liu, P.; Lin, L.; Tan, X. Rapid determination of hemoglobin concentration by a novel ensemble extreme learning machine method combined with near-infrared spectroscopy. *Spectrochim. Acta Part A* **2021**, *263*, 120138. [CrossRef]
6. Xie, L.; Ying, Y.; Ying, T. Classification of tomatoes with different genotypes by visible and short-wave near-infrared spectroscopy with least-squares support vector machines and other chemometrics. *J. Food Eng.* **2009**, *94*, 34–39. [CrossRef]
7. Li, L.; Li, D. A Hybrid Multivariate Calibration Optimization Method for Visible Near Infrared Spectral Analysis. In Proceedings of the 2021 7th International Conference on Condition Monitoring of Machinery in Non-Stationary Operations (CMMNO), Guangzhou, China, 11–13 June 2021; pp. 76–81. [CrossRef]
8. Chen, H.; Xu, L.; Ai, W.; Lin, B.; Feng, Q.; Cai, K. Kernel functions embedded in support vector machine learning models for rapid water pollution assessment via near-infrared spectroscopy. *Sci. Total Environ.* **2020**, *714*, 136765. [CrossRef]
9. Xiao, D.; Liu, C.; Le, B.T. Detection method of TFe content of iron ore based on visible-infrared spectroscopy and IPSO-TELM neural network. *Infrared Phys. Technol.* **2019**, *97*, 341–348. [CrossRef]
10. Tian, H.; Zhang, L.; Li, M.; Wang, Y.; Sheng, D.; Liu, J.; Wang, C. WSPXY combined with BP-ANN method for hemoglobin determination based on near-infrared spectroscopy. *Infrared Phys. Technol.* **2019**, *102*, 103003. [CrossRef]
11. Li, Y.; Ma, B.; Li, C.; Yu, G. Accurate prediction of soluble solid content in dried Hami jujube using SWIR hyperspectral imaging with comparative analysis of models. *Comput. Electron. Agric.* **2022**, *193*, 106655. [CrossRef]
12. Yun, Y.H.; Li, H.D.; Deng, B.C.; Cao, D.S. An overview of variable selection methods in multivariate analysis of near-infrared spectra. *TrAC Trends Anal. Chem.* **2019**, *113*, 102–115. [CrossRef]
13. Zhang, X.; Xu, J.; Yang, J.; Chen, L.; Zhou, H.; Liu, X.; Li, H.; Lin, T.; Ying, Y. Understanding the learning mechanism of convolutional neural networks in spectral analysis. *Anal. Chim. Acta* **2020**, *1119*, 41–51. [CrossRef]
14. Cui, C.; Fearn, T. Modern practical convolutional neural networks for multivariate regression: Applications to NIR calibration. *Chemom. Intell. Lab. Syst.* **2018**, *182*, 9–20. [CrossRef]
15. Mishra, P.; Passos, D. Multi-output 1-dimensional convolutional neural networks for simultaneous prediction of different traits of fruit based on near-infrared spectroscopy. *Postharvest Biol. Technol.* **2022**, *183*, 111741. [CrossRef]
16. Fukuhara, M.; Fujiwara, K.; Maruyama, Y.; Itoh, H. Feature visualization of Raman spectrum analysis with deep convolutional neural network. *Anal. Chim. Acta* **2019**, *1087*, 11–19. [CrossRef]
17. Ng, W.; Minasny, B.; McBratney, A. Convolutional neural network for soil microplastic contamination screening using infrared spectroscopy. *Sci. Total Environ.* **2020**, *702*, 134723. [CrossRef]
18. Tian, S.; Wang, S.; Xu, H. Early detection of freezing damage in oranges by online Vis/NIR transmission coupled with diameter correction method and deep 1D-CNN. *Comput. Electron. Agric.* **2022**, *193*, 106638. [CrossRef]
19. Mishra, P.; Passos, D. A synergistic use of chemometrics and deep learning improved the predictive performance of near-infrared spectroscopy models for dry matter prediction in mango fruit. *Chemom. Intell. Lab. Syst.* **2021**, *212*, 104287. [CrossRef]
20. Huang, Y.; Lin, M.; Cavinato, A.G.; Mayes, D.M.; Rasco, B.A. Influence of temperature on the measurement of NaCl content of aqueous solution by short-wavelength near infrared spectroscopy (SW-NIR). *Sens. Instrum. Food Qual.* **2007**, *1*, 91–97. [CrossRef]
21. Mishra, P.; Biancolillo, A.; Roger, J.M.; Marini, F.; Rutledge, D.N. New data preprocessing trends based on ensemble of multiple preprocessing techniques. *TrAC Trends Anal. Chem.* **2020**, *132*, 116045. [CrossRef]
22. Wiedemair, V.; Langore, D.; Garsleitner, R.; Dillinger, K.; Huck, C. Investigations into the Performance of a Novel Pocket-Sized Near-Infrared Spectrometer for Cheese Analysis. *Molecules* **2019**, *24*, 428. [CrossRef]
23. Lu, Z.; Chen, X.; Yao, S.; Qin, H.; Zhang, L.; Yao, X.; Yu, Z.; Lu, J. Feasibility study of gross calorific value, carbon content, volatile matter content and ash content of solid biomass fuel using laser-induced breakdown spectroscopy. *Fuel* **2019**, *258*, 116150. [CrossRef]
24. Geladi, P.; Kowalski, B.R. Partial least square regression: A tutorial. *Anal. Chim. Acta* **1986**, *185*, 1–17. [CrossRef]
25. Bao, Y.; Liu, F.; Kong, W.; Sun, D.W.; He, Y.; Qiu, Z. Measurement of Soluble Solid Contents and pH of White Vinegars Using VIS/NIR Spectroscopy and Least Squares Support Vector Machine. *Food Bioprocess Technol.* **2014**, *7*, 54–61. [CrossRef]

26. Liu, F.; He, Y.; Wang, L. Comparison of calibrations for the determination of soluble solids content and pH of rice vinegars using visible and short-wave near infrared spectroscopy. *Anal. Chim. Acta* **2008**, *610*, 196–204. [CrossRef]
27. Wu, Q.; Song, T.; Liu, H.; Yan, X. Particle swarm optimization algorithm based on parameter improvements. *J. Comput. Methods Sci. Eng.* **2017**, *17*, 557–568. [CrossRef]
28. Liu, K.; Chen, X.; Li, L.; Chen, H.; Ruan, X.; Liu, W. A consensus successive projections algorithm–multiple linear regression method for analyzing near infrared spectra. *Anal. Chim Acta* **2015**, *858*, 16–23. [CrossRef] [PubMed]
29. Araújo, M.R.C.S.U.; Saldanha, T.C.B.; Galvão, R.K.H.; Yoneyama, T.; Chame, H.C.; Visani, V. The successive projections algorithm for variable selection in spectroscopic multicomponent analysis. *Chemom. Intell. Lab. Syst.* **2001**, *57*, 65–73. [CrossRef]
30. Bjerrum, E.J.; Glahder, M.; Skov, T. Data Augmentation of Spectral Data for Convolutional Neural Network (CNN) Based Deep Chemometrics. *arXiv* **2017**. [CrossRef]
31. Zhu, J.; Sharma, A.S.; Xu, J.; Xu, Y.; Jiao, T.; Ouyang, Q.; Li, H.; Chen, Q. Rapid on-site identification of pesticide residues in tea by one-dimensional convolutional neural network coupled with surface-enhanced Raman scattering. *Spectrochim. Acta Part A* **2021**, *246*, 118994. [CrossRef] [PubMed]
32. Malek, S.; Melgani, F.; Bazi, Y. One-dimensional convolutional neural networks for spectroscopic signal regression. *J. Chemom.* **2018**, *32*, e2977. [CrossRef]
33. Zhu, H.; Yang, L.; Gao, J.; Gao, M.; Han, Z. Quantitative detection of Aflatoxin B1 by subpixel CNN regression. *Spectrochim. Acta Part A* **2022**, *268*, 120633. [CrossRef]
34. Liu, Z.; Cai, W.; Shao, X. Outlier detection in near-infrared spectroscopic analysis by using Monte Carlo cross-validation. *Sci. China Ser. B: Chem.* **2008**, *51*, 751–759. [CrossRef]
35. Sewdien, V.N.; Preece, R.; Torres, J.; Rakhshani, E.; Van, D. Assessment of critical parameters for artificial neural networks based short-term wind generation forecasting. *Renew. Energy* **2020**, *161*, 878–892. [CrossRef]
36. Chen, X.; Yu, R.; Ullah, S.; Wu, D.; Zhang, Y. A novel loss function of deep learning in wind speed forecasting. *Energy* **2021**, *238*, 121808. [CrossRef]
37. Koshoubu, J.; Iwata, T.; Minami, S. Elimination of the Uninformative Calibration Sample Subset in the Modified UVE (Uninformative Variable Elimination)–PLS (Partial Least Squares) Method. *Anal. Sci.* **2001**, *17*, 319–322. [CrossRef]
38. Dixit, Y.; Al-Sarayreh, M.; Craigie, C.R.; Reis, M.M. A global calibration model for prediction of intramuscular fat and pH in red meat using hyperspectral imaging. *Meat Sci.* **2021**, *181*, 108405. [CrossRef]
39. Yun, Y.; Wang, W.; Tan, M.; Liang, Y.; Li, H.; Cao, D.; Lu, H.; Xu, Q. A strategy that iteratively retains informative variables for selecting optimal variable subset in multivariate calibration. *Anal. Chim. Acta* **2014**, *807*, 36–43. [CrossRef]

MDPI AG
Grosspeteranlage 5
4052 Basel
Switzerland
Tel.: +41 61 683 77 34

Sensors Editorial Office
E-mail: sensors@mdpi.com
www.mdpi.com/journal/sensors

Disclaimer/Publisher's Note: The statements, opinions and data contained in all publications are solely those of the individual author(s) and contributor(s) and not of MDPI and/or the editor(s). MDPI and/or the editor(s) disclaim responsibility for any injury to people or property resulting from any ideas, methods, instructions or products referred to in the content.

www.ingramcontent.com/pod-product-compliance
Lightning Source LLC
Chambersburg PA
CBHW041408210926
51850CB00013B/3209
* 9 7 8 3 7 2 5 8 2 3 6 7 3 *